SUPERCONDUCTIVE PARTICLE DETECTORS

Advances in the Physics of Condensed Matter

SUPERCONDUCTIVE PARTICLE DETECTORS

26 – 29 October 1987

INSTITUTE FOR SCIENTIFIC INTERCHANGE

Edited by
ANTONIO BARONE
Dipartimento di Scienze Fisiche
Università di Napoli
and
Istituto di Cibernetica del CNR
Arco Felice, Napoli

World Scientific
Singapore • New Jersey • Hong Kong

Published by

World Scientific Publishing Co. Pte. Ltd.
P.O. Box 128, Farrer Road, Singapore 9128

U. S. A. office: World Scientific Publishing Co., Inc.
687 Hartwell Street, Teaneck NJ 07666, USA

SUPERCONDUCTIVE PARTICLE DETECTORS

ISBN 9971-50-611-4

Printed in Singapore by Utopia Press.

PREFACE

It is general opinion that a Workshop devoted to the potential of Superconductivity, in the field of radiation detection, was particularly timely and important. Indeed such a meeting between outstanding representatives of various scientific communities (Superconductivity, Nuclear Physics, Astrophysics) has proven to be extremely stimulating and fruitful.

The warm atmosphere created during the Workshop has produced a deep interchange of ideas and triggered new common projects in this area. Once again the ISI has offered all the most excellent conditions for the success of such an event.

I wish to thank the President, Prof. Tullio Regge and the Director of the ISI, Prof. Mario Rasetti for providing this opportunity. Thanks are extended to the staff in Villa Gualino, in particular to the coordinator of the ISI Mrs. Tiziana Bertoletti

Many thanks are also due to Dr. Carlo Carmerlingo, Dr. Roberto Cristiano Mrs. A.M. Mazzarella and C. Salinas, of the Istituto di Cibernetica, for their valuable cooperation during the various stages of the organization.

The Chairman
Antonio Barone

The present volume includes a collection of reviews and original contributions presented at the Workshop on "Superconductive Particle Detectors" held at the Institute for Scientific Interchange, Villa Gualino, in Torino, Italy.

The workshop has been organized by the:

INSTITUTE FOR SCIENTIFIC INTERCHANGE, TORINO
ISTITUTO ITALIANO PER GLI STUDI FILOSOFICI, NAPOLI
ISTITUTO DI CIBERNETICA DEL CNR, ARCO FELICE, NAPOLI

and it was held also under the auspices of the:

ISTITUTO NAZIONALE DI FISICA NUCLEARE

CONTENTS

SUPERCONDUCTIVE PARTICLE DETECTORS

NONEQUILIBRIUM SUPERCONDUCTIVITY FOR PARTICLE DETECTORS

K. E. GRAY

Materials Science Division,
Argonne National Laboratory,
Argonne, Illinois 60439, USA

A considerable amount of attention has been devoted to nonequilibrium superconductivity over the last 10-15 years. A fairly complete and quantatitive understanding of the experimental and theoretical aspects of the subject has emerged [1,2]. In this paper, those aspects of nonequilibrium superconductivity which are relevant to a majority of particle detector applications will be reviewed, and new calculations, more specific to actual detector applications, will be presented. The primary focus is on ionizing particles for which the characteristic energy is greater than typical superconducting energy gap values, Δ, of about 1 meV. Thus microwave and far-infrared detection is excluded, although many of the results and consequences may also apply in those cases.

For typical examples of interaction with high energy particles, the initial disturbance creates a state which is very far from equilibrium and spatially localized around the particle interaction or its path through the detector. Such a state will decay towards equilibrium at the bath temperature in a characteristic time during which the disturbance will also diffuse away from the point of origin. Thus, for example, concepts which are valid near-to-equilibrium may have to be re-examined in the present context. In addition, the very interesting and important contribution of charge imbalance effects [3] to nonequilibrium superconductivity need not be considered here since particle interactions will not produce such an imbalance. Finally, a new aspect of the problem involves the very short time period after particle impact during which the local temperature is in excess of the superconducting transition temperature, T_C, and the above considerations are irrelevant.

Thin films are particularly important for detectors and are special because of the large relative surface area through which the excess energy can escape, usually via phonons. This escape process can play a predominant role in the response of these detectors and will be discussed at length.

The measurements of the relevant relaxation rates, i.e., quasi-particle-pair, inelastic quasiparticle scattering and phonon escape, are summarized in the next section, together with a discussion of the quasiparticle diffusion length. Following that, states far-from-equilibrium are discussed including calculations based on various approximations. As a specific example which is particularly useful for particle detector applications, the nonequilibrium state of a thin film irradiated with a minimum ionizing particle is described. This leads into the final sections which discuss two important applications: high temporal and spatial resolution position detectors; and particle detectors with good energy resolution.

CLOSE-TO EQUILIBRIUM

Quasiparticle-pair relaxation rate. The first measurement of the quasiparticle-pair relaxation rate, Γ_R, was done by Taylor [4] using a double tunnel junction, injection-detection experiment. In this experiment, Γ_R is determined by measuring the excess quasiparticle density, ΔN, due to tunnel injection at a known rate I_o, where $\Delta N \equiv N - N(T)$ and $N(T)$ is the density in thermal equilibrium. Subsequent measurements using this technique on a number of materials have been reported and are summarized in [1,2]. The important role of the phonons and their escape through the film surfaces was soon recognized [5], and the relevant rate equations are:

$$\partial N/\partial t = I_o + D\nabla^2 N + 2\Gamma_B N_\omega - RN^2, \qquad (1)$$

$$\partial N_\omega/\partial t = RN^2/2 - \Gamma_B N_\omega - \Gamma_{es}[N_\omega - N_\omega(T)], \qquad (2)$$

where the N^2 term reflects the fact that the quasiparticles must recombine in pairs, and N_ω is the number density of phonons with energy $\hbar\Omega \geq 2\Delta$, which break pairs at a rate Γ_B to create new quasiparticles or escape with a rate Γ_{es}. It can be shown that R is the average rate at which a quasiparticle recombines with any of the other quasiparticles, even though it can only recombine with N/2 of them, i.e., those with opposite spin. In that case $\Gamma_R = RN$, where N is the total density (of both spins). A diffusion term has been added for the more general case of spatially inhomogeneous injection to be dealt with later. Diffusion of phonons in thin films is neglected due to the higher probability of their escape through the film surfaces.

In equilibrium, I_o is zero, and $2\Gamma_B N_\omega(T) = RN(T)^2$, so that the ratio R/Γ_B is determined only by Fermi-Dirac and Bose-Einstein statistics and the densities of states of electrons and phonons. Straightforward evaluation shows that R/Γ_B is a constant at low temperatures.

In the above experiments, I_o is assumed spatially uniform and the junction sizes are considerably larger than the diffusion distance of quasiparticles so the diffusion term is neglected. A second important simplification results from considering the limit of small injection, i.e., $\Delta N \ll N(T)$. The nonlinearity disappears, and Eq. 1 reduces to

$$\partial N/\partial t = I_o + 2\Gamma_B \Delta N_\omega - 2\Delta N \Gamma_R(T). \tag{3}$$

If the nonequilibrium phonons escaped from the films without any interaction, ΔN_ω could be neglected, and steady state measurements of I_o and ΔN would give Γ_R directly. Unfortunately, ΔN_ω is not small and has the same temperature dependence as $\Delta N \Gamma_R(T)$. Solving Eqns. 1 and 2 together, Rothwarf and Taylor [5] have shown that the steady state experiments measure an effective recombination rate $\Gamma_R^* = I_o/2\Delta N$, where

$$\Gamma_R^* = \Gamma_R/F_\omega, \tag{4}$$

$$F_\omega \equiv 1 + \Gamma_B/\Gamma_{es}. \tag{5}$$

The author [6] has shown that Eqns. 4 and 5 are also valid for time dependent experiments. Although the phonon trapping factor, F_ω, is difficult to reliably calculate or measure in most experiments, direct measurements of Γ_R are possible near T_c, so that F_ω can be estimated (see below). There is a critical review of calculations and measurements of F_ω by Kaplan [7].

Experiments on aluminum double junctions have verified that near thermal equilibrium, that Γ_R (and Γ_R^*) are proportional to N(T). However at very low temperatures, Γ_R^* saturates rather than decreasing exponentially with Δ/k_BT, as N(T) does. Experimently, the saturation is caused by recombination in the normal cores of magnetic flux lines due to residual fields or those used to destroy the Josephson effect in one of the junctions. Other possible causes include heating effects, diffusion of quasiparticles out of the detector junction, overinjection and injector saturation [6]. The case of overinjection is well explained by Rothwarf and Taylor (see Eq. 10), and results in the same value [6] of $R^*=\Gamma_R^*/N(T)$ as the above fit to N(T).

Several calculations of R exist, and they are summarized by Kaplan, et al. [8] who also show

$$R = [2\Delta(0)/k_BT_c]^3/[4N(0)\Delta(0)\tau_o] \tag{6}$$

$$\tau_o = [Z_1(0)\hbar/2\pi(k_BT_c)^3]\ \alpha^2F(\Omega)/\Omega^2\ , \tag{7}$$

where $Z_1(0)$ is the real part of the Eliashberg renormalization factor at $\omega=0$, $\alpha^2F(\Omega)$ is the electron-phonon coupling constant times the phonon density of states and τ_o is the time constant characterizing the strength of the electron-phonon interaction in a specific material.

A direct method of measuring R or τ_o is in charge imbalance experiments, since phonon readsorption cannot cause a charge imbalance [3]. Dynes, et al. [9] have shown that Γ_R can be deduced from the lifetime broadening of the energy gap measured in a PbBi tunnel junction in equilibrium, but the technique is restricted to materials for which $\Gamma_R \sim \Delta(0)/\hbar$. Finally, measurements of Γ_R very close to T_c are not affected by phonon trapping because there are no special recombination phonons, i.e., N_ω , of energy $\hbar\Omega=2\Delta \gg k_BT$. Such measurements in aluminum films [10] obey the theoretical dependence calculated by Schmid and Schoen [11] :

$$\Gamma_R = 2.28\,(\Delta/k_BT)/\tau_o, \tag{8}$$

and the resulting τ_o agrees with that measured in charge imbalance experiments [3]. Note that Eq. 8 can be understood in a simple physical picture (see [1]).

<u>Phonon escape rate.</u> The difficulties associated with phonon trapping can be illustrated by considering the limit of Eqns. 1 and 2 when $\Gamma_{es} \ll \Gamma_B$ or $F_\omega \gg 1$. In that case, the measured Γ_R^* is just $\Gamma_{es}(\Gamma_R/\Gamma_B)$, where the ratio in brackets is wholly

determined by statistics. In other words, the information about the strength of the electron-phonon coupling is lost, and Γ_{es} is measured. Since Γ_{es} depends strongly on total film thickness, d, it is convenient to consider $\Gamma_{es}=\eta s/4d$ where s is the average phonon velocity and η is the intrinsic, average phonon transmission probability at the film-substrate boundary.

The above measurements of R and R* in aluminum films [6,10] lead to F_{ω}~30 for double junctions of total thickness of 90 nm on sapphire substrates. Such a large F_{ω} and the measured R* of 1.2 x 10^{-12} cm^3/s implies that $\Gamma_{es}=1.4 \times 10^8$ s^{-1}, assuming transverse phonons dominate. The resulting estimate of η is ~0.017, and this agrees suprisingly well with recent measurements [12] of PbBi films on sapphire substrates, which show η~0.018, if the transverse sound velocity for Pb is used.

Quasiparticle diffusion length. For nonuniform injection of quasiparticles, the diffusion term must be included in Eq. 1. For small deviations from equilibrium the concept of the quasiparticle diffusion length has been introduced [13]. Because of the small Γ_R* at low temperatures, quasiparticles diffuse by a random walk due to elastic collisions with impurities and defects and by inelastic scattering into other quasiparticle states before recombining. The average distance they move is given by

$$\Lambda^* = \sqrt{D/\Gamma_R^*}. \qquad (9)$$

Typical values are ~100 μm for Al at low temperature, and ~1 μm for Sn near T_c. Note that Λ* is not an intrinsic material property due to the dependence on defects and F_{ω}.

Energy dependent inelastic scattering rate. The total inelastic scattering rate for quasiparticles includes the recombination rate Γ_R and the rate of scattering into other quasiparticle states, Γ_S. Both of these have been shown by Kaplan, et al. [8] to be strongly energy and temperature dependent. The first experiment to measure this energy dependence was by Miller and Dayem [14] who used the same double junction as for measurements of Γ_R*. The idea is illustrated in the inset of Fig. 1. Quasiparticles injected into the middle film from the left can tunnel into the larger gap superconductor on the right, but only if they do not lose energy by inelastic scattering in the middle film. This small current in the right-hand junction, which adds to the current from the thermalized injected quasiparticles (curve labeled x1), is readily seen in the (+) curves when the current axis is greatly expanded, as in those labeled x10^2, x10^3 and x10^4. It has all the features of the x1 curve, but is smaller and

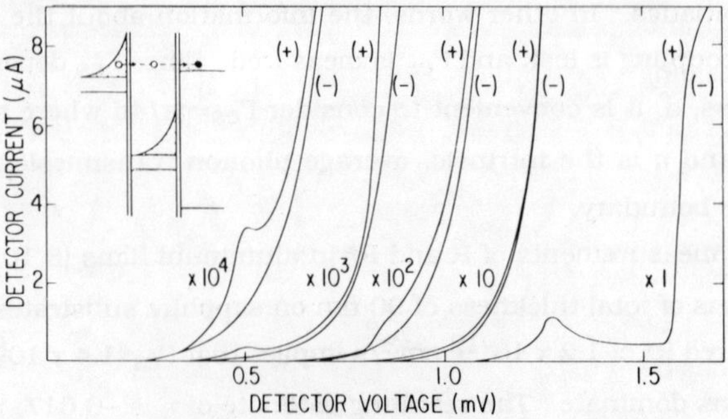

Fig. 1. The I(V) of the large gap junction for intense quasiparticle injection showing the polarity dependent excess current at low voltages (~0.5-1 mV) from which the inelastic scattering rate can be determined.

shifted to lower voltage by exactly the bias on the left-hand junction. It appears predominently in one polarity only because of the charge imbalance associated with the directly injected quasiparticles.

From these measurements [1], the energy dependent inelastic scattering rate, $\Gamma_I(E)=\Gamma_S(E)+\Gamma_R(E)$, and the elastic charge imbalance relaxation rate can be derived. Note that by using the difference between the (+) and (-) curves, phonon readsorption plays no role and it is $\Gamma_R(E)$ and not $\Gamma_R^*(E)$ that is relevant. The $\Gamma_I(E)$ behavior reflects the E^3 dependence expected for electron-phonon scattering due to the Ω^2 density of final (phonon) states, and the magnitude is close to theoretical expectations [8] based on the value of τ_o found above. Earlier measurements on more disordered aluminum films by Kaplan, et al. [15] show reasonably consistent results for the energy dependence but reflect the higher rate of electron-phonon scattering normally found in highly disordered materials [16].

FAR-FROM-EQUILIBRIUM

One of the biggest differences caused by a large perturbation from thermal equilibrium is the quasiparticle relaxation term in Eq. 1 which can no longer be linearized. Special cases can be solved analytically, but the important general case including diffusion and time dependence is insolvable. However progress has also been made by appealing to certain approximations.

Special cases. An exact solution was readily obtained by Rothwarf and Taylor [5] if both diffusion and time dependence are neglected. The result is:

$$N = \sqrt{N(T)^2 + I_o/R^*}, \tag{10}$$

and this equation has been used [6] to determine R* (see above).

If diffusion in one spatial dimension only is included, but time dependence neglected, then solutions related to the Weierstrassian functions can be obtained [17]. However the value of these is restricted to highly specialized experimental configurations.

Coupled quasiparticle-phonon propagation in bulk. In order to understand the propagation of nonequilibrium disturbances in bulk crystals of Pb, Narayanamurti, et al. [18] avoided the nonlinearity associated with the quasiparticle recombination by dealing with an effective quasiparticle temperature T_e instead of N:

$$\Delta T_e = T_e - T_b = \Delta N/\{\partial N(T)/\partial T\}. \tag{11}$$

The complete description includes separate heat flow equations for both phonons and electrons which are coupled through relaxation terms:

$$(\partial/\partial t - \alpha_e^2\nabla^2)T_e = -(T_e - T_p)\Gamma_R, \tag{12}$$

$$(\partial/\partial t - \alpha_p^2\nabla^2)T_p = -(T_p - T_e)\Gamma_B. \tag{13}$$

Here α_e^2 and α_p^2 are the thermal diffusivities of the electrons and phonons and T_e and T_p are their effective temperatures. The solutions are carried out analytically using Fourier transform techniques and are in good agreement with their experimental measurements. Such solutions are inadequate for thin films because they ignore phonon escape, however it would seem possible to add an appropiate phonon escape term, $-(T_p - T_b)\Gamma_{es}$, to their phonon heat flow equation (Eq. 13) and affect a solution in two dimensions.

Distribution function approximations. The quasiparticles obey Fermi-Dirac statistics in thermal equilibrium, and two simple *ad hoc* extensions for nonequilibrium situations have been proposed. In the first, an additional chemical potential μ^* is included for the quasiparticles[19]. In another, the energy distribution is characterized by a temperature T^* which is greater than the bath temperature [20]. Direct measurements of the distribution function in nonequilibrium films of aluminum [21] have shown that both must be included to find a reasonable approximate description of the nonequilibrium caused by intense tunnel injection of quasiparticles. Others [22] have found the T^* distribution is adequate for small injection rates due to laser pulses.

These approximations can be related to the simplification found by using Eq. 11 above. However, since there is no prescription to determine μ^* and T^*, these approximations are primarily useful when the energy distribution of quasiparticles is needed. Note that this is specifically neglected in Eqns. 1 and 2.

Numerical solutions. If the energy distribution of the quasiparticles can be neglected, then the full nonlinear Eqns. 1 and 2 can be numerically integrated to obtain useful results for specific experiments, and this seems the most appropiate approach for many particle detector applications. Although this represents no problem in principle, the wide range of time and distance scales makes the calculation time consuming even for modern computers. Given whatever initial conditions prevail, Eqns. 1 and 2 are used to calculate the changes ΔN and ΔN_ω, as a function of $\mathbf{r}$, for a time step Δt. The derivatives associated with the diffusion term in Eq. 1 are determined from the values of N at neighboring values of $\mathbf{r}$.

However, if the energy distribution is necessary, then the more complicated equations described in [23] must be numerically solved, together with the energy gap equation (see Eq. 14). Note that these are extensions of the kinetic equations considered extensively by Chang [24], except that variations with time and spatial coordinates must be included. These add enormously to a computer program which is already quite time consuming and may make the numerical solution intractable from a practical point of view.

Energy gap. The energy gap, Δ, for a BCS superconductor is determined by the following integration of the quasiparticle distribution function, f(E), times the reduced BCS density of states $\rho(E) \equiv N_S(E)/N_N(E)$:

$$\ln\{\Delta(0)/\Delta\} = \int_0^\infty dE\rho(E)f(E)/E, \qquad (14)$$

where $\Delta(0)$ is the energy gap at zero temperature which is $1.76k_BT_c$ for weakly coupled superconductors. In equilibrium, f(E) is the Fermi-Dirac distribution function and the BCS temperature dependence is found for $\Delta(T)$. For nonequilibrium situations, f(E) must be determined by one of the above methods to calculate Δ. For a study [1] of aluminum films subjected to intense tunnel injection of quasiparticles, the direct measurement of f(E) by tunneling gave a somewhat better Δ than the μ^*,T^* model.

THIN FILM PARTICLE DETECTOR

As an example of the application of nonequilibrium concepts to particle detectors, a description of the interaction of a minimum ionizing particle (MIP) with a thin film will be presented. Some prelimenary results have been given in a previous publication [23] for a 5-MeV α-particle. To avoid the complication of the large energy deposited in the substrate for α-particles, only MIP are considered here. As an example of the importance of various processes, consider a MIP passing through a 1 μm thick superconducting film in a direction normal to the film. It does this in $\sim 3\times10^{-15}$ s, creating high energy quasiparticles as primary excitations. However these excitations are not created uniformly in the superconducting film but rather along the one-dimensional MIP path. The initial electron kinetic energy will be quite large ($\gg k_BT_c$) in the region close to the particle path, and it is necessary to know the diffusion of energy away from this path, both laterally in the film plane and from the film into the substrate and onto the heat sink.

<u>Thermalization to $T_e=T_c$.</u> Time-resolved-spectroscopy has been used recently [25] to measure the thermal relaxation of hot electrons in metals surfaces resulting from ultra-fast laser pulses. After a time of a few ps the electrons exhibited a local equilibrium with the phonons at a somewhat elevated temperature. A slower relaxation process, dominated by heat diffusion, returned the system to ambient temperature. Allen [26] has calculated the electron-phonon relaxation process, which can be described by:

$$dT_e/dt = \Gamma_t(T_p-T_e), \qquad (15)$$

$$\Gamma_t = 3\hbar\langle\Omega^2\rangle\lambda/\pi k_B T_e , \tag{16}$$

where λ is the electron-phonon coupling constant used to calculate T_c. After this equilibrium between electrons and phonons, the energy scale is of order the Debye frequency, ω_D, and further thermalization to the much lower energies associated with superconductivity ($\sim \Delta \sim \omega_D/100$) cannot be described by Eqns. 15 and 16. However for both detector applications considered here, a knowledge of the time and length scales of the normal, "hot-spot" region is of great importance.

The maximum possible size of the hot-spot is found by balancing the energy deposited by the particle with the enthalpy difference between the superconducting state at T_b and the normal state, Δh:

$$r_c = \sqrt{|dE/dx| / \pi \Delta h}, \tag{17}$$

where $|dE/dx|$ is the energy loss per unit length for the particle (typical values range from $\sim 5\times10^{-10}$ j/cm for a 5-MeV α-particle to $\sim 2\times10^{-12}$ j/cm for MIP). However this gives no information about the time scale, and the dynamics are particularily important because the maximum hot-spot size cannot be realized if the excess excitations diffuse further than r_c before thermalization to energies of order $\Delta \sim T_c$. The reason for this is that high energy electronic excitations have relatively little effect on the energy gap (see Eq. 14). In addition, a full dynamical solution gives the time scale of the hot-spot which is a crucial consideration for fast time resolution.

In view of the previously mentioned difficulties of affecting a complete analytical or numerical solution, a different, simplified approach is considered. During the thermalization from ω_D to Δ, the excess energy exchanges many times between electrons and phonons. A great calculational simplification comes by fixing the energy exchange at each process by its average value. For the Debye phonon spectrum, the average phonon energy emitted by an electron of energy E is 3E/4, with the consequent reduction of the electron energy to E/4. When phonons of frequency ω scatter electrons, two excitations are created whose average energies are $\omega/2$. Thus from one phonon gerneration to the next, the energy decreases as $\omega_{n+1}=3\omega_n/8$. Since $\omega_D \sim 100\Delta$, it takes only about 5 phonon generations to achieve average energies of $\sim\Delta$.

In order to be quantitative, the magnitudes and ratios of the phonon-electron and electron-phonon scattering rates at energy E are needed from [8]. (The latter can be thought of as the ratio of the electron to phonon specific heats at a temperature of E/k_b). For Nb, the ratio[8] is approximately one for $E \sim \Delta$, and decreases greatly at higher energy. Therefore the time scale is determined predominantly by the phonons, and is of order 10 psec. During this time the electrons will diffuse about 0.2 μm for a diffusion constant of 25 cm^2/sec, and the last phonon generation will travel furthest, but only about 0.01 μm. However for Al, the ratio of scattering rates is significantly greater than one for $E \sim \Delta$, due to the large ratio of ω_D/Δ, and thus the low energy electrons determine both the time and distance scales. The time scale is of order 2 nsec which is much longer then Nb due to the weak electron-phonon coupling in Al. For the same diffusion constant the electrons diffuse about 2 μm and the last phonon generation moves only about one-third of that ballistically.

For comparison, the values of r_c determined from Eq. 17, assuming a MIP and a reduced temperature of 0.9, are 0.05 and 0.5 μm respectively for Nb and Al. Thus a smaller diffusion constant, i.e., shorter electron mean-free-path, may be necessary in this example to take full advantage of the deposited energy to produce a large normal region, i.e., a hot-spot. Note that in the presence of fields or transport currents the effective T_c, and therefore Δh can be greatly reduced, leading to larger r_c.

A final important comment. Because of the very short ballistic propagation distance of phonons and the low phonon escape probability, phonon escape can be reasonably neglected during the above stages of thermalization.

Numerical solution below T_c. It can be reasonably supposed that the initial state is given by a Gaussian distribution of quasiparticles around the particle path with a maximum value of $N(T_c)$ at r=0, and a characteristic width given by 2 r_c Since the proper calculation involving the full kinetic equations is impractical, several assumptions will be made in order to affect a tractable solution; these show how the general trends of the solutions depend on various parameters, including τ_o, Δ, Γ_{es}, etc. First, assume the substrate is in thermal equilibrium with the bath. Although this is not very realistic for α-particles, it is acceptable for completely absorbed radiation like x-rays or for MIP if the substrate thermal conductivity is much better than the film. Next the energy dependence of N is neglected so that Eqns. 1 and 2, rather than the full kinetic equations, need to be numerically integrated.

It is instructive to first find the results for uniform injection, i.e., neglecting diffusion. This calculation has been made using parameters which are relevant to ~50 nm films of aluminum on sapphire substrates, and a reduced temperature of T_b/T_c~0.4. At this temperature, $N(T_b)$ is about 1/40 times $N(T_c)$. Initially the rate of drop in N is high since $\Gamma_R^*=R^*N$ is large, but for longer times, N approaches $N(T_b)$ and the rate approaches the equilibrium value of $2R^*N(T_b)$, where the factor of 2 reflects the pair-wise recombination, i.e., the distribution changes at twice the rate of the individual quasiparticles (see [6]).

If diffusion is included the computer calculation outlined above requires a significantly longer time due to the wide range of distance scales needed. However, a feeling for the nature of the solution can be obtained more simply. Again using the example of aluminum films, the distance scale r_c was found to be about 0.5 μm for a MIP because of the low specific heat at T_c of about 1.4 K. Assuming a diffusion constant, D, of 25 cm^2/sec, the diffusion radius becomes ~5 μm after a time of only $1/1000\Gamma_R^*$, during which less than 0.3% of the initial quasiparticles have recombined. However, during this time diffusion has diluted the excess quasiparticle density, ΔN, by almost 100-fold and the recombination rate becomes approximately the thermal equilibrium one (i.e., $2R^*N(T_b)$ in this example. Thus the diffusion is so fast that the high initial recombination rate is never realized.

For higher T_c materials like Sn and Nb, as was calculated above, r_c is smaller and Γ_R^* larger. These have opposite effects on the diffusion radius calculation, but do not cancel entirely. Estimates for Nb indicate a larger fraction of the quasiparticles may be lost during the initial diffusion period.

In conclusion, in the limit that very few quasiparticles are lost during the initial diffusion period when Γ_R^* is large, diffusion can be dealt with very simply.

PARTICLE POSITION DETECTOR

The primary purpose of these detectors is to locate in space the position of a particle path with high spatial and temporal resolution. In some applications the particle may be adsorbed in the detector, e.g., x-rays. Superconducting detectors consisting of Josephson tunnel junctions have been proposed in [23], and thin film strip detectors in [27]. Switching times of 10 ns have been observed in thin film strips and Josephson junctions can switch faster than 100 ps. In each case the interaction of the particle heats the superconductor locally, thereby reducing the critical current, I_c, to a value below the applied current, I. Because of inevitable variations in I_c, due to inhomogeneities in the junction barrier or the film thickness and/or width, the value of I cannot be too close to I_c. Thus the perturbation of the particle must be a reasonable fraction of the junction area or the film strip width. It is clear that in order to achieve switching for a given $\varepsilon=1-I/I_c$, the strip width can be larger than the linear dimension of a square Josephson junction by a factor of $\sqrt{4/\pi\varepsilon}\sim3.6$ for $\varepsilon=0.1$. A discussion of thin film strip detectors can be found elsewhere in this book [28].

In addition to inhomogeneities in the tunnel barrier, thermal fluctuations can cause premature switching in Josephson junctions. If I/I_c must be well below one, say 0.9, due to barrier inhomogeneities, then the fluctuation switching has been shown not to be a problem[29]. Careful analysis for the case of $T_b=1.3$ K, $r_c=0.25$ μm and a junction area of $10\pi r_c^2$has shown that fluctuation switching will be the equivalent to a flux of 100 particles/cm^2/s for a Josephson critical current density of 1500 A/cm^2. For 2500 A/cm^2 the equilivalent flux is perhaps ten orders of magnitude lower. Thus uniform barriers should be the major concern toward increasing the junction size.

TUNNEL JUNCTION PARTICLE ENERGY DETECTOR

There has been considerable interest in this possibility since the pioneering work of Wood and White [30]. The hope is to improve the energy resolution over existing semiconducting devices since the superconducting gap is much smaller and thus more excitations can be created for a given particle energy. Recent work has realized this possibility [31,32]. See also other papers in this book.

In this detector application the tunnel junction is generally between two superconducting films and is biased at a voltage below the sum of the two energy gaps. For the case of two identical superconducting films, the current, I(T), in thermal equilibrium is proportional to N(T). In a nonequilibrium situation the current will depend not only on N, but also the energy distribution of quasiparticles, f(E). The detailed equations which cover this dependence are given in [23]. However, as noted above, the calculation of f(E) as a function of time and spatial location is not practical, so again in this case some simplifying assumptions are needed. The easiest is to ignore the energy dependence and use the numerical solutions for $N(\mathbf{r},t)$ to obtain ΔI by integrating $N(\mathbf{r},t)$ over space and time.

Note that if few quasiparticles are lost before rapid diffusion occurs, then only the time integral is needed (see above). For example, because the total number, not density, of quasiparticles created by the particle interaction diffuses so rapidly, the density $N(\mathbf{r},t) \sim N(T_b)$ for all $\mathbf{r}$ and t, and the signal seen from the tunnel junction decays exponentially with a rate $2\Gamma_R^*(T_b)$ from an initial value of $\Delta I(0)=\pi r_c^2 I(T_b)/A$, where A is the junction area. Thus the integrated charge is $\pi r_c^2 I(T_b)/2A\Gamma_R^*(T_b)$.

Experimental considerations. Since the quasiparticles which are created by the interaction will still diffuse a distance of order Λ^* before recombining, those ionizing particles which interact within a distance Λ^* of the junction edge can have a different sensitivity. Unless the film is terminated there, the quasiparticles can leak out. Use of an island geometry [33] is helpful to minimize this effect.

Another consideration which can effect the resolution is that of nonuniformities in the junction barrier, the phonon escape rate or the film thickness. This can be seen by writing down the additional junction current, ΔI, due to the particle interaction[1]:

$$\Delta I = \int \Delta I(\mathbf{r})\, d\mathbf{r}, \tag{18}$$

$$\Delta I(\mathbf{r}) = 2e\Gamma_T(\mathbf{r})N_p(\mathbf{r}), \tag{19}$$

$$\Gamma_T(\mathbf{r}) = g_{nn}(\mathbf{r})/4ed(\mathbf{r})N(0), \tag{20}$$

where $\Gamma_T(\mathbf{r})$ is the average tunneling rate for a quasiparticle, $N_p(\mathbf{r})$ is the number of quasiparticles per unit of film area at $\mathbf{r}$ (the two-dimensional coordinate in the film plane) created by the particle interaction, $g_{nn}(\mathbf{r})$ is the normal conductance of the junction per unit area, N(0) is the density of states at the Fermi energy and the time dependence has been suppressed since it is irrelevant to this arguement. Now $g_{nn}(\mathbf{r})$ depends on the local barrier thickness and height, $\Gamma_T(\mathbf{r})$ also depends on the local film thickness $d(\mathbf{r})$, and $N_p(\mathbf{r})$ depends on the local phonon escape rate through $\Gamma_R^*(\mathbf{r})$ and Eqns. 4 and 5. Note that N_p is inversely related to $\Gamma_R^*(\mathbf{r})$, so the additional current per unit area in Eq. 19 is seen to depend on the ratio $\Gamma_T(\mathbf{r})/\Gamma_R^*(\mathbf{r})$. Note that since $\Gamma_R^*(\mathbf{r})$, is proportional to $1/d$ in the usual case of $F_\omega \gg 1$, ΔI is independent of d in that limit.

However in the usual case of two active films in the tunnel junction the situation is far more complicated (see [34]). In either case, the total signal will be proportional to the local film thickness at the path of particles which traverse the entire film since the energy deposit is $|dE/dx|d$.

Quasiparticle trapping. In order to increase the interaction probability very thick films are sometimes required. An ingenious technique to enhance the measured signal has been proposed by Booth [35]. He shows that the sensitivity can be increased by up to a factor of 100 or more by trapping the quasiparticles in another superconductor of lower gap, on which the tunnel junction is formed. The idea is that quasiparticles from the large gap superconductor diffuse into the small gap region where they can lose some energy by scattering. It is then energetically impossible for them to diffuse back to the large gap superconductor and they are effectively trapped. Because the gap is smaller, the number of quasiparticles is greater than originally created in the large gap superconductor. While one does not necessarily create more excitations than if the small gap superconductor were directly irradiated, the excess quasiparticles are trapped in a potentially smaller volume to enhance the tunneling probability (see Eq. 20). Booth [35] has also proposed cascading a series of tunnel junctions to achieve active gain. After tunneling into the large gap part of each electrode, a gain in quasiparticle number occurs upon diffusing into the small gap region, which is positioned for tunneling into the next composite electrode. The energy is supplied by the voltage bias.

Superconducting transistor. A second method to achieve active gain does not require additional junctions. If the tunnel junction has a sufficiently high g_{nn}, then Γ_T can be larger than Γ_R^*, and each quasiparticle created by the particle interaction can cause more than one electron to cross the tunnel barrier in a cyclical process described in [36]. Values of gain up to 4 have been measured in aluminum double junction devices [36]. Unfortunately, Γ_T cannot be indefinitely large and is limited to $\sim\Gamma_I$ (see below), and the ratio Γ_I/Γ_R is predicted [8] to be ~2 for typical transistor operation, so that the maximum gain [1] of this transistor effect is $\sim\Gamma_I/2\Gamma_R^*=F_\omega$.

Maximum tunneling conductance. In a tunnel junction, the tunneling rate, Γ_T, can never be greater than the total inelastic scattering rate, Γ_I, because a second quasiparticle cannot tunnel into an occupied state [1]. This idea is consistent with the experimental observation of a maximum g_{nn} for tunnel junctions which scales with $1/\tau_0$ for various materials. For example, in 30 nm thick aluminum films, if

$g_{nn}>3\times10^5$ S/cm^2, the current-voltage characteristic, I(V), departs from ideal behavior. At low temperatures this corresponds to $\Gamma_T\sim10^7$ s^{-1} which is similar to the Γ_I predicted by [8] for the characteristic time, τ_o, found experimentally to be 90 ns. Note that for thicker films or for smaller junction areas, such that diffusion out of the junction occurs, higher values of g_{nn} are possible for Al. In the case of tin junctions, good I(V) are found for $g_{nn}\sim10^7$ S/cm^2, reflecting the smaller τ_o (~2.3 ns), while in Nb, for which $\tau_o\sim0.15$ ns, values of $\sim10^8$ S/cm^2 have been achieved.

ACKNOWLEDGEMENTS

This work supported by the U.S. Department of Energy, Division of Basic Energy Sciences-Materials Sciences under contract #W-31-109-ENG-38.

REFERENCES

[1] K.E. Gray, in Nonequilibrium Superconductivity, Phonons and Kapitza Boundaries, edited by K.E. Gray, (Plenum, New York, 1981), p. 131.

[2] Nonequilibrium Superconductivity, Phonons and Kapitza Boundaries, edited by K.E. Gray, (Plenum, New York, 1981); NonequilibriumSuperconductivity, edited by D.N. Langenberg and A.I. Larkin, (Elsevier, Amsterdam, 1986).

[3] J. Clarke, in Nonequilibrium Superconductivity, Phonons and Kapitza Boundaries, edited by K.E. Gray, (Plenum, New York, 1981), p. 353; and Nonequilibrium Superconductivity, edited by D.N. Langenberg and A.I. Larkin, (Elsevier, Amsterdam, 1986), p. 1.

[4] B.N. Taylor, Thesis, Univ. of Pennsylvania, 1963.

[5] A. Rothwarf and B.N. Taylor, Phys. Rev. Lett. **19**, 27 (1967).

[6] K.E. Gray, J. Phys. F:Metal Phys. **1**, 290 (1971).

[7] S.B. Kaplan, J. Low Temp. Phys. **37**, 343 (1979).

[8] S.B. Kaplan, C.C. Chi, D.N. Langenberg, J.J. Chang, S. Jafarey and D.J. Scalapino, Phys. Rev. **B14**, 4854 (1976).

[9] R.C. Dynes, V. Narayanamurti and J.P. Garno, Phys. Rev. Lett. **41**, 1509 (1978).

[10] I. Schuller and K.E. Gray, Phys. Rev. Lett. **36**, 429 (1976).

[11] A. Schmid and G. Schoen, J. Low Temp. Phys. **20**, 207 (1975).

[12] R. Gross, U. Klass and H. Bender, Extended Abstracts of 1987 International Superconductivity Electronics Conference, Tokyo, Japan, August 28-29, 1987, p. 345.

[13] A.B. Pippard, J.G. Sheperd and D.A. Tindall, Proc. Roy. Soc. (London) **A324**, 17 (1971).

[14] B.I. Miller and A.H. Dayem, Phys. Rev. Lett. **18**, 1000 (1967).

[15] S.B. Kaplan, J.R. Kirtley, and D.N. Langenberg, Phys. Rev. Lett. **39**, 291 (1977).

[16] A. Schmid, Z. Physik **271**, 251 (1974).

[17] K.E. Gray, Thesis, Univ. of Cambridge, 1969.

[18] V. Narayanamurti, R.C. Dynes, P. Hu, H. Smith and W.F. Brinkman, Phys. Rev. **B18**, 6041 (1978).

[19] C.S. Owen and D.J. Scalapino, Phys. Rev. Lett. **28**, 1559 (1972).

[20] W.H. Parker, Phys. Rev. **B12**, 3667 (1975).

[21] H.W. Willemsen and K.E. Gray, Phys. Rev. Lett. **41**, 812 (1978).

[22] A.D. Smith W.J. Skocpol and M. Tinkham, Phys. Rev. **B21**, 3879 (1980).

[23] A. Barone, S. de Stefano and K.E. Gray, Nucl. Inst. and Meth. **A235**, 254 (1985).

[24] J.J. Chang, in Nonequilibrium Superconductivity, Phonons and Kapitza Boundaries, edited by K.E. Gray, (Plenum, New York, 1981), p. 263.

[25] R.W. Schoenlein, W.Z. Lin, J.G. Fujimoto and G.L. Eesley, Phys. Rev. Lett. **58**, 1680 (1987).

[26] P.B. Allen, Phys. Rev. Lett. **59**, 1460 (1987).

[27] N.K. Sherman, Phys. Rev. Lett. **8**, 438 (1962); E.C. Crittendon, Jr. and D.E. Spiel, J.Appl. Phys. **42**, 3182 (1971).

[28] R.G. Wagner and K.E. Gray, this book.

[29] P. Silvestrini, O. Liengme and K. E. Gray, to appear Phys. Rev. **B**, (1987).

[30] G.H. Wood and B. White, Appl. Phys. Lett. **15**, 237 (1969).

[31] D. Twerenbold, Europhys. Lett. **1**, 209 (1986), and this book.

[32] H. Krauss, T. Peterreins, F. Probst, F.v. Feilitzsch, R.L. Mossbauer and E. Umlauf, Europhys. Lett. **1**, 161 (1986), and this book.

[33] A. Barone, G. Darbo, S. De Stefano, G. Galinaro, A. Siri, R. Vaglio and S. Vitale, Nucl. Instr. and Meth. **A234**, 61 (1985).

[34] D. Twerenbold, Phys. Rev. **B34**, 7748 (1986).

[35] N.E. Booth, Appl. Phys. Lett. **50**, 293 (1987), and this book.

[36] K.E. Gray, Appl. Phys. Lett. **32**, 392 (1978).

SUPERCONDUCTING DETECTORS FOR NUCLEAR PHYSICS, PARTICLE PHYSICS AND ASTROPHYSICS

N. E. BOOTH

Department of Nuclear Physics,
University of Oxford,
Oxford, OX1 3RH, U.K.

1. INTRODUCTION

Some ideas on configurations of superconducting detectors for particle tracking and calorimetry are presented. The quasiparticle trapping mechanism is discussed and how it can be implemented in bulk detectors. Recent results on the fabrication and performance of single-crystal indium detector prototypes are presented. Possible applications of the quasiparticle multiplier are discussed.

One might well ask why superconducting particle detectors are of any interest. Working in the necessary cryogenic environment is expensive, laborious, and demanding. There are at least two areas where the extra effort is worth it. One of these areas encompasses the special cases where very good energy resolution is required. The smallness of the superconducting energy gap has the result that it takes only about $\epsilon = 10^{-3}$ eV to produce an excitation [1] whereas it takes about 1 eV in a semiconductor and 25 eV in a gas. The energy resolution is proportional to $\epsilon^{1/2}$ and in principle is more than an order-of-magnitude better in a superconductor as compared with any conventional device. Another important property of superconductors is that they span a wide range of atomic number.

Thus, for example, superconducting Sn with $Z = 50$ could be used for X-ray astronomy; a thickness of only about 10 μm would have excellent efficiency up to about 10 keV X-ray energy. For γ-ray spectroscopy and low-level γ-ray detection the high photo-peak efficiency and good energy resolution potentially provided by high atomic number superconductors is exceedingly attractive. Table 1 shows a comparison of some superconductors with the widely-used Ge detector.

Table 1

Comparison of some superconductors to Ge for γ-ray detection.

	Ge	Nb	Ta	Pb
Atomic number	32	41	73	82
Atomic weight	72.60	92.91	180.95	207.21
Density, g cm^{-3}	5.32	8.57	16.6	11.3
Superconducting transition temp., K	–	9.25	4.48	7.20
Debye temp., K	400	250	230	88
Linear photo-electric absorption coefficient at 511 keV, cm^{-1}	0.020	0.077	0.89	0.91
Energy gap, meV	670	1.53	1.2	1.37
ϵ, meV	2,960	2.6[a]	2.0[a]	2.3[a]
Fano factor [b]	0.12	0.2[a]	0.2[a]	0.2[a]
Resolution (FWHM) at 511 keV, keV	1.0[c]	0.038[b]	0.034[b]	0.036[b]

[a] Calculated. See ref.1.

[b] Energy resolution = $2.35\,(\epsilon F E)^{\frac{1}{2}}$ where F is the Fano factor.

[c] Typical value.

A second area where superconducting detectors offer a solution is the need for large mass, totally sensitive detectors for solar neutrinos and for the hypothetical weakly-interacting massive particles (WIMPS) which are perhaps responsible for the dark matter problem. Here the requirement apart from large mass is sensitivity to very small energy depositions. There appear to be two possible methods to achieve this using superconductors. One is to use the quasiparticles produced in a bulk superconductor. This is discussed in detail later for the specific case of indium, but the principles apply to other superconductors. Another way is to detect the phonons produced in a single-crystal insulator or semiconductor (or possibly even a superconductor) with superconducting transition-edge bolometers or superconducting tunnel junctions. Arrays of series-connected junctions of a low-gap superconductor (e.g. Al) offer the possibility of large area coverage and high efficiency for detection of the phonons.

Turning now to accelerator particle physics we might consider the hypothetical high energy physics experiment of the future shown in Fig.1. One storage ring with superconducting magnets is already in operation and others are planned. There are several superconducting solenoid magnets in operation. What are the possibilities for making the other items in Fig.1, the electronics, the tracking detector, and the calorimeter, also with superconductors? Josephson junction logic circuits and very fast ADC's and also superconducting transmission lines have been produced but with little application so far [2]. SQUIDs have been used as amplifiers, and another amplifier, the quasiparticle multiplier, will be discussed later. The superconducting computer may never be commercially viable, but superconducting electronics could well play an important role in high energy physics in the future.

The traditional tracking detector, the gaseous proportional drift chamber, has been supplemented in recent years by silicon strip detectors which have better spatial resolution and act as vertex detectors. There are two problems with these; the silicon detectors are thicker than one would like from the point-of-view of multiple Coulomb scattering, and their performance deteriorates with radiation dose [3]. Superconducting strips or series-connected arrays of tunnel junctions can have spatial resolution limited only by state-of-the art lithography and the number of electronic channels one is prepared to pay for. They are also much more radiation resistant. The superconductors can be very thin; if they are on a substrate similar in thickness to the silicon detectors then they will also be sensitive to the phonons produced by the passage of a particle through the substrate. Otherwise there is the possibility of depositing the superconductor on a much thinner substrate, such as a film of kapton.

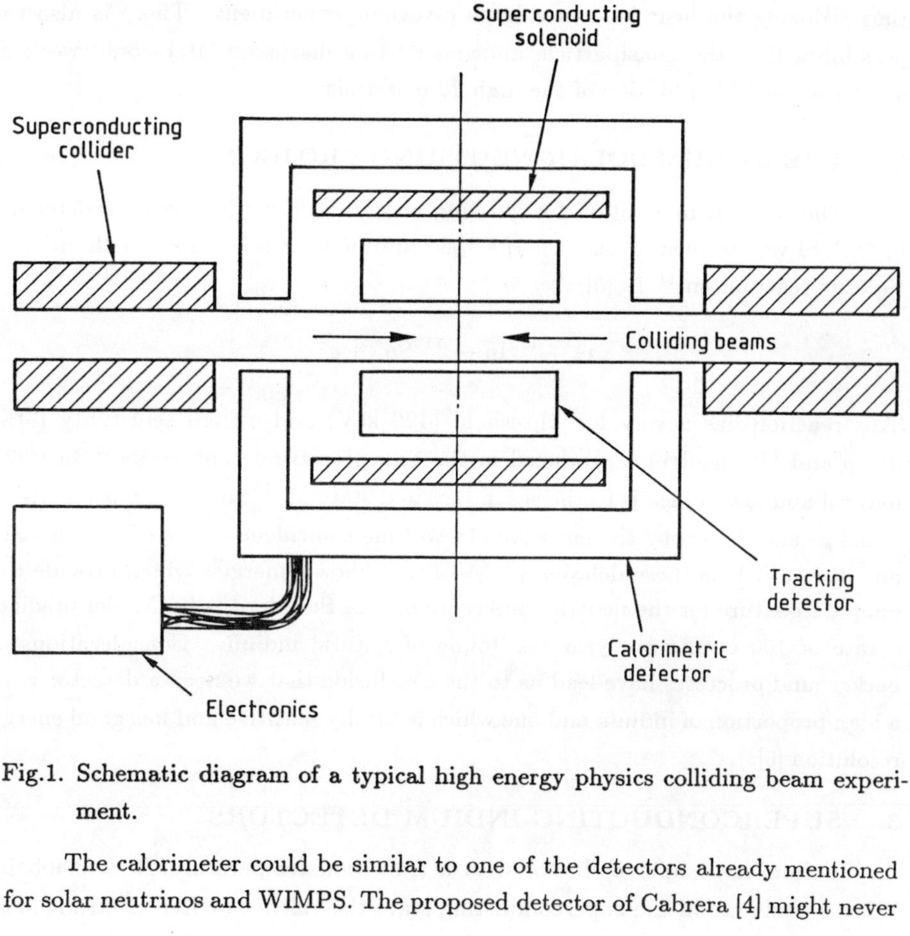

Fig.1. Schematic diagram of a typical high energy physics colliding beam experiment.

The calorimeter could be similar to one of the detectors already mentioned for solar neutrinos and WIMPS. The proposed detector of Cabrera [4] might never succeed in detecting solar neutrinos because of background problems, but I see it as the natural successor to the solid-state drift chamber.

Before I stop speculating and get back to reality let me mention one or two things about the new high-temperature superconductors. My first reaction is one of disgust because the new materials have, at least temporarily, diverted the efforts of the Cambridge part of our collaboration. On the positive side they offer the possibility of superconducting transmission lines between 4 K and 77 K, thus reducing the heat load in a large cryogenic experiment. There is also the possibility that the quasiparticle multiplier to be discussed later could work at 4 K by using NbN plus one of the high T_c materials.

2. THE INDIUM SOLAR NEUTRINO PROJECT

The original motivation for working on superconducting particle detectors in Oxford was to measure the energy spectrum of solar neutrinos via the inverse β-decay reaction on ^{115}In [5],

$$\nu_e + {}^{115}\mathrm{In} \rightarrow {}^{115}\mathrm{Sn}^* + e^- .$$

This reaction has a very low threshold (128 keV) and a high sensitivity t the $p-p$ and ^{7}Be neutrinos produced in the Sun. Neutrino capture in ^{115}In (96% natural abundance) leads to the second excited state of ^{115}Sn which has a half-life of 3.3 μs and decays by the emission of two time-coincident γ-rays of energies 116 and 497 keV. It is these delayed γ-rays of well-known energies which provide the unique signature for the neutrino interaction. The Standard Solar Model predicts a rate of 100 events per year per tonne of natural indium. Considerations of background processes have lead us to the conclusion that we need a detector with a high proportion of indium and one which is totally sensitive and has good energy resolution [6].

3. SUPERCONDUCTING INDIUM DETECTORS

Indium is a superconductor with a transition temperature $T_c = 3.408$ K. When metals are in the superconducting state the valence or free electrons near the Fermi level (energy E_F) are bound in Cooper pairs (binding energy $2\Delta(T)$) due to the special nature of the interactions between electrons and phonons. At any temperature $T < T_c$ Cooper pairs are occasionally broken by phonons, and the excitations, called quasi-electrons and quasi-holes, or collectively quasiparticles, can recombine and emit a phonon. The number-density of quasiparticles $N(T)$ is proportional to the Boltzmann factor $\exp(-\Delta(T)/kT)$ and so is strongly temperature dependent. The recombination rate follows a similar dependence.

What happens when an ionizing particle, such as an electron from a neutrino interaction, deposits its kinetic energy in a superconductor? The processes involved are shown schematically in Fig.2 [7]. In the first 10^{-12} s many electrons are excited to energies well above E_F. Within this time they lose energy by interacting with other electrons and by radiating numbers of phonons. Those phonons with energy $\Omega \geq 2\Delta(T)$ interact mainly by breaking Cooper pairs. After about 10^{-10} s there remains an excess of quasiparticles, of order 10^9 per MeV of energy deposition, due to Cooper-pair breaking, and an excess of phonons of energy $\Omega < 2\Delta(T)$.

In a pure superconductor the recombination lifetime can be made long by reducing $N(T)$, i.e. by operating at a temperature well below T_c. To give an idea of the magnitude of the excitation caused by an ionizing interaction let us consider an indium crystal of volume 1 cm^3 at 0.3 K, a temperature easily achievable with a dilution refrigerator; an energy deposition of 1 MeV, creating 10^9 excess quasiparticles, will double the number density. This is equivalent to raising the quasiparticle temperature by 10 mK in this 1 cm^3 volume.

The best way to detect such an excitation is to use a superconducting tunnel junction. This consists of the detecting film or crystal, a tunnel barrier consisting of a thin insulating layer (often the natural oxide), and a counter-electrode film of the same or a different superconductor. We [5] and other groups [8–12] have made such devices using superconducting thin films, and detected pulses due to α-particles and X-rays.

To do the solar neutrino experiment we must have the indium in pieces of

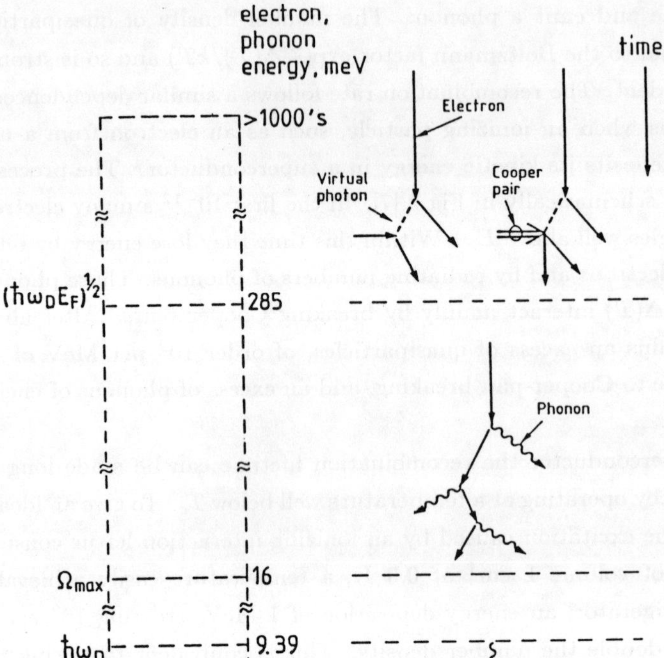

Fig.2. Schematic representation of the energy scale and processes involved after the interaction of a nuclear particle with a superconductor.

bulk material, at least 1 g each, and not in the form of thin films. Moreover we need sufficiently good time resolution to resolve the electron emitted in the neutrino capture process from the delayed 116-keV γ-ray which will almost always interact in the same piece. Further, we need to solve the technical problems of making junctions on the surface of bulk indium, electrical contacts, etc.

In Fig.3(a) let S_1 represent a piece of indium of volume V with a junction of area A_J on one surface. We will assume that the volume of the counterelectrode S_3 is negligible compared to that of the single crystal detector S_1, so that all the ionizing events of interest break Cooper pairs in S_1. For simplicity we assume however that S_3 is the same superconductor as S_1. An ionizing event at $t = 0$ creates n_0 excess quasiparticles via Cooper-pair breaking. These excess quasiparticles although created locally will diffuse rapidly in high-purity single crystal material, and after a few transit times across the crystal will be uniformly distributed throughout the volume V. For the present discussion we will not consider the signal current during these first few transit times since it depends on the location of the ionizing event, on the geometry, and on the quasiparticle diffusion coefficient, and requires a detailed calculation. With these simplifications the operation of the detector may be envisaged by analogy with the radioactive decay scheme shown in Fig.3(b). Here $n_1(t)$ is the excess number of quasiparticles in S_1, $n_d(t)$, the number which have tunnelled (rate constant λ_1) and $n_l(t)$ the number which have been lost due to recombination (rate constant $2\lambda_{1R}$, the factor 2 being due to the fact that 2 quasiparticles are lost in a recombination event) and due to out-diffusion into the electrical lead to S_1 (rate constant λ_{1D}). The differential equations corresponding to Fig.3(b),

$$\frac{dn_1}{dt} = -(\lambda_1 + \lambda_{1D} + 2\lambda_{1R})n_1 \qquad (1)$$

and

$$\frac{dn_d}{dt} = \lambda_1 n_1 \qquad (2)$$

are easily solved to give the signal current

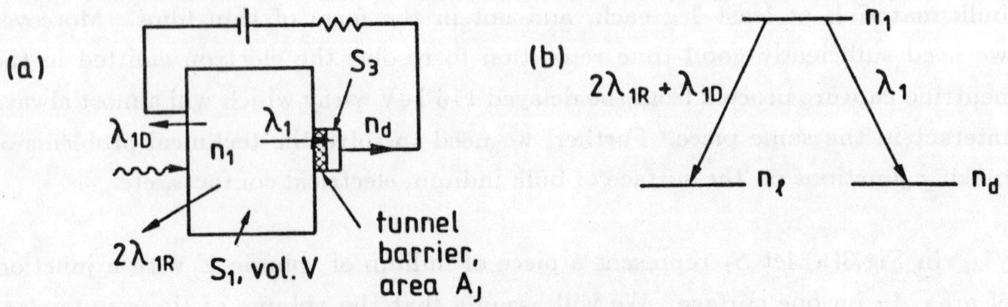

Fig.3. Tunnel junction on the surface of a superconducting crystal of volume V, (a) schematic arrangement, (b) radioactive decay analogue.

$$i_s = |e|\frac{dn_d}{dt} = \lambda_1 n_0 \exp(-\Lambda_1 t) \tag{3}$$

where

$$\Lambda_1 = \lambda_1 + 2\lambda_{1R} + \lambda_{1D} \tag{4}$$

The total charge that can be collected is

$$q(\infty) = |e|n_0\lambda_1/\Lambda_1 \tag{5}$$

For good operation we want $\lambda_1 \gg \lambda_{1D}, \lambda_{1R}$. The out-diffusion rate can be minimized by geometrical considerations of the electrical contact to S_1. We will show later how it can be totally eliminated. The recombination rate can be made arbitrarily low by operating at a low temperature, but care must be taken to avoid any normal-metal regions or magnetic impurities.

The tunnelling rate constant λ_1 is given by

$$\lambda_1 = \frac{1}{2e^2 N_1(0)}\frac{G_{nn}}{V} \tag{6}$$

where $N_1(0)$ is the single-spin density of states at the Fermi surface of S_1 (13.6×10^{21} eV^{-1} cm^{-3} for In) and G_{nn} is the normal state conductance of the junction. Unfortunately the signal current is inversely proportional to the volume of the crystal. Making A_J large increases the junction capacitance C_J ($\simeq 4\mu$F cm^{-2} for indium oxide) and this increases the signal rise-time. The factor to optimize is the specific conductance g_{nn}. The highest value achieved experimentally is about $10^8\ \Omega^{-1}\,\mathrm{cm}^{-2}$. Inserting this and $A_J = 1\,\mathrm{mm}^2$ into (3) gives, for $n_0 = 10^9$ (about 1 MeV energy deposition) in a crystal of volume 1 cm^3, $i_s = 0.04\,\mu$A. Assuming we have a load $R_L = 50\ \Omega$ and that the dynamic resistance of the junction, R_D, is much larger than R_L we expect a maximum signal voltage of 2 μV. This is very small, but is detectable with a reasonable signal-to-noise ratio with low-noise semiconductor electronics. However, the latter sets a limit on the energy resolution which can be achieved. Moreover, the high value of g_{nn} which we have used in the calculation may not be routinely achievable. Although the 10^9 excess quasiparticles may represent a large deviation from thermal equilibrium, the basic problem is that the tunnel junction acts like a semi-permeable membrane—it takes a long time for the signal to get out, 4 ms in this example, and consequently the magnitude of the signal current is very small.

As a result of thinking about how to circumvent this problem we came up with the idea of quasiparticle trapping [13]. Suppose that between the indium crystal and the junction we have a small volume v of another superconductor S_2 (in practice aluminium) which has a smaller gap than indium. Over the distance of a few superconducting coherence lengths the indium gap Δ_1 will change to the aluminium gap Δ_2 as shown in Fig.4. Then quasiparticles created in the indium can diffuse into the aluminium where they can lose energy by phonon emission and drop down to the gap edge of aluminium. They now have insufficient energy to get back into the indium and are trapped in the smaller volume of aluminium where they hit the tunnel barrier much more frequently, thereby increasing the signal current. The additional rate constants λ_2 (for trapping), and λ_3 (for tunnelling out of the trap) are introduced in Fig.5(a). Figure 5(b) shows the radioactive decay analogue. The trap opens up another decay channel which can be much faster than the direct process with rate constant λ_1. We have also allowed for recombination in S_2 via λ_{2R}. How well such a device works depends upon the scattering time $\tau_s(E/\Delta_2)$ for a quasiparticle of energy E ($\approx \Delta_1$) in S_2 to relax to close to the gap edge of S_2 by phonon emission. The energy dependence is known theoretically [14] and several measurements have been made for aluminium [15,16]. From these results we can estimate $\tau_s = 5$ ns for the indium-aluminium combination.

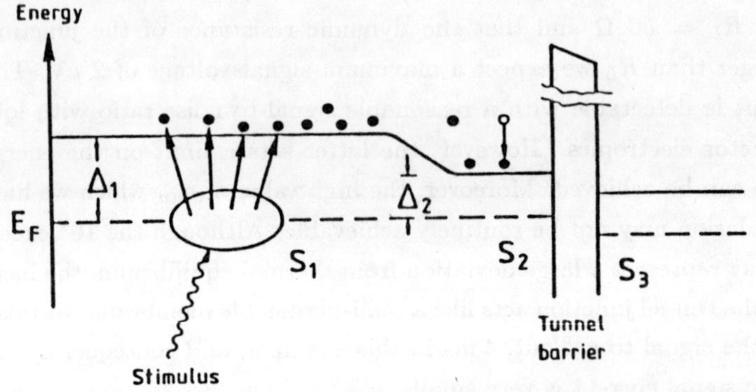

Fig.4. Energy diagram showing trapping of excess quasiparticles produced in superconductor S_1 by an adjacent superconductor S_2 of lower gap.

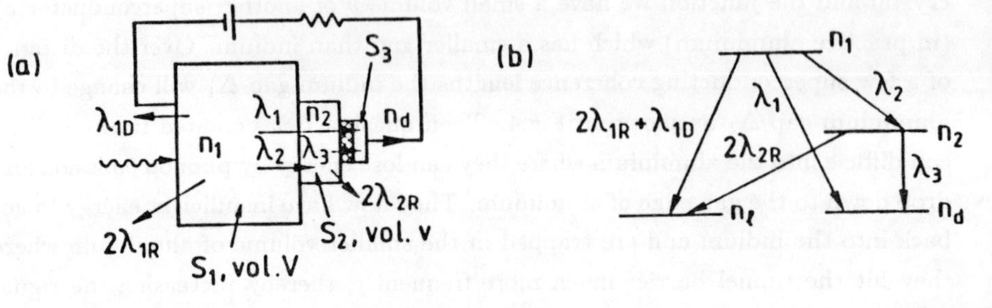

Fig.5. Tunnel junction detector incorporating a trap, (a) schematic arrangement, (b) radioactive decay analogue.

Assuming that the fraction of time the excess quasiparticles spend in S_2 is just the volume fraction $v/(V+v) \simeq v/V$ we have

$$\lambda_2 = \frac{v}{V} \frac{1}{\tau_s(\Delta_1/\Delta_2)} \tag{7}$$

and

$$\lambda_3 = \frac{1}{2e^2 N_2(0)} \frac{G_{nn}}{v} \qquad (8).$$

Neglecting λ_{2R}, the optimum situation occurs for $\lambda_3 = \lambda_2 \gg \lambda_1$, which gives

$$\left(\frac{v}{V}\right)^2 = \frac{\tau_s(\Delta_1/\Delta_2)}{2e^2 N_2(0)} \left(\frac{G_{nn}}{V}\right) \qquad (9).$$

Figure 6(a) shows the current pulse shape for $G_{nn}/V = 10^6 \Omega^{-1}$ cm^{-3} and various trap volumes. The signal current can be increased by a factor of more than 100. Figure 6(b) shows how the maximum signal current and the optimum trap size depend upon G_{nn}/V, for the indium-aluminium combination and also for the niobium-aluminium combination. In the latter case $\tau_s(\Delta_1/\Delta_2)$ is only about 0.15 ns and this results in an increase of signal current by a factor of 10^4.

Our programme of proving experimentally that trapping works, and of implementing it to make indium particle detectors has been curtailed in recent months by the fact that our colleagues in Cambridge have become very actively engaged in work on the new high-temperature superconductors. However we have some experimental results which we can report which are consistent with the trapping idea.

Figure 7(a) shows a drawing of a partially completed device with eight Nb base electrodes which are covered with insulating anodized aluminium except for the contact pads at the periphery and the tunnel junction areas at the centre. The junctions are Nb–AlOx–Al. On one such device, after ion-beam precleaning, we deposited a Nb film over all eight junction mesas to give the device shown in Fig.7(b). On a second device we deposited an indium film and then subsequently grew an indium crystal over the junction mesas in an inert atmosphere. The result is shown in Fig.7(c). In both devices the Al mesas act as traps. For the first device with the top Nb film we compute a trapping time and a tunneling time from the trap both of less than 1μs. For the indium crystal device the tunneling time from the trap is estimated to be about 1μs but the trapping time is several ms because the traps are relatively much smaller in volume. Future devices will have close to the optimum trap size and a configuration more like that illustrated schematically in Fig.8.

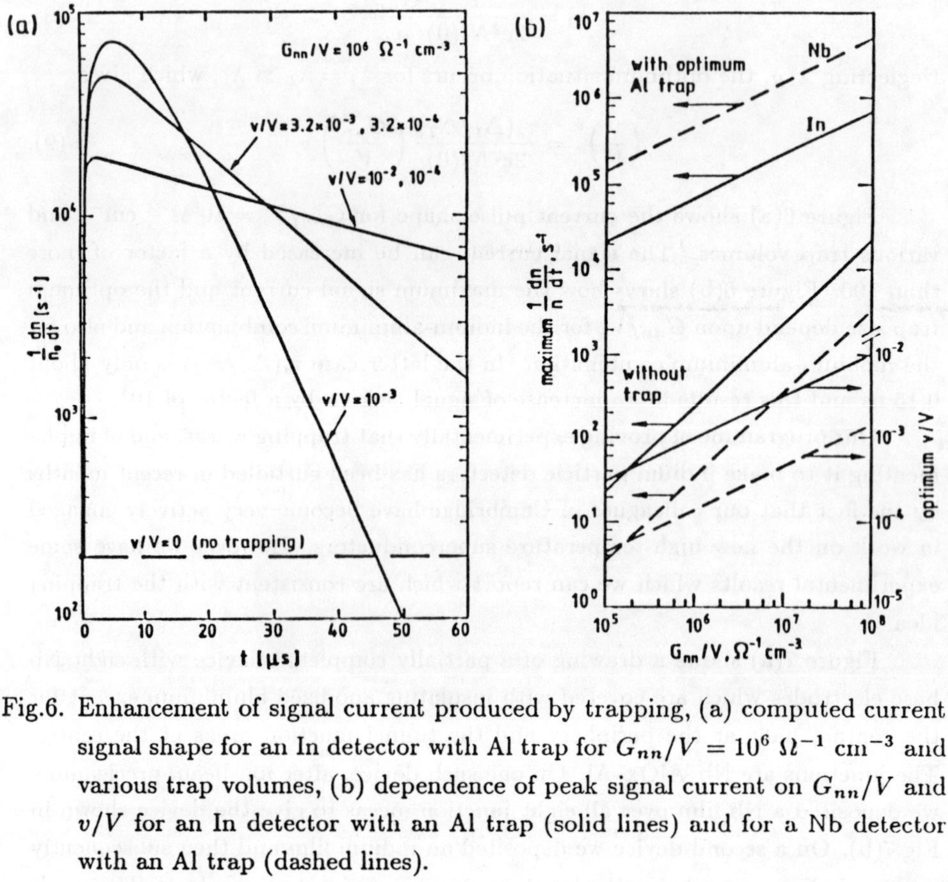

Fig.6. Enhancement of signal current produced by trapping, (a) computed current signal shape for an In detector with Al trap for $G_{nn}/V = 10^6\ \Omega^{-1}\ \mathrm{cm}^{-3}$ and various trap volumes, (b) dependence of peak signal current on G_{nn}/V and v/V for an In detector with an Al trap (solid lines) and for a Nb detector with an Al trap (dashed lines).

The experiment we did with the devices shown in Fig.7(b) and (c) was to pulse one junction beyond the sum-of-the-gaps edge of the I/V characteristic via a transformer. This injects quasiparticles into the top Nb film or indium crystal. We then looked for a signal current on one of the other junctions which was voltage biased close to $(\Delta_{\mathrm{Nb}} - \Delta_{\mathrm{Al}})/e$. Tracings of oscilloscope photographs are shown in Fig.9. In the case of the Nb film the signal current closely follows $|V_{in} - 2\ \mathrm{mV}|$, as shown in Fig.9(a), indicating that the tunneling time is short.

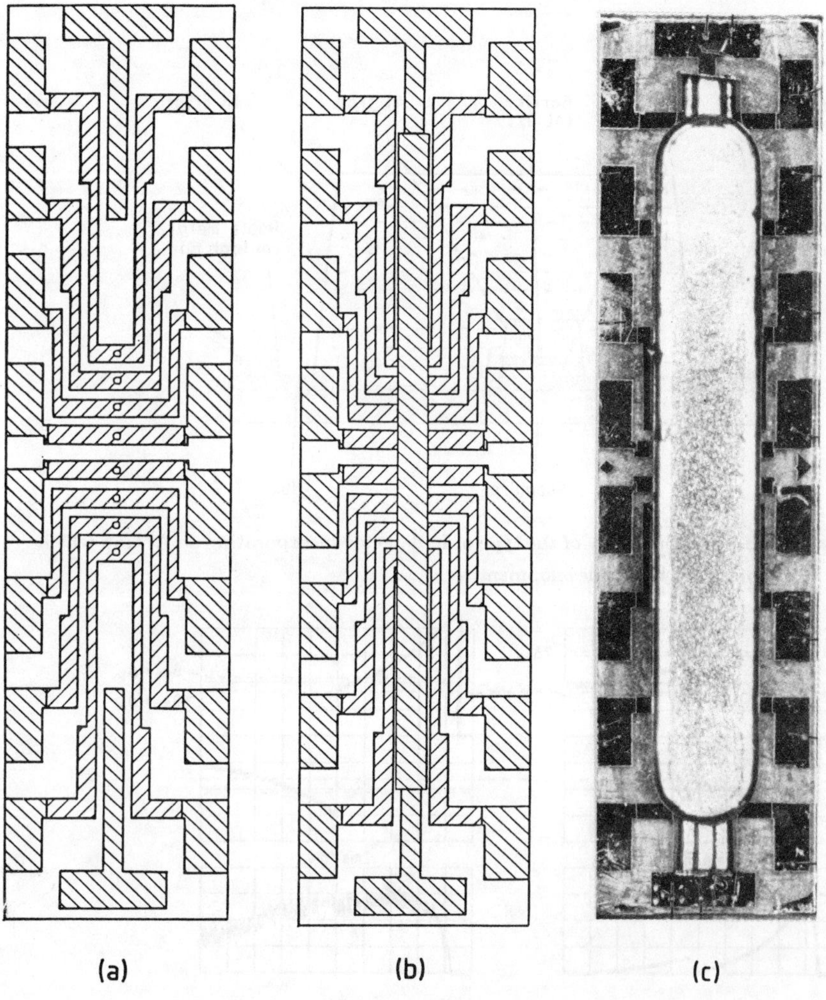

Fig.7. Test devices for the trapping mechanism, (a) drawing of partially completed device on a sapphire substrate, (b) drawing of a similar device after deposition of a Nb counterelectrode film, (c) photograph of a device with an indium crystal grown over the junctions.

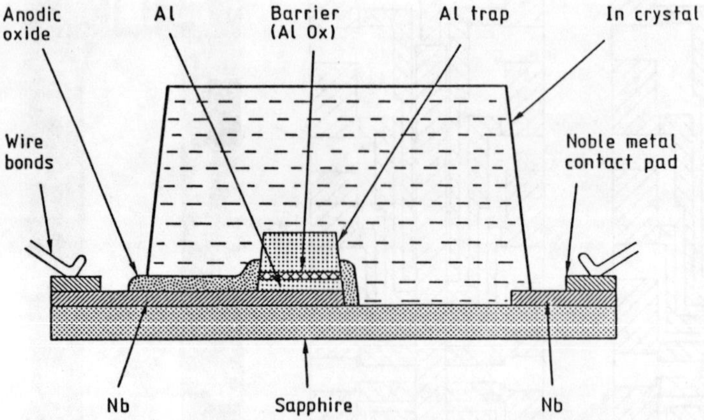

Fig.8. Cross-sectional view of the type of In detector incorporating an Al trap which is presently under development.

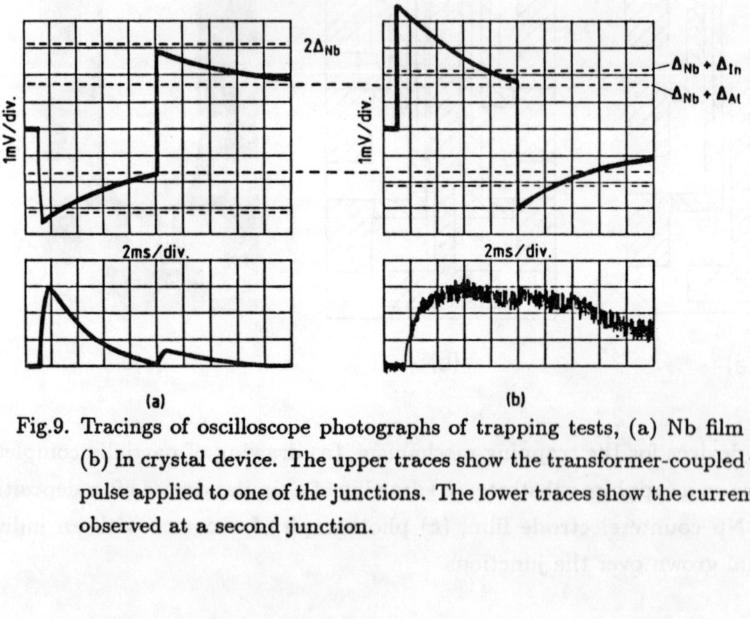

Fig.9. Tracings of oscilloscope photographs of trapping tests, (a) Nb film device, (b) In crystal device. The upper traces show the transformer-coupled voltage pulse applied to one of the junctions. The lower traces show the current signal observed at a second junction.

The value of $V_{in} - 2$ mV was the threshold for seeing any current signal and lies between $(\Delta_{Nb} + \Delta_{Al})/e$ and $2\Delta_{Nb}/e$. In the case of the In crystal the rise time and tunneling time are much longer as expected, as shown in Fig.9(b). Note that in both cases the current signal is independent of the polarity of the voltage pulse. Although these results do not constitute a proof of the trapping scheme they are consistent with it. A second piece of evidence comes from the dependence of the signal current on the detector junction bias. In our Nb–AlOx–Al junctions the Al gap for reasons we don't understand in detail [17], but which has been observed before [18], was only 0.14 meV. This has the result that $(\Delta_{In} - \Delta_{Al}) - 2\Delta_{Al}$ is a positive number (0.11 meV). This means that relaxation phonons emitted in the trapping process have sufficient energy to break a Cooper pair in the Al (the multiplication process to be discussed in the next section) and to excite one of the quasiparticles to 0.11 meV above the gap edge. We did indeed observe a peak in the signal current at a bias

$$V = [(\Delta_{Nb} - \Delta_{Al}) - (\Delta_{In} - 3\Delta_{Al})]/e = 1.29 \text{ mV}.$$

Unfortunately there was some leakage current in these junctions and we did not pursue these measurements further.

4. THE QUASIPARTICLE MULTIPLIER

As mentioned previously, when the ratio of the gaps Δ_1/Δ_2 is greater than 3, a multiplication process can occur because some of the phonons emitted in the relaxation process have sufficient energy to break a pair in S_2 creating more quasiparticles. This results in additional amplification of the current signal. It also leads to the concept of the quasiparticle multiplier [13], the superconductor equivalent of the photomultiplier. Figure 10(a) shows the processes involved in the quasi-particle multiplier, and Fig.10(b) suggests one of many possible configurations for the multiplier and a possible bias arrangement. Each junction should be biased in the range $(\Delta_1 - \Delta_2) < |eV| < (\Delta_1 + \Delta_2)$, but either bias polarity can be used, and if an injector junction is used it can also operate with either polarity.

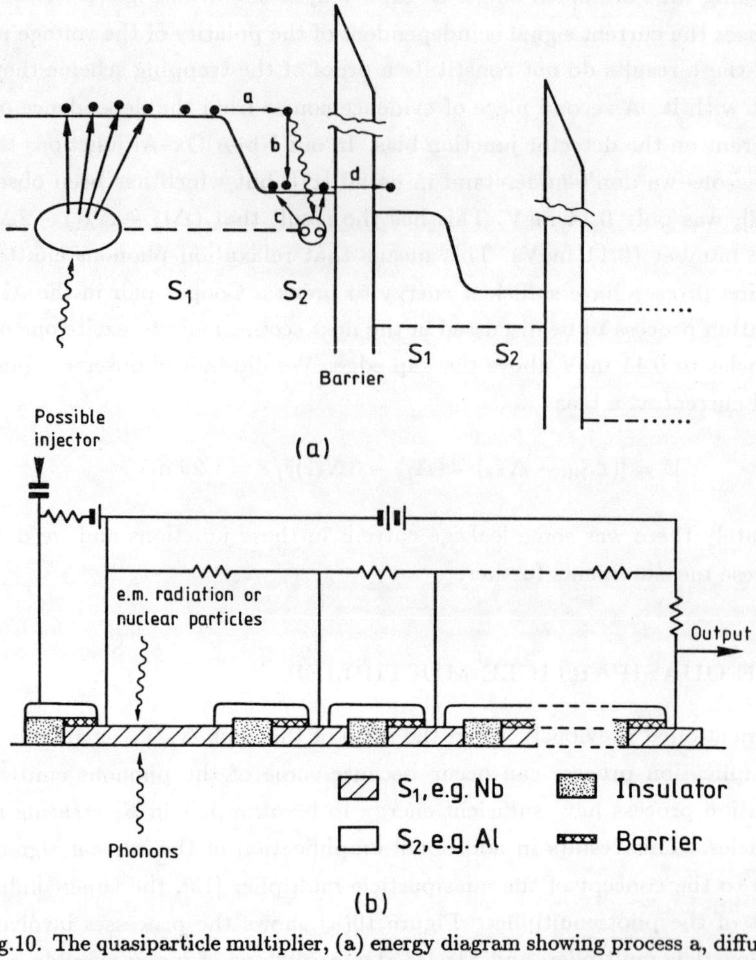

Fig.10. **The quasiparticle multiplier, (a) energy diagram showing process a, diffusion from S_1 into S_2; process b, relaxation to Δ_2 with phonon emission; process c, pair-breaking by phonon of energy $\geq 2\Delta_2$; process d, tunneling, (b) one of many possible configurations for a quasiparticle multiplier.**

The gain of the multiplier has been estimated by J.L.Lloyd and is about 4.5 per stage for the Nb–Al system [19]. Phonon losses which depend on the particular geometrical arrangement will of course reduce this. Fluctuations in the gain introduce only an additional 4.5% error above the statistical error in the initial number of quasiparticles.

There appear to be a number of applications of the quasiparticle multiplier. The one that interests us most is to use it as a preamplifier for the indium detectors. It would be fabricated on the substrate at the same time as the detection junction structure.

Another application is as a very sensitive detector. In principle it should be sensitive to a single photon or phonon of energy a few times $2\Delta_1$, just as a photomultiplier can detect a single optical photon.

Another aspect of the quasiparticle multiplier pointed out by J.L.Lloyd is that it has a structure somewhat analogous to a charge-coupled device. Provided the effective recombination lifetime is sufficiently long it should be possible to store quasiparticles in one of the stages by reducing the bias. Thus one could do all the exciting things that can be done with CCD's and have internal amplification as well.

The main drawback of the multiplier is that it has to operate well below the T_c of S_2. This is not a problem for most particle detector applications because we really need a dilution refrigerator anyway. For other applications it would be advantageous to operate at 4 K. This may be possible if S_1 is one of the new high T_c materials, and a conventional high T_c material, such as NbN, is used as S_2.

We are currently making the first prototypes of the multiplier.

5. SUMMARY

Although we don't yet have realistic superconducting particle detectors, much of the pioneering work has been done, and I think we have a good understanding of the problems. The potential applications in nuclear physics, particle physics, astrophysics, and other areas are very wide. The quasiparticle trapping scheme, although needing quantitative experimental proof, overcomes some of the past problems and makes bulk superconducting detectors feasible, in particular indium detectors for solar neutrinos. Results presented at this Workshop strongly indicate that bulk single-crystal detectors of insulators and semiconductors equipped with superconducting sensors are also feasible [4,20].

Development work on the quasiparticle multiplier has just started. If it really works we have the possibility of a low-noise preamplifier, a very sensitive detector, and CCD-like applications.

I cannot stress enough the importance of the fabrication technology. In order to make real progress we have to have available the kind of technology which the semiconductor industry has, with only a few subtle differences.

Finally I wish to thank and congratulate Professor Barone and the other organizers of this Workshop. It has stimulated interest in this field and involved more physicists. As a direct result I predict we will make much faster progress in this very exciting field.

ACKNOWLEDGEMENTS

I am indebted to D.J. Goldie, B.M. Hawes, J.L. Lloyd, C. Patel and G.L. Salmon from Oxford, and J.E. Evetts, J.H. James, J.M. Lumley, G.W. Morris and R.E. Somekh from Cambridge, whose active participation has made this paper possible.

REFERENCES

[1] The value of ϵ for superconducting Sn has been estimated by M. Kurakado and H. Mazaki, *Nucl. Instr. and Meth.* **185**, 141 (1981); M. Kurakado, *ibid.* **196**, 275 (1982).

[2] For general reviws on the status of superconducting electronics see the special issue on Josephson devices of IEEE Trans. Elec. Dev., October 1980, and B.R. Barnard, "Applications of the Josephson Junction", *GEC Jour. Res.* **2**, 205 (1984).

[3] G. Charpak and F. Sauli, "High-Resolution Electronic Particle Detectors", *Ann. Rev. Nucl. Part. Sci.* **34**, 285 (1984).

[[4] B. Cabrera, "Silicon crystal acoustic detectors, particle detectors utilizing ballistic phonons", these Proceedings.

[5] N.E. Booth, G.L. Salmon and D.A. Hukin, in *Solar Neutrinos and Neutrino Astronomy*, edited by M.L.Cherry, W.A.Fowler and K.Lande, *AIP Conf. Proc.* No.**126**, (American Institute of Physics, New York, 1985) p.216.

[6] N.E. Booth, "The Indium Solar Neutrino Project", *Sci. Prog. Oxf.* (to be published).

[7] J.J. Chang and D.J. Scalapino, *J. Low Temp. Phys.* **31**, 1 (1978); C.C. Chi, M.M.T. Loy and D.C. Cronemeyer, *Phys. Rev.* **B23**, 124 (1981).

[8] G.H. Wood and B.L. White, *Can. J. Phys.* **51**, 2032 (1973).

[9] M. Kurakado, *J. Appl. Phys.* **55**, 3185 (1984).

[10] A. Barone, G. Darbo, S. de Stefano, G. Gallinaro, A. Siri, R. Vaglio and S. Vitale, *Nucl. Instrum. Meth.* **A234**, 61 (1985).

[11] H. Kraus, Th. Peterreins, F. Pröbst, F. v. Feilitzsch, R.L. Mössbauer, V. Zacek and E. Umlauf, *Europhys. Lett.* **1**, 161 (1986).

[12] D. Twerenbold, *Europhys. Lett.* **1**, 209 (1986).

[13] N.E. Booth, *Appl. Phys. Lett.* **50**, 293 (1987).

[14] S.B. Kaplan, C.C. Chi, D.N. Langenberg, J.J. Chang, S. Jafarey and D.J. Scalapino, *Phys. Rev.* **B14**, 4854 (1976); **B15**, 3567(E) (1977).

[15] C.C. Chi and J. Clarke, *Phys. Rev.* **B19**, 4495 (1978).

[16] M.V. Moody and J.L. Paterson, *Phys. Rev.* **B23**, 133 (1981).

[17] For a discussion of this effect see J.J. Chang, in "Nonequilibrium Superconductivty, Phonons, and Kapitza Boundaries", edited by K.E.Gray (Plenum, New York, 1981), Chap.9.

[18] A.F. Cattell, A.R. Long, A.C. Hanna and A.M. MacLeod, *J. Phys. F: Met. Phys.* **13**, 855 (1983).

[19] J.L. Lloyd, private communication.

[20] Th. Peterreins, F. Pröbst, F. von Feilitzsch and H. Krauss, "Detection of phonons created by α-particle absorption in a Si single crystal with superconducting tunnel diodes", these Proceedings.

MEASURING THE ENERGY OF IONIZING RADIATION WITH GIAEVER-TYPE SUPERCONDUCTING TUNNELING JUNCTIONS

DAMIAN TWERENBOLD

Astrophysics Division
European Space Research and Technology Center
Postbus 299, 2200 AG. Noordwijk,
The Netherlands

1. INTRODUCTION

Superconductivity, being a coherence phenomena on the meV energy scale, is expected to response sensitively to the deposition of ionizing energy. This has been demonstrated in the last few years[1-8]. The ionizing energy can either be deposited by the energetic photoelectron emitted after the absorbtion of an x-ray photon, or by a minimal ionizing particle passing through the superconductor. Along the trajectories of the energetic particle atoms will be ionized. The electrons with an energy of the order of eV will 'cool' by electron-electron and electron-phonon interactions and ultimately break Cooper pairs. The result is an excess number of quasiparticles in the superconductor. Information on quasiparticle concentrations in superconductors can be obtained by measuring the subgap current in superconducting tunneling junctions[9].

In the following section, the basic experimental results of using Sn/Sn-ox/Sn junctions as x-ray detectors are summarized. The nonequilibrium physics of these detectors is discussed in section 3. Specifically, the rate equations for nanosecond processes are presented. Quasiparticle diffusion is treated 2-dimensionally. The quadratic self-recombination term of the excess quasiparticles yields the nonlinear effects observed in the preceding work[5-8].

2. EXPERIMENTAL RESULTS WITH SN-JUNCTIONS

The superconducting tunneling junctions used to detect the Mn-K_α (E=5.89 keV) and the Mn-K_β (E=6.49 keV) x-rays of a ^{55}Fe source consisted of two thin Sn films (typically 0.2-0.7 μm, evaporated on fused silica substrates) separated by a very thin natural oxide insulating barrier of the order of a few 10 Å. The areas varied between 50 μm x 50 μm and 150 μm x 150 μm. A magnetic field (~ 50-100 Gauss) was applied parallel to the insulating barrier to suppress the DC Josephson current (the tunneling of the Cooper pairs across the barrier). The operating temperature was 0.4 K in order to reduce the recombination of excess quasiparticles with thermally excited quasiparticles, the density of the latter declining exponentially with temperature as $\exp(-\Delta/kT)$. The junctions were DC biased at a voltage of about 0.7 Δ/e and AC coupled to a charge sensitive amplifier. The results are summarized as follows (for the K_α line):

accumulated charge	: $Q = 2\cdot 10^6$ e
minimal ionizing energy	: $\varepsilon = 2.5$ meV
energy resolution (with electronic noise)	: $\Delta E = 65$ eV
energy resolution intrinsic	: $\Delta E = 40$ eV

The latter number was obtained by a method discussed in ref[6,8]. With the minimal ionizing energy of 2.5 meV and an assumed Fano factor of 1 the intrinsic resolution ought to be 9 eV. The cause for the discrepancy is still an open question. There are indications that it could be owing to local fluctuations in the films (e.g granularity).

In any realistic application, the electronic noise is a problem owing to the large capacity of these devices. In a charge sensitive circuit, the dominant noise source is the intrinsic series noise of the input FET, amplified by the capacitance of the junction. The required signal-to-noise ratio sets a limit on the maximal allowed area of the STJ-detectors[10]. This is especially of concern when one wishes to make arrays of junctions in order to cover a reasonable detecting area[11]. This problem can be approached from two sides: by reducing the intrinsic noise of the FET, as then the number of detectors, and hence of preamplifiers, reduces linearly. Owing to the relatively low bandwith of the STJ-detector (< 1MHz) the 1/f noise of the FET has to be very low, leaving JFETs as the only candidates. The noise figures will have to be less than 1 nV/$\sqrt{Hz}$. Secondly, junctions with a smaller specific capacity have to be fabricated. This can be obtained by using semiconductor films having lower barrier heights. Then, the thickness of the insulating film can be increased. But, unfortunately, this barrier height will have to be decreased quadratically for a same good performance of the STJ-detector[10].

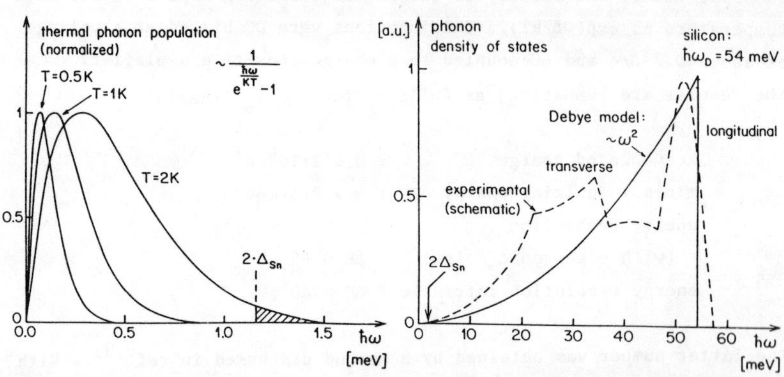

Figure 1: (a) (fused silica) : the assumed normalized thermal phonon population at the three temperatures 2 K, 1 K and 0.5 K. (b) (silicon) : the phonon density states in silicon assumed to be populated by ballistic phonons.

An interesting additional effect was observed when a single crystal silicon was used as a substrate. When exposing Sn-junctions evaporated on highly doped single crystal Si substrates to 6 keV x-rays, an order of magnitude more signals were detected than in the case of Sn-junctions on fused silica substrates. This was interpreted to be due to the transmission of energetic phonons ($\hbar\omega > 2\Delta$) from the substrate into the superconducting films, where excess quasiparticles were created by phonon pair breaking. The largest pulse heights of these signals were of the same order as in the case of x-rays absorbed in the superconducting film directly. No signals were seen from x-rays absorbed in the fused silica. This cannot be due to a smaller phonon transmission coefficient from fused silica into Sn. Quite contrary, the numbers are[12]:

		longitudinal	transverse
$\eta_{Sn - SiO_2}$	:	0.44	0.15
$\eta_{Sn - Si}$	:	0.26	0.088

A possible reason for the strikingly different behaviour is the different energy distribution of the x-ray induced phonons: In the case of fused silica (Fig.1a), all the x-ray energy thermalizes on a time scale much faster than the tunneling time, which is of order 1 μsec. Virtually no phonons exist with an energy larger than 2Δ at temperatures less than 1 K. In the case of silicon, however, it is a well known fact that high energy phonons have a long life time and propagate ballistically at low temperatures[13]. The density of states of these ballistic phonons is indicated schematically in Fig.1b. The cutoff energy of the phonons in the Debye modell for Silicon is 54 meV. The relative position of 2Δ (the threshold energy of the detector) is strikingly different in the two cases.

From these observations, one may hence propose[13,6,11] to detect and measure, maybe with high resolution, the energy of x-rays absorbed in suitably fabricated crystal substrates which serve in the same time as substrates for superconducting tunneling junctions. The decisive question is, of course, what determines the resolution of this STJ/crystal detector. Most probably this will be given by the statistics in the phonon production following the absorbtion of the x-ray. Competing energy deposition mechanisms are:

- electron-hole production in the crystal with recombination times longer than 1 μsec (beyond the band width of the STJ).
- photo production, escaping the crystal.
- collective excitation modes with energies less than 2Δ.
- production of crystal defects with long life times.

However, the high resolution of 17 eV at 6 keV of the Goddard x-ray bolometer[14] already suggests that these processes cannot be dominant. But one has to be aware of the fact that in the Goddard x-ray bolometer all phonon energies are measured and no low energy cutoff exists. In the case of a STJ-detector, the low energy cutoff is $\hbar\omega = 2\Delta$.

Concerning the superconducting tunneling junctions as directly absorbing x-ray detectors, the question emerges what limits the observed intrinsic resolution of these detectors. The fluctuation in local (~ 1μm) properties of the superconducting films could be one cause. This view is supported by the following section. Experimentally, this could be investigated by exciting the STJ-detector locally either by using a very small pinhole mask or by exposing the detector to a highly collimated electron beam with electron energies of the order of 10 keV, as is proposed by the Tuebingen group in this workshop.

3. SUB-NANOSECOND NONEQUILIBRIUM PHYSICS IN THE STJ-DETECTOR

When an x-ray is absorbed in a superconducting film, the quantities describing the superconductive behaviour - Cooper pairs, quasiparticles and phonons - deviate from their equilibrium values. In contrast to most nonequilibrium experiments with lasers or heat sources[9], an x-ray deposits a well defined energy, localized both in volume (of the order of 1 μm^3) and in time. A variety of different phenomena were observed and have been discussed in detail in ref[6-8]. The observations can be readily described by a set of 4 Rothwarf-Taylor equations[15] coupling the quasiparticle numbers and the 2Δ-phonon numbers of each of the two films. The physics is described by the following rates (the values are for the Sn-junctions):

- R : quasiparticle recombination coefficient ($2.5\cdot10^{-19}$ m^3/sec)
- τ_B : phonon pair breaking time (~ 0.1 nsec)
- t_γ : phonon escape time (~ 0.1 - 1 nsec)
- λ : quasiparticle mean free path (~ 0.01 - 1 μm)
- τ_{tun} : quasiparticle tunneling time (~ 1 - 5 μsec)

Owing to the backtunneling effect between two superconducting films[16], the tunneling current is at any time the sum of the electrons flowing from the lower film into to the upper film and from the upper into the lower, respectively[5-8].

With the rate equations of ref[7,8] the dependence of the detector signal on temperature could be explained by the recombination of excess quasiparticles with thermally excited quasiparticles. Owing to the specific geometry of our detectors, quasiparticles diffusing from x-rays absorbed in the long (superconducting) current leads yielded a longer rise time of the signal, a fact which could also described well with the equations. In addition, the back tunneling had to be taken into account to properly fit the two simultaneously obtained pulse shapes from x-rays absorbed in either of the two films (one set of parameters has to describe both pulses). The physics described occured in the time scales from a few nsec to several µsec. In the following, these considerations will be extended to the sub-nanosecond regime.

To study the processes which occur at these early time scales in a STJ-detector, and to discuss its implications on the operation as a detector, we will reduce the R-T equations to the following two radial equations (2-dim polar coordinates are used, and axial symmetry is assumed):

$$\begin{aligned} \frac{d}{dt} q(r,t) &= 2\tau_B^{-1} p(r,t) - \frac{R}{V(r)} q(r,t)^2 \\ \frac{d}{dt} p(r,t) &= -(\tau_B^{-1} + \tau_\gamma^{-1}) p(r,t) + \frac{1}{2}\frac{R}{V(r)} q(r,t)^2 \end{aligned} \qquad (1)$$

where $q(r,t)$: number of quasiparticles at time t in the segment between r and r+dr

$p(r,t)$: number of 2Δ-phonons at time t in the segment between r and r+dr

$V(r) = h\cdot\pi\cdot((r+dr)^2 - r^2)$: volume of the segment

h : film thickness

The temperature is assumed to be sufficiently low that thermal recombination is negligable. Quasiparticle tunneling is neglected as it becomes effective only at much larger times (only one film of the junction is hence investigated).

To solve these equations we assume the quasiparticles to diffuse by the following radial diffusion law in two dimensions:

$$n(r,t) = \frac{1}{4\pi Dt} e^{-\frac{r^2}{4Dt}} \qquad (2)$$

where $D = 0.5 \cdot v \cdot \lambda$ is the diffusion constant
$v = 5 \cdot 10^5$ μm/μsec : the velocity of quasiparticles
λ = quasiparticle mean free path

To solve the rate equations numerically the following approximative scheme is used: at each time step t the distributions are set as follows:

$$\begin{aligned} q(r,t) &= Q(t)\, n(r,t)\, r\,dr \\ p(r,t) &= P(t)\, n(r,t)\, r\,dr \end{aligned} \qquad (3)$$

where Q(t) : total number of quasiparticles at time t
P(t) : total number of 2Δ-phonons at time t

The total quasiparticle number Q(t+Δt) and the total 2Δ-phonon number P(t+Δt) is obtained by firstly calculating the change Δq(r,t) and Δp(r,t) at a given radius r with help of the rate equations, i.e.

$$\begin{aligned} \Delta q(r,t) &= \Delta t \cdot 2\tau_B^{-1} p(r,t) - \Delta t \cdot \frac{R}{V(r)} q(r,t)^2 \\ \Delta p(r,t) &= -\Delta t \cdot (\tau_B^{-1} + \tau_\gamma^{-1}) p(r,t) + \frac{\Delta t}{2} \frac{R}{V(r)} q(r,t)^2 \end{aligned} \qquad (4)$$

These contributions are then integrated over the 2-dim area and added to yield the new total numbers:

$$Q(t+\Delta t) = Q(t) + 2\pi \sum_r \Delta q(r,t)$$
$$P(t+\Delta t) = P(t) + 2\pi \sum_r \Delta p(r,t) \qquad (5)$$

For simplicity, the radial phonon distribution is assumed to follow the radial quasiparticle diffusion. For short times, i.e. $t < \tau_B$, this can be justified by the fact that phonons are much slower than quasiparticles. Actually, the energy stored in the phonon system lags behind the quasiparticle system and the nonlinear effects we are interested in would be enhanced. At later times, i.e. $t > \tau_B$, the phonons nearly all are converted to quasiparticles owing to the fast phonon pair breaking time.

The energy deposited by the ionizing radiation is exchanged between the quasiparticles and the the 2Δ-phonons owing to the recombination factor R/V(r). Conversely, phonon pair breaking τ_B turns the stored excitation energy from 2Δ-phonons into quasiparticles. The self-recombination term shows nonlinear effects owing to the quadratic $q(r,t)^2$ term in equation (1). The ultimate decay channel for the stored ionizing energy is the phonon escape process.

Solutions of the rate equations (1) are illustrated in Fig 2. The time evolution of the total quasiparticle numbers Q(t) and the 2Δ-phonon numbers P(t) are shown for the two cases: (a) all the ionizing energy is initially in the quasiparticle system and, (b) all the energy is initially in the 2Δ-phonon system. The phonon pair breaking time is chosen to be 0.1 nsec and the phonon escape time 0.3 nsec. To enhance the quadratic effect, a thin film of 0.1 μm and a short quasiparticle diffusion length of 10 Å has been used. In case (a), the quadratic recombination term produces phonons maximaly at the time scale of 10 psec. Owing to phonon escape these phonons than vanish. This is reflected in the time evolution of the total energy (the phonon numbers are multiplied by a factor of 2 because they carry an energy of 2Δ) In case (b), the phonon pair breaking is comparably slow compared to the relevant time of the quadratic effect which is determined by the temporal evolution of the localization of the excess quasiparticles. The FWHM of the diffusion is 1 μm at 1 nsec, which

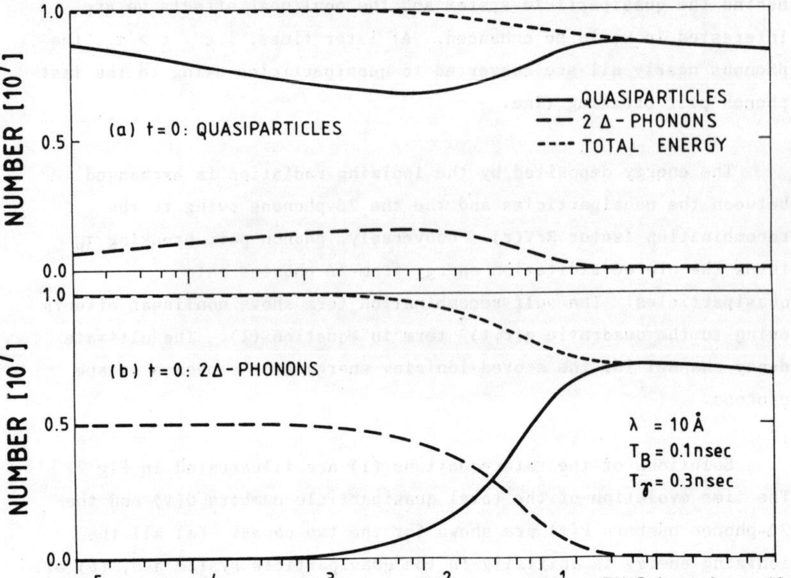

Figure 2: The time evolution of the quasiparticle number, the 2Δ-phonon number and the stored excitation energy in a Sn film of 0.1 μm thickness. (a): At t=0 the total energy of 10^7 Δ is in the quasiparticle system. (b): At t=0 the total energy is in the 2Δ-phonon system.

shows that the effects are very local indeed. In case (b), the total energy stored at 10 nsec is less because there were more phonons at the pair breaking time scale than in case (a). For all times larger than 1 nsec, most of the energy is stored in the quasiparticles. But as phonon escape is a fast process at these larger times, the energy leaks out of the superconductor in spite of the very small number of phonons present.

The quasiparticle mean free path λ has been varied by 10 % to observe how a local variation of λ (e.g. grain size) could determine

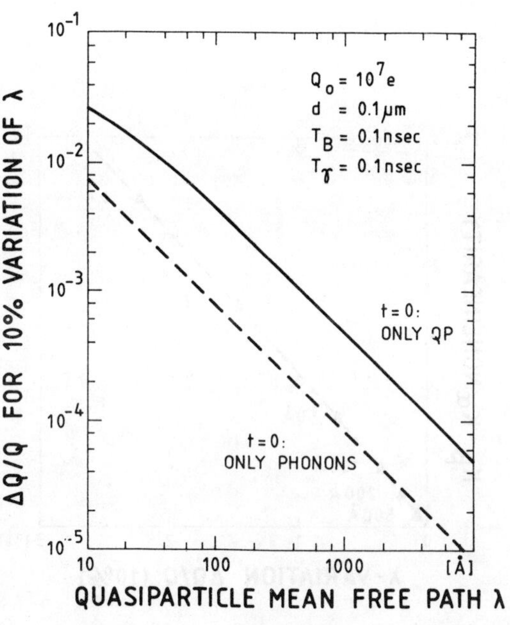

Figure 3: The relative variation ΔQ/Q in the quasiparticle number for a 10 % variation of λ is plotted against the quasiparticle mean free path λ.

the pulse height of the detector. The relative variation in the residual quasiparticle number (at 100 nsec), defined by

$$\frac{\Delta Q}{Q} = \frac{Q(\lambda + 0.1\lambda) - Q(\lambda)}{Q(\lambda)} , \qquad (6)$$

is plotted against λ in Fig.3. Again, the two cases (a) and (b) of Fig.2 have been included. From Fig.3 it is apparant that one can achieve a fluctuation of the pulse height on the 1 % level by this nonlinear effect (an intrinsic resolution of ΔE = 40 eV at the energy of 6 keV is 0.7 %). If one wishes to achieve a good intrinsic energy resolution in STJ-detectors the quasiparticle mean free paths have either to be large or at least should not fluctuate locally by less than a few percent.

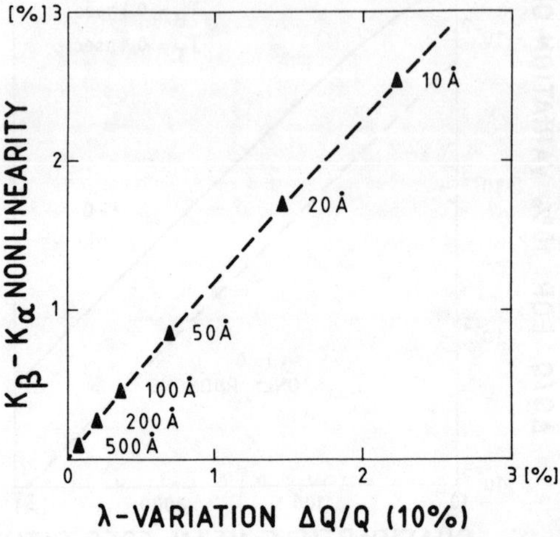

Figure 4: The K_β - K_α nonlinearity is plotted against the relative variation ΔQ/Q for various quasiparticle mean free paths λ.

In the high resolution spectra of the Sn-junctions, a $K_\beta - K_\alpha$ nonlinearity defined as

$$\left(\frac{E_{K_\beta}}{E_{K_\alpha}}\right) - \left(\frac{Q_{K_\beta}}{Q_{K_\alpha}}\right)_{Obs.} \quad (7)$$

was observed[6-8]. Experimentally this was 2 %. The same effect can be reproduced in the numerical solution of the model described above. If one solves the rate equations for two initial numbers 10^7 (K_α) and $1.1 \cdot 10^7$ (K_β) a smaller ratio than 1.1 is observed at times later than 1 nsec. In Fig.4, the deviation of this residual ratio (at 100 nsec) is plotted against the relative variation $\Delta Q/Q$ of Fig.3 for various quasiparticle mean free paths λ. The relation is surprisingly linear.

This suggests that the residual intrinsic resolution in the STJ-detector and the $K_\beta - K_\alpha$ nonlinearity are connected by the same nonlinear process, namely the quadratic self recombination term. This term is important between 1 psec and 1 nsec where the excess quasiparticles are concentrated in a volume of the order of $h \cdot 1\ \mu m^2$. For a larger quasiparticle mean free path the diffusion is faster, which means that a less number of recombination phonons, leaving the system by phonon escape, are produced in the relevant time scale.

In the Sn-junctions, a striking granularity could be observed under a scanning electron microscope[6,8]. It appeared that films with larger grains showed a better energy resolution than in those with smaller grains. But, as these films were grown under an oxygen pressure of $2 \cdot 10^{-5}$ mbar, the films are probaly highly disturbed yielding a quasiparticle mean free path λ less than the observed size of the granularity. For future improvement, high quality films are required with large quasiparticle meen free paths in order to suppress the nonlinear effect.

4. CONCLUSION

Solving the 2-dim Rothwarf-Taylor equations in a quasiparticle diffusion model shows a nonlinear response of the superconducting tunneling junction x-ray detector. The nonlinearity, which shows up in a smaller calculated pulse height ratio of the K_β to K_α line, is linearily correlated with the relative fluctuation of the pulse height $\Delta Q/Q$ when the quasiparticle mean free path λ is varied. The latter could be the reason for the observed intrinsic resolution of 0.7 % at 6 keV in Sn junction STJ-detectors. High quality superconducting films with large quasiparticle mean free paths would hence both reduce the nonlinearity and increase the intrinsic energy resolution of the detector.

REFERENCES

1) G.H.Wood and B.L.White, Appl.Phys.Letters 15 (1969) 237
2) M.Kurakado and H.Mazaki, Phys.Rev B 22 (1980) 168
3) A.Barone et al., Nucl.Instr.and Meth. A234 (1985) 61
4) H.Kraus, Th.Peterhans, F.Pröbst, F.v.Feilitzsch, R.L.Mössbauer, V.Zacek, and E.Umlauf, Europhys. Lett., 1 (1986) 161
5) D.Twerenbold, Europhys. Lett., 1 (1986) 209
6) D.Twerenbold and A.Zehnder, J.Appl.Phys. 61 (1987) 1
7) D.Twerenbold, Phys.Rev B, 34 (1986) 7748
8) D.Twerenbold, Ph.D.thesis, Diss.ETH No. 8038, (Eidgenössische Technische Hochschule,Zürich,1986)
9) for a review on nonequilibrium superconductivity refer to e.g. 'Nonequilibrium Superconductivity, Phonons, and Kapitza Boundaries', ed. K.E.Gray, Plenum Press, New York, 1981

D.N.Langenberg, in 'Low Temperature Physics LT14', eds H.Krusius and H.Vuorio, Volume V 223, North Holland, Elsevier, 1975

10) D.Twerenbold, Nucl.Instr.and Meth. A260 (1987) 430

11) D.Twerenbold, to be published in the Proceedings of the London Conference on Position Sensitive Detectors (London, 1987) (special issue of NIM section A)

12) S.B.Kaplan, J.Low.Temp.Phys., 37 (1979) 343

13) B.Cabrera, Proceedings of the VI Moriond Workshop on Massive Neutrinos in Particle Physics and Astrophysics, Tignes, 1986, ed. J.Tran Thanh Van (Editions Frontieres, Gif sur Yvette, France, 1986)

14) D.McCammon, M.Juda, J.Zhang, S.S.Holt, R.L.Kelley, S.H.Moseley and A.E.Szymkowiak, Proc. 18th Conf. on Low Temperature Physics, Kyoto, 1987, Japanese Journal of Applied Physics, 26 (1987) Supplement 26-3

15) A.Rothwarf and B.N.Taylor, Phys.Rev.Letters 10 (1967) 14

16) K.E.Gray, Appl.Phys.Letters 32 (1978) 392

QUASIPARTICLE DIFFUSION IN SUPERCONDUCTING TUNNELING JUNCTIONS FOR X-RAY DETECTORS

W. ROTHMUND, and A. ZEHNDER

Schweizerisches Institut fur Nuklearforschung
CH-5234 Villigen, Switzerland

1. INTRODUCTION

We report on the recent experimental results on high resolution X-ray detection with superconducting $Sn/SnO_x/Sn$ tunneling junctions. An earlier used preparation method of the junctions using a stencil mask and the best obtained results so far are briefly reviewed. The quasiparticle diffusion in the superconducting films was a severe limitation for the resolution of the detector. Further improvement could then be obtained by using a standard photolithographic lift-off technique, which yielded a very well-defined geometry of the junctions. As a next step, we investigated the influence of the microscopic structure of the films on the diffusion of the excess quasiparticles. We found that the size of the grains in the films is of great influence on the resolution of the detector and on the number of accumulated charge. Finally we present a diffusion model for the quasiparticle-phonon system. It turned out that we had to add an empirical loss term for quasiparticles to the rate equations to explain our experimental data.

Already in 1969, G. H. Wood et al.[1] pointed out that a superconducting tunneling junction is a powerful tool for the detection of ionizing radiation. This arises from the fact that the quasiparticle excitations of a superconductor are separated from the superconducting ground state by an energy gap of the order of 1 meV, which is three orders of magnitude smaller than the gap of a semiconductor. Hence thousand times more free charge carriers will be produced by the absorption of radiation in a superconductor than in a semiconductor. Moreover, the excited phonons in a superconductor are not necessarily lost, since the Debye energy (~ 10 meV) is considerably greater than the superconducting energy gap. The ideal resolution of the $Sn/SnO_x/Sn$ junctions considered here would be 5 eV FWHM at 5.89 keV (Mn-K-lines), assuming a Fano factor of 1. The best resolution we achieved so far is 41 eV FWHM.

2. EXPERIMENTAL RESULTS

2.1 Stencil Mask Preparation Technique

Tin was evaporated thermally and deposited through a stencil mask onto the substrates. The lower film was oxidized by glow discharge (oxygen pressure $8 \cdot 10^{-2}$ mbar, oxidation time 4-8 min, current 20 mA). The typical junction geometry was:

Junction area	: $50 \cdot 100\ \mu m^2$
Width of current leads	: 10-15 μm
Thickness of films	: 0.15 μm (lower film), 0.6 μm (upper film)
Barrier thickness	: 10-20 Å

With this shadow technique, no reproduceable geometry was feasible, because the mask was not in an intimate contact to the substrate. For the same reason the edges of the films were not very sharp.

This sample was then mounted in a ^{3}He-cryostat and cooled down to 0.4 K. A charge sensitive preamplifier (designed by P. Jokinen, SIN) was used to collect the produced excess charge carriers tunneling across the oxide barrier. The output signal of the preamplifier was calibrated by means of a testpulse, which was fed onto a small capacitor (1 pF) at its input. The testpulse width provided also an estimate of the electronic noise contribution. The transmission frequency of the preamplifier was around 5 MHz, and the lower bandwidth was given by the RC-time of the feedback loop (50-100 μs).

Fig. 1 shows the best obtained results with this kind of detector[2)]. As can be seen, the signals are asymmetric. The lower energy part of the peaks is broadened due to quasiparticle diffusion from the tunneling region into the current leads. A gaussian can be fitted to the high energy side of the signals, which yields a resolution of 65 eV FWHM for the K_α-peak. The overall resolution is 90 eV in this case.

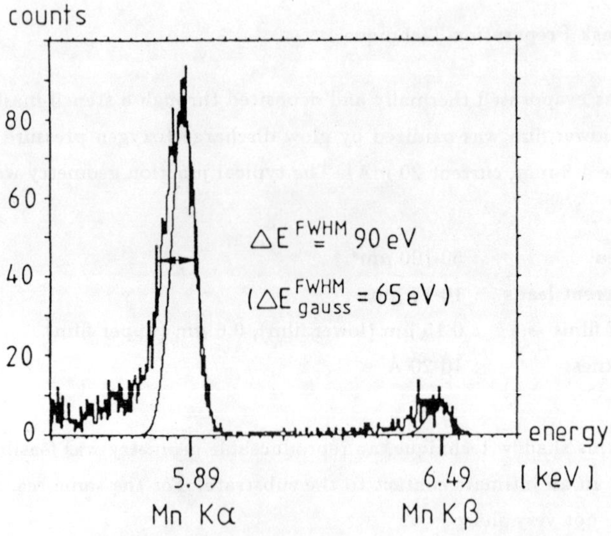

Figure 1 Resolution of the MnK_α peak, measured with a junction prepared by stencil mask evaporation. The low energy broadening of the peaks is due to the diffusion of quasiparticles away from the tunneling region[2].

The different contributing signals can clearly be distinguished in the scatter plot of fig. 2, where the risetime is plotted versus the pulseheight[2]. Events taking place in either film are separated from one another in risetime due to the different film thicknesses. The K_α- and K_β-peaks of each film can be seen at the high energy end of the spectrum. The pulses in region A vary in pulseheight but have constant risetime. These are events which take place in the region of overlap, showing only a partial deposition of the ionizing energy (photo- or Auger-electron escape). The pulses in region B are due to X-rays absorbed in the current leads. Here the excess quasiparticles first have to diffuse into the tunneling region before they can cross the barrier, leading to increasing risetime and energy loss with increasing distance.

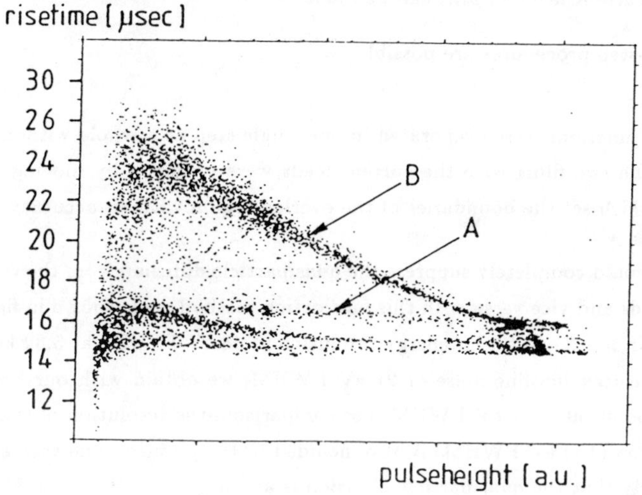

Figure 2 Scatter plot (risetime vs pulseheight) of a junction deposited through a stencil mask, exhibiting the different characters of events in a junction (for details see text)[2)].

As mentioned above, this were the best results that we obtained with junctions prepared by stencil mask technique. In most cases, the diffusive broadening of the peaks was much stronger than in this example. We even operated a junction with a very clean, leakage current free I-V characteristic, but observed no X-ray pulses at all. After reducing the width of the current leads mechanically and cooling down again, the signals appeared, but were rather distorted due to quasiparticle diffusion.
Consequently, we tried to suppress this diffusive broadening by using a photolithographic lift-off process to prepare clean and reproduceable junctions with thin current leads. A description of this technique and the results are given in the following section.

2.2 Photolithographic Lift-Off Technique

Preparing the junctions by means of photolithographic lift-off has the following advantages:

- Reproduceable geometry with sharp film edges
- Possibility of easily varying the junction geometry

- Thin current leads (4 μm) can be made
- Multi-step procedures are possible

Our junctions were evaporated in one single step as a whole without breaking the vacuum. The two films with the current leads were obtained by moving a slit over the substrate which set the boundaries of the overlap region with an accuracy of 20-30 μm.

We could completely suppress the quasiparticle diffusion from the tunneling region into the leads and vice versa with this fabrication method, as is shown in fig. 3 and fig. 4. The peaks in fig. 3 exhibit an energy resolution of 48 eV FWHM at 5.89 keV. Correcting for the measured biasline noise of 24 eV FWHM, we obtain with our best detector an intrinsic resolution of 41 eV FWHM. For comparison, the resolution of the best available SiLi-detectors (150 eV FWHM) is also included in the picture. The scatter plot in fig. 4 shows clearly that the quasiparticle diffusion is absent.

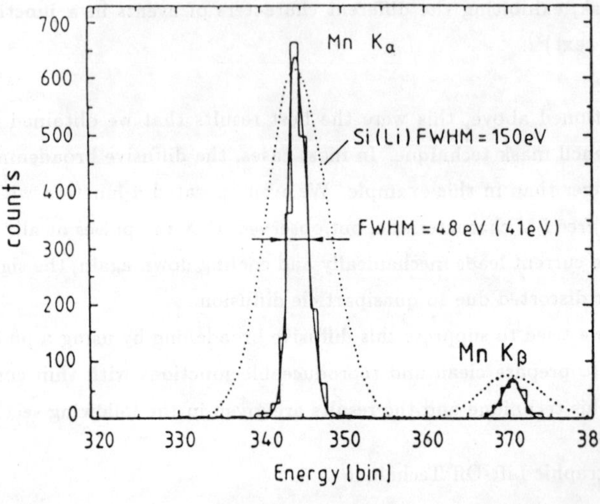

Figure 3 MnK$_\alpha$ and MnK$_\beta$ peaks of our best detector prepared by photolithographic lift-off. The diffusive broadening is totally suppressed. The resolution of a SiLi detector is included for comparison.

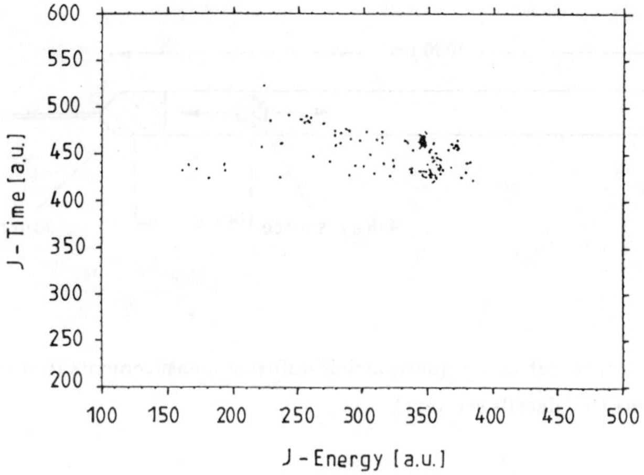

Figure 4 Scatter plot of the signals of the same junction as in figure 3. In contrast to figure 2 the diffusion bands A and B are absent.

However, with the thin current leads alone the diffusive broadening of the peaks could not be suppressed in all cases. Many junctions still showed quasiparticle diffusion. We therefore investigated the influence of the microscopic film structure on the quasiparticle diffusion and hence on the detector response.

2.3 Quasiparticle Diffusion Measurments In Films With Different Granularities

To investigate the diffusion of excess quasiparticles in thin superconducting films with different microscopic structures, we operated $Sn/SnO_x/Sn$ junctions with a wide current lead, which could be illuminated by a moveable ^{55}Fe X-ray point source. Fig. 5 shows the experimental set-up.

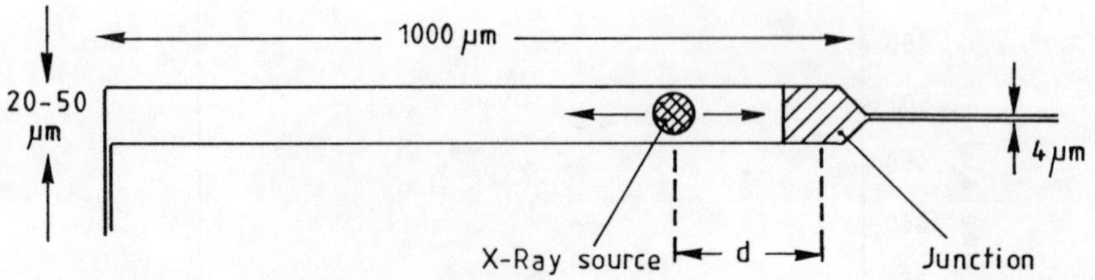

Figure 5 **Experimental set-up for quasiparticle diffusion measurements in superconducting films (for details see text).**

The risetime and the pulseheight of the signals could be obtained as a function of the distance d from the source to the region of overlap. The source could be driven along the upper current lead from outside the cryostat with a micrometerscrew. We were able to adjust the distance d reproduceable with an accuracy of ~20 μm. The area exposed to the X-rays had a diameter of ~100 μm.

We performed these measurments for different temperatures and junctions with different granularities. With increasing temperature (0.45 K - 0.65 K), the pulseheight of the signals as well as the risetime decreased slightly, indicating the growing importance of quasiparticle recombination.

Fig. 6 and fig. 7 show the pulseheight and the risetime versus distance d for two junctions with different granularity. One junction (junction 1) had large grains (~1 μm) with only a few links between them, the other one (junction 2) had small grains (~0.25 μm) with many links. As can be seen, the junction with the large grains collects more charge in the region of overlap, but exhibits a steep decrease of the pulseheight with increasing d. The risetime of the signals is higher for junction 1 and increases much stronger with increasing d than for junction 2. The numerically calculated curves, which are also included in the plots, are discussed in section 3.

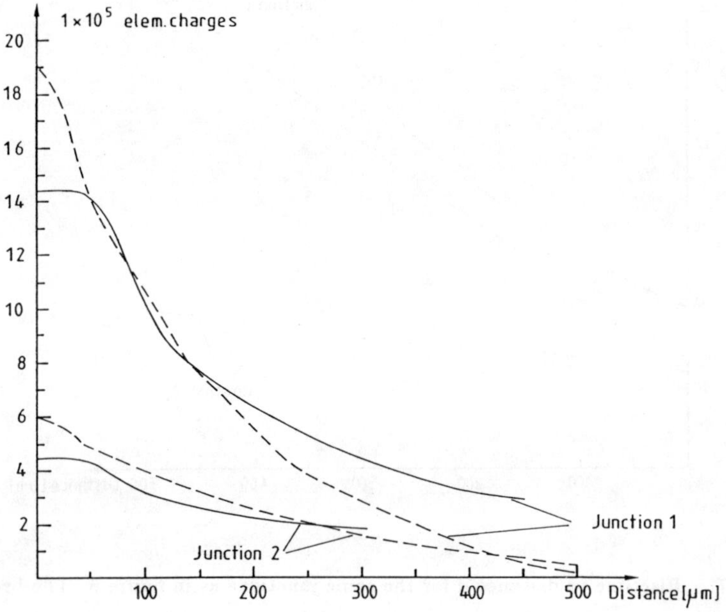

Figure 6 Pulseheight of the X-ray signals vs distance d from the source to the tunneling region for two junctions with different granularities. The theoretical curves (broken lines) have been calculated numerically using the model explained in section 3.

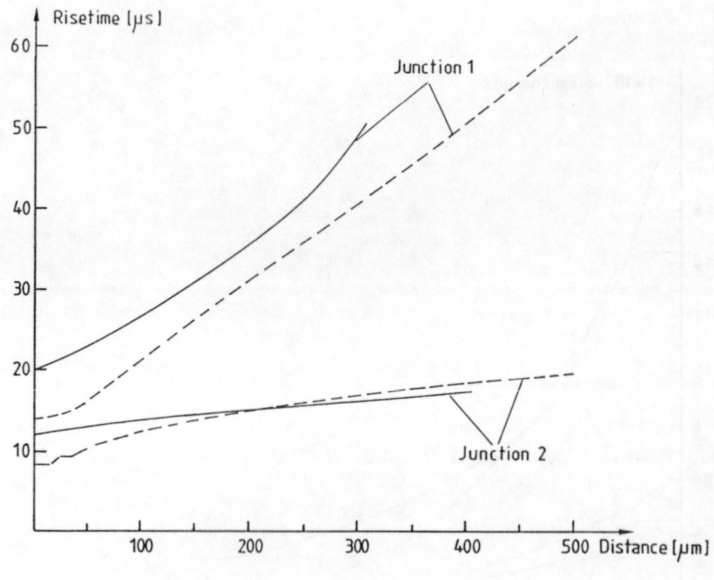

Figure 7 Risetime vs distance d for the same junctions as in figure 6. The broken lines represent the numerically calculated curves.

3. THEORY AND DISCUSSION

3.1 Absorption Of X-Rays In Superconductors: Typical Time Scales

When a 6 keV X-ray is absorbed in a superconducting film, a photoelectron is produced which ionizes atoms along its trajectory, leading to electrons with energy of several eV. Then inelastic scattering processes distribute the energy among more electrons in a time of the order of picoseconds[3)] until the mean energy per electron is e few meV. Finally, we get a distribution of excess quasiparticles with a mean energy of the order of the gap and phonons with energies smaller than twice the gap within the first nanosecond. At temperatures $T/T_c < 0.2$ the quasiparticle recombination time is of the order of several hundred μs[4)]. Since our junctions show tunneling times of typically 10 μs, the quasiparticles cross the oxide barrier and contribute to the collected charge before they decay.

3.2 Rate Equations For Quasiparticles And Phonons

The dynamical behaviour of the quasiparticle-phonon system in a superconductor can be described by the Rothwarf-Taylor equations for the number densities of the respective quantities[5].

$$\begin{aligned}
\frac{d}{dt}(n_{th}+n_{ex}) &= 2\tau_B^{-1}(n_{th}^*+n_{ex}^*) - R(n_{th}+n_{ex})^2 - \tau_T^{-1}n_{ex} \\
\frac{d}{dt}(n_{th}^*+n_{ex}^*) &= -\tau_B^{-1}(n_{th}^*+n_{ex}^*) + \frac{1}{2}R(n_{th}+n_{ex})^2 - \tau_\gamma^{-1}n_{ex}^*
\end{aligned} \qquad (1)$$

n : Number density of (thermal and excess) quasiparticles

n^* : Number density of (thermal and excess) phonons

τ_B^{-1} : Phonon pair breaking rate

R : Recombination coefficient

τ_T^{-1} : quasiparticle tunneling rate

τ_γ^{-1} : Phonon escape rate

Evaluating the quadratic terms, one can subtract the contribution of the thermal number densities in both equations, which yields an expression for the remaining excess densities. To account for the quasiparticle diffusion, we multiply these equations with a time dependent volume V(t), which is given by the film thickness and the FWHM of a gaussian describing the spreading of the quasiparticle cloud with time. This gives two equations for the total numbers of excess quasiparticles and phonons for each film of the junction:

$$\begin{aligned}
\frac{d}{dt}N_1 &= 2\tau_B^{-1}N_1^* - \frac{R}{V_1(t)}N_1^2 - (\tau_R^{-1}+A_1(r_0,t)\tau_{T12}^{-1})N_1 + A_2(r_0,t)\tau_{T21}^{-1}N_2 \\
\frac{d}{dt}N_1^* &= \frac{1}{2}\tau_R^{-1}N_1 + \frac{1}{2}\frac{R}{V_1(t)}N_1^2 - (\tau_B^{-1}+\tau_{2W}^{-1}+\tau_{\gamma 1}^{-1}+\tau_{\gamma 12}^{-1})N_1^* + \tau_{\gamma 21}^{-1}N_2^* \\
\frac{d}{dt}N_2 &= 2\tau_B^{-1}N_2^* - \frac{R}{V_2(t)}N_2^2 - (\tau_R^{-1}+A_2(r_0,t)\tau_{T21}^{-1})N_2 + A_1(r_0,t)\tau_{T12}^{-1}N_1 \\
\frac{d}{dt}N_2^* &= \frac{1}{2}\tau_R^{-1}N_2 + \frac{1}{2}\frac{R}{V_2(t)}N_2^2 - (\tau_B^{-1}+\tau_{2W}^{-1}+\tau_{\gamma 2}^{-1}+\tau_{\gamma 21}^{-1})N_2^* + \tau_{\gamma 12}^{-1}N_1^*
\end{aligned} \qquad (2)$$

$N_i(N_i^*)$	: Total number of quasiparticles (phonons) in film i
$\tau_R^{-1} = 2Rn_{th}$	: Thermal recombination rate
$\frac{R}{V_i(t)} \cdot N_i$	: Self recombination rate
τ_{Tij}^{-1}	: Tunneling rate from film i into film j
$A_i(r_0, t)$	: Fraction of quasiparticles in tunneling volume of film i
r_0	: Space coordinate of initial event
$\tau_{\gamma i}^{-1}$	: Phonon escape rate from film i into surroundings
$\tau_{\gamma ij}^{-1}$	: Phonon transmission rate from film i into film j
τ_{2W}^{-1}	: Break up rate of 2 Δ phonons

The tunneling rate is determined by the normalconducting resistance of the junction R_N (slope of I-V characteristic at $V > 2\Delta/e$) and the film thickness t[6]:

$$\tau_T^{-1} = (R_N e^2 N_o A t)^{-1} \tag{3}$$

Here N_o is the single spin density of states at the Fermi energy, e is the electron charge and A is the area of the junction. The temperature dependent recombination rate τ_R^{-1} of the quasiparticles was determined by Kaplan et al.[4].

$$\tau_R^{-1}(T) = \sqrt{\pi(\frac{2\Delta}{K T_c})^5} \sqrt{T/T_c} \, e^{-\Delta/KT} \cdot \tau_o^{-1} \tag{4}$$

(low temperature limit)

Δ is the energy gap of the superconductor at T = 0 K, T_c is its critical temperature and K is the Boltzmann constant. τ_o is a material dependent characteristic time. For tin at T = 0.4 K, the calculated recombination time is (τ_o = 2.3 ns):

$$\tau_R(Sn, T = 0.4\,K) = 3.7\,ms \tag{5}$$

The geometric properties of the junction are included in the coefficients $A_i(r_0, t)$. The FWHM of the diffusion gaussian is given by $\sqrt{<v_F> \lambda\, t}$, with $<v_F>$ and λ being the thermal average of the Fermi velocity[7] and the quasiparticle mean free path, respectively.

3.3 Discussion Of The Quasiparticle Diffusion Measurments

An X-ray of 5.89 keV energy creates in tin approximately $6 \cdot 10^6$ charge carriers[8], but we only collect ~20% of the initial charge. We evaluated the rate equations (2) numerically in order to reproduce the experimental results (see fig. 6 and 7) by adjusting the quasiparticle mean free path.

We defined a diffusion time τ_D given by

$$\tau_D = \frac{L^2}{<v_F> \lambda} \quad (6)$$

where L is the length of the tunneling region. The influence of the quasiparticle diffusion on the pulseheight and the risetime of the signals can be pointed out by considering two contrary situations:

1. $\tau_D \ll \tau_T$: Fast diffusion limit

   If the quasiparticle diffusion is much faster than the tunneling time, they will extend over the total area (junction and lead) before the onset of tunneling. Since the recombination time, which describes the only loss mechanism for quasiparticles, is in the temperature range of $T/T_c < 0.2$ much longer than the tunneling time (see (5)), the quasiparticles in the leads are not lost. In this case the time τ_C, which is needed to collect the charge with the preamplifier, will be

   $$\tau_C = \tau_T \frac{A_{tot}}{A_j} \quad (7)$$

   where A_{tot} and A_j is the total area and the junction area, respectively. The preamplifier will collect the whole amount of initially produced charge, but if τ_C becomes comparable with the RC-time τ_F of the preamplifier feedback loop, the pulseheight of the signals depends on this feedback-time τ_F, which in our case could be calibrated by a pulser with adjustable rise- and falltime. However, pulseheight and risetime will be independent of the location of the X-ray absorption.

2. $\tau_D \gg \tau_T$: Slow diffusion limit

   If the X-ray is absorbed in the tunneling region, the charge collection time τ_C will be given by the tunneling time

   $$\tau_C = \tau_T \quad (8)$$

   and the total initial charge will be collected. If the absorption takes place in the current leads, the charge collection time will be dominated by the slow diffusion of the quasiparticle cloud, increasing rapidly with increasing distance d. The pulseheight of the signals will then again depend on τ_F, for the same reason as mentioned above.

The risetime slope of junction 1 in fig. 7 indicates, that the junction is close to the slow diffusion limit, thus exhibiting a small quasiparticle mean free path. Therefore, the charge collection time in the tunneling region should be close to the tunneling time. We adjusted the quasiparticle mean free path in our calculations to reproduce the amount of collected charge and the risetime of the pulses over the junction region. However, we could not account for the steep decrease of the pulseheight with increasing distance d.

From the dependence of the risetime on d (see fig. 7) we see, that junction 2 is closer to the fast diffusion limit. The quasiparticle mean free path has therefore to be set larger than for junction 1. Thus the charge collection time in the region of overlap should be much larger than the tunneling time. But from fig. 7 it becomes apparent that junction 2 shows even a faster risetime than junction 1. With this model, the experimental data of junction 2 could not be reproduced.

To fit our data satisfactory, we had to insert an empirical quasiparticle loss term into the rate equations (2), which is characterized by a loss rate representing an additional free parameter in the numerical calculations. The so calculated curves with the assumed loss times can be seen in fig. 6 and 7.

Some possible explanations for this yet unknown loss mechanism could be:

- Location dependent energy gap in the superconducting films. Quasiparticles could be trapped in regions with a lowered energy gap and be lost if they are not adjacent to the tunneling barrier.
- Impurity aided quasiparticle recombination.

Further work has to be done to clarify this behaviour.

4. CONCLUSIONS

In our work with superconducting $Sn/SnO_x/Sn$ tunneling junctions as X-ray detectors we could show, that for high resolving power two parameters have to be optimized:

1. The quasiparticle diffusion from the tunneling region into the current leads and vice versa has to be suppressed.
2. The grain structure has to be optimized to lead the lowest possible loss rate.

A yet unknown loss mechanism had to be inserted into the rate equations (2) describing the quasiparticle-phonon system to obtain agreement with the experimental measurments. We observed that increasing the grain size leads to the collection of more charge and hence improves the resolution.

We could suppress the quasiparticle diffusion by preparing the junctions by means of photolithographic lift-off, which provides very thin leads and a clean geometry, resulting in an improved energy resolution.

Further investigations will have to be done on the identification of the quasiparticle loss mechanisms to get some insight into the influence of the microscopic film structure on the detector response.

ACKNOWLEDGEMENTS

We are very grateful to Prof. Dr. Melchior and his collaborators at ETH Zürich, namely R. Widmer, who helped us with the photolithographic preparation techniques and provided us with the major equipment, and to P. Jokinen, who supplied the necessary electronics.

REFERENCES

1) Wood, G.H. and White, B.L., Appl. Phys. Lett. 15, 237 (1969).
2) Twerenbold, D. and Zehnder, A., J. Appl. Phys. 61,1 (1987).
3) Chi, C.C., Loy, M.M.T. and Cronemeyer, D.C., Phys. Rev. B23, 124 (1981).
4) Kaplan, S.B., Chi, C.C. and Langenberg, D.L., Phys. Rev. B14, 4854 (1976).
5) Rothwarf A. and Taylor, B.N., Phys. Rev. Lett. 19, 27 (1967).
6) Ginsberg, D.M., Phys. Rev. Lett. 8, 204 (1962).
7) Narayanamurti, V., Dynes, R.C., Hu, P., Smith, H. and Brinkmann, W.F., Phys. Rev. B18, 6041 (1978).
8) Kurakado, M., Nucl. Instr. and Methods 196, 275 (1982).

LOCATION SELECTIVE IRRADIATION OF SUPERCONDUCTING TUNNEL JUNCTIONS WITH LOW ENERGY X-RAYS

H. KRAUS, F. PROBST, F.V. FEILITZSCH, TH. PETERREINS

Technische Universitat Munchen,
D-8046 Garching, Germany

We present experimental results on high resolution X-ray detection using superconducting $Sn/Sn0_x/Sn$ tunnel junctions.

Our earlier experiments and calculations indicate that the energy resolution in our superconducting tunnel junctions is limited by quasiparticle diffusion. Consequently, we performed experiments to investigate quasiparticle diffusion in the junction. We report on experiments where the junction was scanned by X-rays and pulse height spectra for various locations of irradiation were recorded. These experiments reveal detailed information on quasiparticle dynamics. Diffusion and other loss processes are discussed.

With our experimental information on loss processes we believe that considerable improvements in the energy resolution can still be achieved. Up to now, we demonstrate an energy resolution of 88eV (FWHM) at 5.89keV.

1. Introduction

Since a few years the properties of superconducting tunnel junctions as high resolution X-ray detectors are investigated [1,2]. As the energy necessary to create a single charge carrier in the superconductor is less by a factor of app. 10^3 as compared to a semiconductor, one expects the possible energy resolution to be enhanced by more than one order of magnitude. This would lead to a statistical limitation of the energy resolution of 7eV (FWHM) for 5.89keV X-rays. Up to now, an energy resolution of 88 eV (FWHM) at 5.89keV is demonstrated [3].

2. Motivation Of Quasiparticle Trapping

To understand the results of our earlier experiments, we developed a set of equations that describes the dynamics of quasiparticles and phonons in the superconducting tunnel junctions. This model is based on the Rothwarf-Taylor [4] equations extended for our purposes. As a complete discussion of the rate equations would be beyond the scope of this article we list only the basic processes that have to be included:

1. tunneling of quasiparticles from one film into the other and vice versa
2. diffusion of quasiparticles and phonons within the films
3. recombination of quasiparticles by emission of 2Δ-phonons and pair breaking by reabsorption of these phonons
4. exchange of phonons between the films and their escape into the substrate.

In our experiments, we rendered the thermal recombination of quasiparticles meaningless by lowering the operating temperature of the cryostat to $\frac{1}{10}$ of the superconductor's critical temperature. The use of different geometries for our tunnel junctions gave us first estimations of the extent of the quasiparticle diffusion.

We performed calculations for a junction where the only process to reach the equilibrium state after energy absorption is the escape of phonons from the superconductor into the substrate. The results of these calculations yielded:

1. The time the excess quasiparticles spend in the tunnel junction should be prolonged considerably in comparison to a junction with losses due to quasiparticle diffusion allowed. Thus the total amount of charge carriers crossing the tunnel barrier increases which might largely improve the signal to electronic noise ratio.

2. The energy spectrum is unambiguous. If the lifetime of the excess density is short or comparable to the tunnel time constant, the absorption of equal energies in the two layers of the tunnel junction leads to different signal amplitudes. In the case of the junction assumed above, the excess density lifetime is very long compared to the tunnel time constants of our tin junctions. Thus we expect only one pulse height for both events of energy absorption in the top film as well as in the bottom film.

3. Experiments Employing Quasiparticle Trapping

3.1 Experimental Set-Up

The diffusion losses can be eliminated if the quasiparticles created by energy deposition in the tunnel junction are confined to a volume close to the tunnel barrier. One method to do this is proposed in references [5,6]. The leads to the junction are formed of a superconductor that has a larger energy gap than the superconducting material used to build the junction. Therefore quasiparticles trying to exit the junction are repelled due to an increase of the energy gap at the transition to the lead's superconductor.

We used $Sn/SnO_x/Sn$ junctions with leads made of lead and a geometry depicted in fig. 1. Sizes for the tunnel area varied from 100μm by 100μm up to 300μm by 350μm.

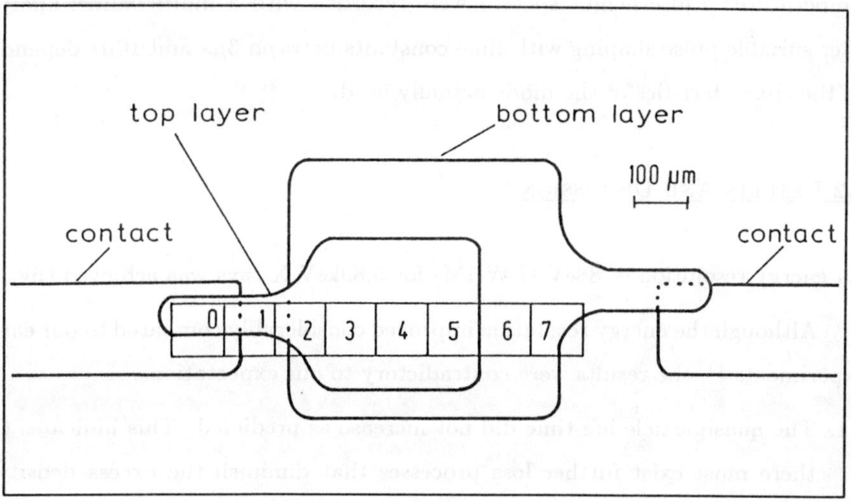

Figure 1: Geometry of the $Sn/SnO_x/Sn$ tunnel junction with leads made of lead. The squares labeled 0 ... 7 denote the areas where the junction was irradiated. These numbers correspond to the spectra in figure 3.

The tunnel junctions were produced by vacuum evaporation of tin through suitable masks and barrier formation by glow discharge oxidation.

The samples were mounted in a vacuum chamber inside a ^{3}He-cryostat. Temperature was measured with a Ge-resistance thermometer attached onto a cold finger close to the junction. A weak magnetic field (20 G - 100 G) parallel to the surface of the junction served to suppress the d.c. Josephson current [7].

Using an ^{55}Fe radioactive electron capture source, we irradiated the detector with 5.89 keV (88%) and 6.49 keV (12%) X-rays. In order to adjust the X-ray flux and to absorb Auger electrons emitted by the source, a mylar foil was inserted between source and detector.

A d.c. bias was supplied with a current source. A 50Ω coaxial cable connected the tunnel junction to a charge sensitive preamplifier operated at room temperature. Pulse height spectra were recorded with a multichannel analyzer after suitable pulse shaping with time constants between $3\mu s$ and $10\mu s$ depending on the characteristics of the diode actually used.

3.2 Results And Discussion

An energy resolution of 88eV (FWHM) for 5.89keV X-rays was achieved (fig. 2).

Although the energy resolution improved considerably compared to our earlier experiments [1)], the results were contradictory to our expectations.

1. The quasiparticle life time did not increase as predicted. This indicates that there must exist further loss processes that diminish the excess density of quasiparticles much faster than expected from thermal recombination alone.
2. Consequently, the peaks resulting from the absorptions in the two layers of the junction can still be resolved and multiple tunneling [8)] of quasiparticles is negligible. Hence, the signal-to-noise ratio did not improve significantly.
3. The lower sides of the observed peaks still have a larger half-width compared to the upper sides. This is a hint that quasiparticle diffusion might still contribute to the finite energy resolution of the detector. This point will be discussed more thoroughly below.

It is apparent from the signal's timing that there is at least one additional loss process that was not taken into account up to now. This or these loss processes are of nearly the same importance as the losses of quasiparticles by diffusion out of the tunneling zone in our earlier experiments. They were then occluded by the diffusion losses and appear now as diffusion losses are abolished.

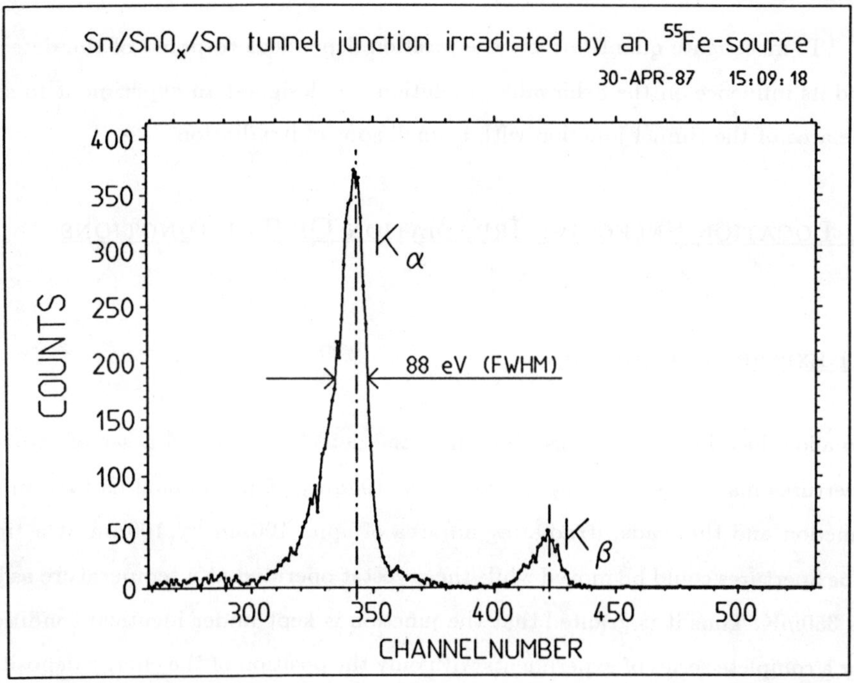

Figure 2: Energy spectrum for the exposure of the complete tunnel junction. The leads to the region where the tunnel barrier is located were covered with sheet copper to prevent irradiation. The energy resolution of 88 eV at 5.89 keV was determined from the well known energy difference of the K_α- and the K_β-energy. The figure shows the pair of peaks resulting from energy absorption in the top layer only.

The asymmetry of the peaks in the spectra indicate that the response of the detector depends on the location of X-ray absorption.

With these results, one has to assume that not quasiparticle diffusion out of the tunnel junction is the resolution limiting factor, but one or more yet unknown loss processes in combination with a finite diffusion constant for quasiparticles inside the films (see section 4.2).

To investigate quantitatively the position dependent response of the detector and its influence on the achievable resolution, we designed an experiment to scan the area of the tunnel junction with a small spot of irradiation.

4. Location Selective Irradiation Of Tin Junctions

4.1 Experimental Details

To allow location selective irradiation of the detector, we used a set of movable apertures made of sheet copper (thickness: 100μm). This enabled us to scan the junction and the leads, irradiating an area of app. 100μm by 100μm at a time. The apertures could be moved while the cryostat operated at a temperature as low as 350mK. Thus it is granted that the junction is kept under identical conditions for a complete series of experiments with only the position of the energy deposition varying. The aperture-mechanism is at a distance of 200μm in front of the junction. Therefore, the boundary of the exposed region is not sharp according to the finite size of the radioactive source.

The operating parameters for the whole experiment were kept unchanged during the eight runs at different positions. The time of measurement for each run was 1000 sec at an average count rate of 4 per second.

The quasiparticle diffusion constant, necessary to calculate the mean propagation length of the quasiparticles, was determined from the conductivity of stripes that were evaporated simultaneously with the layers of the tunnel junction.

4.2 Results And Discussion

The junction geometry is shown in figure 1.

The numbered positions in figure 1 with areas of 100μm by 100μm each correspond to the equally labeled spectra in fig. 3. These areas were selected for the irradiation experiments. It is possible to irradiate both layers simultaneously (positions 2,3,4 and 5), only the bottom layer (pos. 6 and 7) or the leads (pos. 0 and 1).

Spectra 3,4 and 5 are clearly from irradiation of the tunnel barrier region, while spectrum 2 corresponds to a relatively undefined spot at the beginning of the tunnel volume and the transition to the lead. The pair of peaks (K_α, K_β) around pulse height 370 corresponds to absorption events in the top (smaller) layer while the peak around pulse height 180 results from events in the bottom (larger) layer [1)].

The resolution of the pair of peaks resulting from energy absorption in the top layer is best for pos. 4 and 5. It worsens as we approach the rim of the tunnel volume. For the spectra at position 2 and to some extent for pos. 3, one can infer that a pronounced density of excess quasiparticles diffuses into the lead of the top layer. Due to an additional yet unknown loss process, the excess density of quasiparticles has decayed before it is distributed homogeneously over the layer's volume.

Spectra 6 and 7 result from irradiation of the bottom film only. The pulse height decreases with increasing distance from the tunnel barrier region. During the life time of the quasiparticles, most of them in spectrum 5 are in the tunnel area, as they are created inside and spread outwards, with the center of density inside the tunnel area. This creates the peak at channel number 150 with its lower end at channel 100. In the opposite case (spectrum 6) the quasiparticles are created outside the tunnel area and hence must first diffuse into it. Thus the pulse

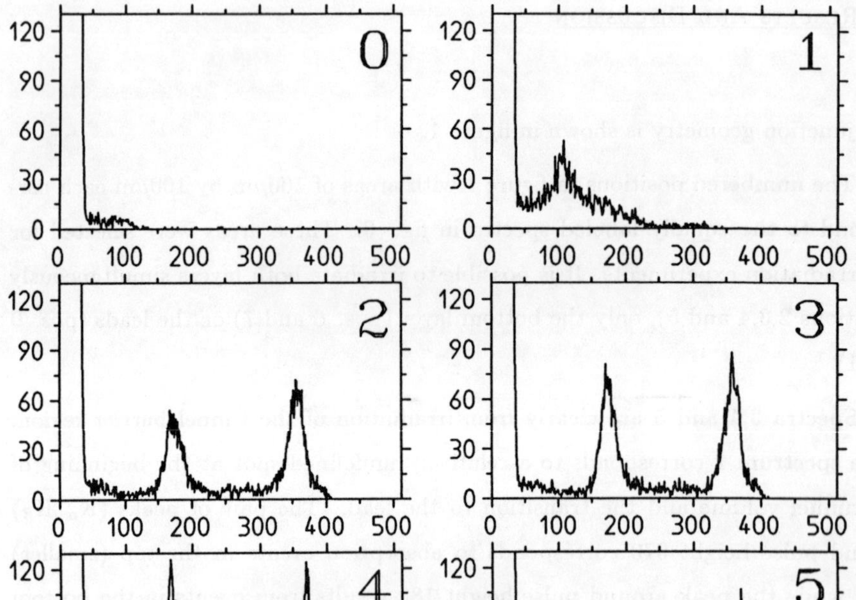

Figure 3: **Energy spectra for location selective irradiation of a junction with geometry depicted in fig. 1. The numbers in the upper right corner of each spectrum corresponds to the denoted locations in fig. 1.**

height drops suddenly from spectrum 5 to spectrum 6, forming a peak at channel number 70. In spectrum 6, some X-rays still hit the top layer (counts at pulse height 370).

Spectrum 7 results from absorption at still farer distances from the tunnel volume producing lower pulse heights.

In spectra 0 and 1 X-ray absorption in the lead lead contributes to the observed signals. It is not easy to distinguish clearly between the different events leading to the pulse height spectra obtained. Especially for spectrum 1 one has to take into account that due to the different energy gaps in tin and lead the number of quasiparticles created is different for the same amount of energy absorbed in the two materials.

Although we used a mylar foil to shield off the auger-electrons emitted from the source, there are two background processes that remain:

1. The mean attenuation averaged with the relative intensities for the X-rays after passing the 350μm mylar foil is only 19450 in a 100μm copper sheet. Thus, the junction as whole is always irradiated with $\frac{1}{19450}$ intensity compared to the intensity in the desired region (See spectrum 0 and 7 in fig. 3).
2. While traversing the two 100μm-wide and 100μm-thick apertures a not negligible amount of X-rays creates secondary electrons due to photo effect along the copper boundary. Due to the narrow distance between the aperture and the junction, these electrons cannot be shielded. They impinge on the junction with an energy distribution from 0 to 6.49 keV and therefore lead to a background of almost equally scattered pulse heights. (See spectra 2 thru 6 in fig. 3).

Summing up all spectra (0...7), that is, simulating a homogeneous exposure of the whole junction leads to the spectrum in fig. 4. The dip at pulse height 40 would vanish if a spectrum resulting from irradiation at a location beyond position 7 was added. The intensified count rate at almost all channels compared to the spectrum in fig. 2 is due to the stray electrons from the aperture now present eight fold. This is especially apparent for channels 210 thru 320. One has to recall that this spectrum (fig.4) is summed up from eight single spectra. Each irradiation involved a set of apertures causing stray electrons mentioned above. This explains the intensified count rate at all channels compared to the spectra resulting from irradiation of the whole junction without apertures.

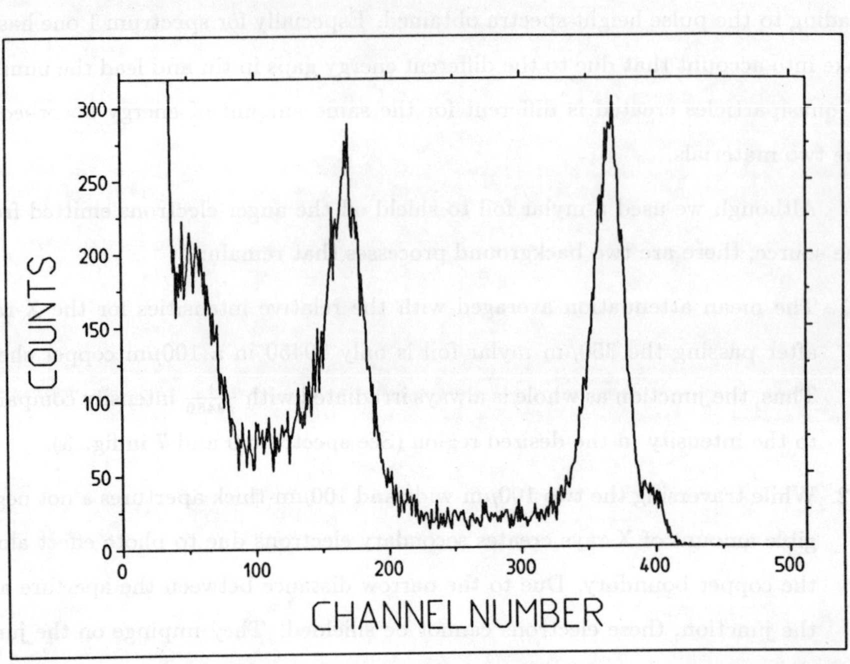

Figure 4: Energy spectrum resulting from superposing (summing counts at equal channelnumbers) the eight spectra of fig. 3. This simulates a homogeneous irradiation of the whole detector.

The rise time of the pulses at the output of the charge sensitive preamplifier was only $10\mu s$, indicating a rapid decay of the quasiparticle excess density.

The lifetime of the excess quasiparticle density is significantly determined by the loss process with the shortest time constant. In our earlier experiments [1]) using junctions without contacts of lead, we measured quasiparticle lifetimes of $5\mu s \ldots 10\mu s$. With the experimental data of timing, geometry and the quasiparticle diffusion constant we concluded that the excess density relaxes via diffusion out of the tunnel volume into the leads. Common to all experiments employing quasiparticle trapping are yet unknown additional loss processes that impose a maximum excess quasiparticle lifetime of $5\mu s \ldots 20\mu s$, depending on the particular junctions and their film properties. Possible reasons for such loss processes could be:

inhomogeneous, location dependent energy gap in the superconductor

film structure (granularity and impurities)

trapped magnetic flux

From the spectra shown in fig. 3, following data are taken:

Table 1: The pulse heights for the absorption of the ^{55}Mn-K$_\alpha$ energy in the top and bottom layer of the junction and their ratios for the locations 2 thru 5 (junction area).

Position:	2	3	4	5
pulse height (top):	357	361	369	373
pulse height (bottom):	168	169	165	143
$\frac{\text{pulseheight (top)}}{\text{pulseheight (bottom)}}$:	2.12	2.13	2.24	2.61

The processes responsible for the observed signals are the tunnel processes [3] (Eq. 1 a,b):

$$I_{T_{top\to bottom}} = e\gamma_{T_{top\to bottom}} N_{top} \tag{1a}$$

$$I_{T_{bottom\to top}} = e\gamma_{T_{bottom\to top}} N_{bottom} \tag{1b}$$

$$\gamma_{T_{top\to bottom}} = \frac{1}{e^2 R_\infty A \cdot 4 \cdot n(0)} \cdot \frac{\Delta + eU}{\sqrt{(\Delta + eU)^2 - \Delta^2}} \cdot \frac{1}{d_{top}} \tag{2a}$$

$$\gamma_{T_{bottom\to top}} = \frac{1}{e^2 R_\infty A \cdot 4 \cdot n(0)} \cdot \frac{\Delta + eU}{\sqrt{(\Delta + eU)^2 - \Delta^2}} \cdot \frac{1}{d_{bottom}} \tag{2b}$$

N_{top}, N_{bottom} are the number of quasiparticles in the top and bottom film of the junction. The tunneling probabilities $\gamma_{T_{top\to bottom}}$, $\gamma_{T_{bottom\to top}}$ are inversely proportional to the thicknesses d_{top}, d_{bottom} of the top and bottom superconducting film. R_∞ is the tunnel barrier resistance taken from the I-V characteristic at a voltage $U \gg 2\frac{\Delta}{e}$, $n(0)$ is the single-spin density of states at the Fermi surface and A is the junction area. In thermal equilibrium both channels equally contribute to the electrical tunnel current, whereas the net quasiparticle flow cancels. An extra tunnel current resulting from the excess number of quasiparticles produced by absorption of ionizing radiation in one of the films is also given by (1a,b).

For the junction actually used in this experiment, $R_\infty = 165\mathrm{m}\Omega$. With the relation $\tau_T = \frac{1}{\gamma_T}$ the tunnel time constants are determined.

$$\tau_T(top \to bottom) = 22.7\mu s$$

$$\tau_T(bottom \to top) = 28.9\mu s$$

As the quasiparticle excess density lifetime is short by a factor of ≈3 compared to the tunnel time, the multiple tunneling of quasiparticles is negligible. Thus the observed pulse height should be dependent only on the quasiparticle density in the film where the energy absorption took place.

The initial number of quasiparticles created by an energy deposition of 6 keV is N $\approx 6 \cdot 10^6$, assuming an effective energy necessary to create a single charge carrier of $\Delta_{\text{eff}} = 1.7 \cdot 58\text{meV}$[9]. The creation takes place at a very narrow point-like spot from which the quasiparticle excess density spreads according to quasiparticle diffusion. $\bar{l} = \sqrt{6D\tau}$ is the distance the quasiparticles travel within the time τ in a material with the quasiparticle diffusion constant D.

The ratio of the signal's pulse heights resulting from equal energy deposition in the two films is predictable for two limits of the quasiparticle's propagation:

1. The propagation of quasiparticles is slow ($\bar{l} \leq 50\mu m$). In both, the small (top) as well as in the large (bottom) layer the quasiparticles don't notice the boundaries of the layers as they have decayed before they reach the rim. They are confined only by the thicknesses of the two layers. Thus the ratio of the observed pulse heights for events in the two layers should comply with the inverse ratio of the film thicknesses.
2. The propagation of quasiparticles is fast ($\bar{l} \to \infty$). This is equal to a quasiparticle excess density that is distributed almost instantly and homogeneously over the junction volume before a significant amount has decayed. N_{6keV} is the number of quasiparticles created from energy deposition of 6 keV. This number of quasiparticles has to be used for N_{top} in equation 2a) to calculate the signal for energy absorption in the top film. The number N_{bottom} to use in 2b) for energy absorption in the bottom film is only a fraction of N_{6keV}, namely $N_{bottom} = N_{6keV} \cdot \frac{A_{top}}{A_{bottom}}$. Therefore, the ratio of the pulse heights should approach the inverse ratio of the film volumes.

The ratios of the layer's thicknesses and volumes are:

$$\frac{V_{bottom}}{V_{top}} = 3.18; \qquad \frac{d_{bottom}}{d_{top}} = 1.27$$

It can be seen in table 1 that the measured ratios for the pulse height lie between the above limits. This is confirmed by the mean propagation length of the quasiparticles computed from the quasiparticle diffusion constant and the quasiparticle life time.

The diffusion constant was calculated to be D= $8.36 \cdot 10^{-3} \frac{m^2}{s}$ after measuring the film resistance above the critical temperature. This diffusion constant is short by a factor of 9 in comparison to our earlier experiments [1)]. Assuming a time τ of $10\mu s$ as the effective quasiparticle lifetime, we estimate a mean propagation length of $\bar{l} = \sqrt{6D\tau} = 708\mu m$ for the quasiparticles.

With a diffusion length of $700\mu m$, neither of the limits discussed above applies (compare again the ratios in table 1).

The pulse height resulting from energy deposition in the top layer varies only by 4.4% over the inspected region of the tunnel area. Taking only spectra 4 and 5, the variation is only 1.08%. The pulse height decreases from the innermost position (5) to the position closest to the contact area (2), where some quasiparticles diffuse out of the tunnel zone. The size of the top layer is $300\mu m$ by $350\mu m$. The area the quasiparticles would cover after $10\mu s$ is – as pointed out above – app. a circle of $1400\mu m$ - diameter. This and the weak position dependence of the pulses lead to the conclusion that the quasiparticle density is distributed almost homogeneously over the top layer's volume.

A much stronger dependance of the pulse height on the location can be observed for signals from absorption in the bottom layer.

The volume of the bottom layer is larger by a factor of 3.18 compared to the top layer volume. This means that a quasiparticle distribution of the homogeneity as in the top film cannot be achieved within the quasiparticle lifetime observed. Furthermore, the boundary of the volume in which the quasiparticles are confined is not the tunnel volume. So, quasiparticles can easily diffuse out of the tunnel region. Here, the largest pulse height is obtained for positions 2 and 3, the locations where the quasiparticle density spends the most time in the tunnel region while filling the whole volume of the bottom film. But before a distribution as homogeneous as in the top film can be established, a large part of the excess density has decayed as can be derived from the timing of the signals. The propagation and the decay of the quasiparticle density inside the superconductor can be derived from spectra 5,6 and 7.

5. Conclusions

In our earlier experiments [1] we irradiated junctions with layers of equal sizes but various thicknesses and different sizes of contact leads as a whole.

With the results of our new experiments we must conclude, that the diffusion of quasiparticles out of the tunnel volume is not the only resolution limiting process but at least to the same extent some unknown processes inside the superconducting layers that trap and thus diminish the number of the quasiparticles. These loss processes have to be abolished before the method of quasiparticle trapping can reveal its full potential power.

It is also shown that with junctions of the quality available now, the sensitive area should be limited to obtain better resolution. The sample discussed in this work showed a shift in pulse height of 1.08% or 64 eV for the 5.89 keV X-rays when the location of irradiation was moved over a distance of $100\mu m$.

Our future efforts will concentrate on the identification of the remaining loss processes. By increasing the diffusion constant we hope that the problem of location dependent detector response will lose its importance.

This work was supported by Bundesministerium für Forschung und Technologie.

6. References

1.) Kraus, H., Peterreins, Th., Pröbst, F., v. Feilitzsch, F., Mößbauer, R.L., Zacek, V. and Umlauf, E. : Europhys. Lett. 1, 161 (1986)

2.) Twerenbold, D. and Zehnder, A. : J. Appl. Phys. 61, 1 (1987)

3.) v. Feilitzsch, F., Hertrich, T., Kraus, H., Peterreins, Th., Pröbst, F., Seidel, W. : "Low Temperature Calorimeters", in : Proceedings of the International Conference on Solar Neutrino Detection with ^{205}Tl and Related Topics, Dubrovnik 1986, to appear in Nucl. Instr. and Meth.

4.) Rothwarf, A., Taylor, B.N. : Phys. Rev. Letters 19, 27 (1967)

5.) Kraus, H. : Diplomarbeit, Technische Universität München 1985

6.) Booth, N. : Appl. Phys. Lett. 50, 293 (1987)

7.) Barone, A., Paternó, G., Russo, M., Vaglio, R. : "Measurements of the Maximum Josephson Supercurrent vs. Magnetic Field as a Probe of Tunnel Barrier Uniformity", in
Nonequilibrium Superconductivity, Phonons and Kapitza Boundaries, ed. by K.E. Gray (Plenum Press, New York, 1981) p.673

8.) Gray, K.E. : "Tunneling: A Probe of Nonequilibrium Superconductivity", in Nonequilibrium Superconductivity, Phonons and Kapitza Boundaries, ed. by K.E. Gray (Plenum Press, New York, 1981) p.131

9.) Kurakado, M. : Nucl. Instr. and Methods 196, 275 (1982)

NEUTRINO MASS DETERMINATION BY β DECAY SPECTRUM: PRESENT AND FUTURE SITUATION

F. CELANI, F. CARDONE,
INFN-LNF 00044 Frascati, Italy

S. PACE
Dip. di Fisica, Univ. di Salerno
84100 Salerno, Italy

We report and discuss the results of the recent esperiments on neutrino $\nu_e(\nu_e)^*$ mass, which used the electron spectrum from tritium β decay.

Up to now all the experiments (not only by β decay) performed on the determination of $m(\nu_e)$ gave no solid proof of $m(\nu_e) \neq 0$.

We analize the energy resolution we need to determinate the eventual mass of ν_e in tritium β decay by the measurement of the end-point in the Kurie plot.

The present experimental apparatuses based on Si-Li detectors have a resolution of the order of 200 eV and in the near future people hope to reach resolutions of about 10 eV by improvements in the present techniques. But it needs to reach resolutions less than 10 eV.

Actually the only way to reach resolutions of the order of eV is by using superconducting junction in the tunnel regime with materials having enaugh low value of the energy gap (0.2 ÷ 2 meV).

* This writing is only to remember the possibility of Dirac or Majorana neutrinos.

1. INTRODUCTION

First of all we have to start saying that up to now there is no definitive evidence of massive neutrinos. We will take in account two ideas generally admitted and invoked as "common sense":

A - An a priori hierarchy among neutrino masses, in this scheme $m(\nu_e)$, $m(\nu_\mu)$, $m(\nu_\tau)$ are in the ratio of m(e), $m(\mu)$, $m(\tau)$. This means that ν_τ is at least 20 times heavier than ν_μ, itself 200 times heavier than ν_e.

B - A bound on the sun of neutrino masses that the astrophysicists gave us which corresponds to $\sim 50 \div 100$ eV.

With these two inputs it is easy to derive sensible guesses for the masses: $m(\nu_\tau)$ of order 10 eV, $m(\nu_\mu)$ of order 1 eV, $m(\nu_e)$ of order 10^{-2} eV or less.

We will not speak about the mixing as a possible direct consequence of the non zero neutrino mass but we will consider the reverse situation: to use the mixing or rather the neutrino oscillation to have limits on neutrino mass.

With these a priori expectations we see that the laboratory experiments are at present hopeless.

Now, infact, the experiments have an energy resolution of about 200 eV and in the near future should reach the level of $\sim$ 10 eV accuracy.

But to improve on this it will be extremely difficult with the present techniques. So we hope to reach in the future the 1 eV level by the new superconducting tunnel junction.

The present results (1986-87) from tritium β decay give us:

Moscow ITEP	20 eV <	$m(\nu_e) < 40$	eV[1]
Zürich		$m(\nu_e) < 18$	eV[2]
Los Alamos		$m(\nu_e) < 29$	eV
Tokyo INS		$m(\nu_e) < 32$	eV[3]

Anyway people prefer to remember that all these results are compatible with 0 mass.

From the double β decay experiments with Ge^{76} we have:

$$m(\nu_e) < 1 \div 2 \text{ eV} \qquad (4), (5)$$

This seems a very stringent limit, on order of magnitude better than the tritium case. From the recent experiments on neutrino oscillations[6] we have that they exclude masses down to ~ 0.1 eV.

Obviously the experiments on neutrino oscillations give us only the $m_2^2 - m_1^2$, but admitting the hierarchy, as we said before, we can set limit on neutrino mass too. The technique of the time of flight used in the case of the burst (s) from SN1987A[7] gives us a limit on $m(\nu_e)$ of order 10 eV, at the level of the tritium measurements. At the end we would like to recall the results about ν_μ and ν_τ masses.

$$m(\nu_\mu) < 270 \text{ keV}^{(8)}$$
$$m(\nu_\tau) < 70 \text{ MeV}^{(9)}$$

2. THEORETICAL BACKGROUND

The electron energy spectrum in nuclear β decay is affected by finite ν_e mass as shown in Fig. 1 where we report the classical Kurie plot showing deviation from straight line for non zero neutrino mass.

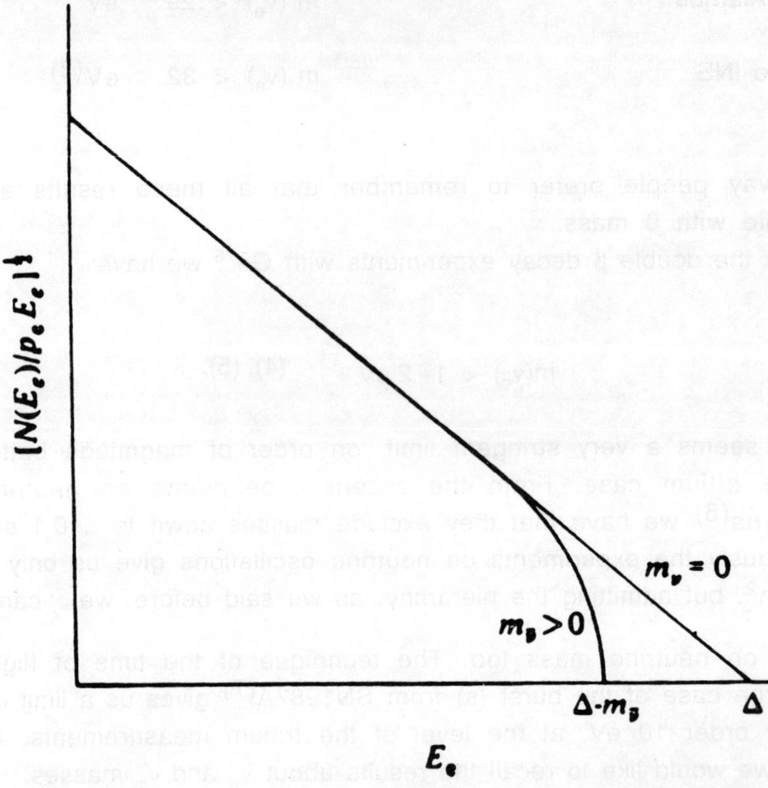

FIGURE 1

The differential transition probability for allowed β decay, averaged over nuclear spin and integrated over e^- and ν_e directions, is:

$$dW = (4G_F^2 \cos\theta_1)/2\pi^3 \; \xi F(Z,E) p_e^2 \, dp_e \, p_\upsilon^2 \, dp_\nu \cdot \delta(\Delta - E - E_\nu) \qquad (1)$$

Including the possibility of finite neutrino mass

$$m_\nu = (E_\nu^2 - p_\nu^2)^{1/2} \qquad (2)$$

Writing $p_\nu^2 \, dp_\nu = p_\nu E_\nu dE_\nu$, and integrating over E_ν, we obtain the electron energy spectrum:

* $$N(E) = G_F^2/2\pi^2 \cos^2\theta_1 \xi F(Z,E) p_e E \cdot (\Delta - E)\sqrt{(\Delta - E)^2 - m_\nu^2} \qquad (3)$$

However, the final atom may be in an excited state i, since each such state i is associated with a different energy Δ_i, (3) must be replaced by

$$N(E) = G_F^2/2\pi^3 \cos^2\theta_1 \xi F(Z,E) p_e E \sum_i W_i(\Delta_i - E)\sqrt{(\Delta_i - E)^2 - m_\nu^2} \qquad (4)$$

where $W_i = | < \psi^f_{atom} | \psi^i_{atom} > |$.

If the sum dominated by a single term and $m_\nu = O$, a plot of $\sqrt{N(E)/pE}$ versus E (kurie plot) is a straight line with negative slope, which intercepts the E axis at Δ (see FIG. 1). A finite mass causes the line curve downward near the endpoint and to intersect the axis with infinite negative slope as shown in FIG. 1.

The existence of other final atomic states usually cause the line to curve the other way, obscuring the effect of finite mass.

* Note: $\theta 1 = \theta_{Cabibbo}$ according the convention of the Kobayashi-Maskawa matrix.

In the three-body β decay we talked about we have the situation:

$$N_1 \rightarrow N_2 + e^- + \nu_e$$

for massive mixed neutrinos such a decay will consist of n components with maximal electron energies (end point energies)

$$\Delta^i_{max} = [M_1^2 + m_e^2 - (M_2 + m_{\nu i})^2] / 2M_1 \qquad i = 1,n \qquad (5)$$

where M_1 is the initial mass of the nucleus and M_2 is the final mass. The corresponding Kurie plot $\sqrt{N/E_e p_e}$, neglecting for a moment the Coulomb correction factor, will have, as said before, a kink at each energy E^i_{max}, but up to now due to poor energy resolution of the present detectors, as we will discuss later, it was not possible to see this effect.

The method of neutrino mass determination based on three-body decays is more sensitive to small neutrino masses, since there is a linear mass dependence of the end point, the accuracy of the neutrino mass determination is essentially equal to the energy resolution of the electron spectrometer.

The best studied case among the three-body β decay is the tritium one.

$$H^3 \rightarrow He^3 + e^- + \nu_e$$

A finite antineutrino mass m_ν will cause deviations from the straight Kurie plot at distances $\sim m_\nu$ from the energies determinated by (5).

In particular we recall that, at least theoretically, the Kurie plot will have an infinite slope at highest end point but in practice, due to the finite resolution of the electron spectrometer, the slope of the Kurie plot is decreased near the end point. So, with a finite neutrino mass the Kurie plot near the end point will be somewhat steeper than expected from the performance of the measuring device assuming m_ν = O see FIG. 2 remembering that Δ = end point energy. The full curves are the expected behavior for $m_\nu = 0$ and $m_\nu = 1$ unit of energy, for the indicated effective resolution width W, the dashed lines represent straight line extrapolation

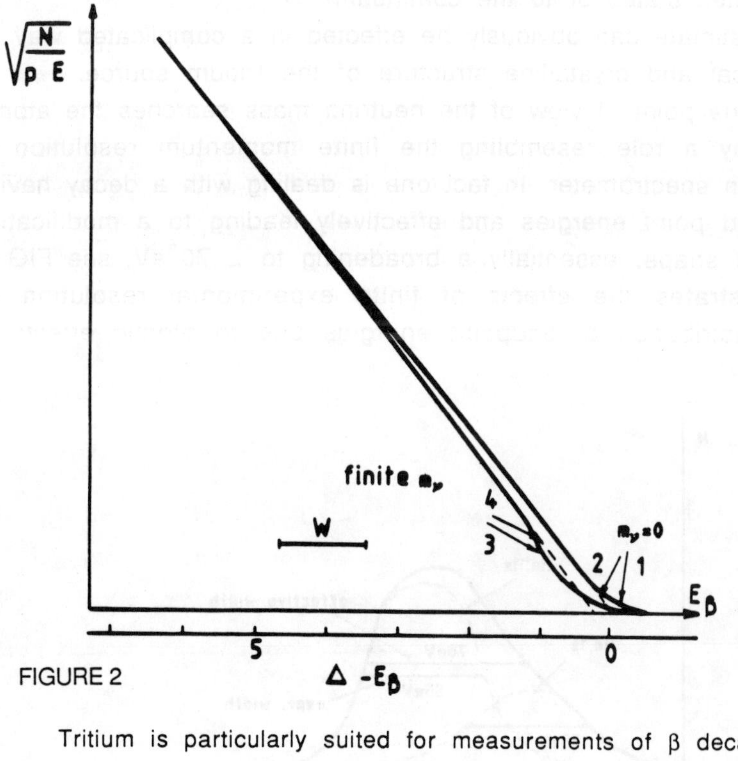

FIGURE 2

Tritium is particularly suited for measurements of β decay due

to its low end point energy Δ = 18.6 keV, low Z value, and convenient half life $T_{1/2}$ = 12.3 year. But the energy resolution of present detectors is poor as we said and so at this level of accuracy an additional difficulty arose.

The initial state for a free tritium atom has one bound electron with binding energy E_B = 13.6 eV during the beta decay nuclear charge suddenly changes by a factor of two. The initial bound spectator (atomic) electron can be shaken up to the $2S_{1/2}$ or to some other excited state of the final He^3 atom.

For a free atom the probability of the final $1S_{1/2}$ state is 70%, of the $2S_{1/2}$ state 25%, and the remaining 5% of the electrons go to higher excited states or to the continuum(10).

This estimate can obviously be effected in a complicated way by the chemical and crystalline structure of the tritium source.

From the point of view of the neutrino mass searches the atomic effects play a role resembling the finite momentum resolution of the electron spectrometer. In fact one is dealing with a decay having several end point energies and effectively leading to a modification of the line shape, essentially a broadening to ~ 70 eV, see FIG. 3 which illustrates the effects of finite experimental resolution 55 eV and distribution of endpoint energies due to atomic effects in tritium(11).

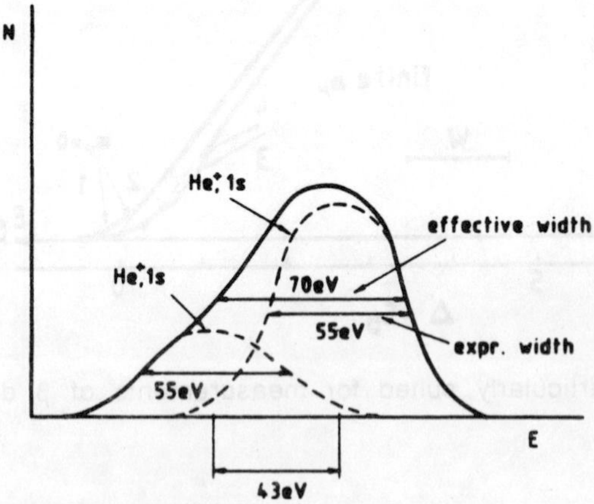

FIGURE 3

So it is possible to perform an experiment with a relatively strong tritium source combined with a good energy resolution but the atomic final state effects represent a serious obstacle to a further improvement of accuracy.

Summarizing it is opportune to have a β source with a low Δ and an hallife reasonably short in order to have an high β emission rate as we said for the tritium $\Delta = 18.6$ keV and $T_{1/2} = 12.3$ years and in the sensible zone of 100 eV the β particles fraction is $\sim 10^{-6}$ whereas in the sensible region of 10 eV is $\sim 10^{-9}$.

Infact remembering:

$$N_\beta \propto F(E,Z)\, p_e\, E\, (\Delta - E) \sqrt{(\Delta - E)^2 - m_\nu^2} \qquad (6)$$

we can see that the stronger change occurs in an interval ΔE below Δ which is at least few times the neutrino mass and the fraction of the number of β particles having $E > \Delta - \Delta E$ (for $m_\nu = 0$) is proportional to $(\Delta E/\Delta)^3$ with proportional constant ≅ 7 assuming that F (E,Z) = constant, as said before the finite resolution of the detector has the effect to deform the "tail" of β spectrum (Kurie plot) in the apposite verse with respect the effect of neutrino mass.

The resolution, as FWHM, was ever greater than neutrino mass, almost in the experiment performed up to now, so we need to know very well the resolution function to calculate the neutrino mass.

The theoretical β spectrum and Kurie plot have to be calculated, at the experimental energy points, by convoluting the resolution function with the theoretical β spectrum:

$$N(E) = C\, F(E,Z)\, p_e\, E\, (\Delta - E) \sqrt{(\Delta - E)^2 - m_\nu^2} \qquad (7)$$

The constant C could be determinated by an opportune fit of the data for particular values of Δ over the β kinetic enrgy range from 10 to 17 keV; also it is possible to use e modified non relativistic coulomb correction factor F (E,Z) according[12] we have to take into account that the main effect of the resolution function R is to "force" the data in interval of width $R \cong$ FWHM.

So in a region of energy R ($R>m_\nu$) below the estrapolated end point Δ there will be a counting number

$$A(m_\nu) \propto \int N_\beta(E)\, dE \propto (R^2 - m_\nu^2)^{3/2} \qquad (8)$$

assuming that the factor $F(E,Z)\, p_e E$ is constant in the integration region.

However if R is over estimated of a quantity ε then $A(m_\nu)$ can be calculated assuming a neutrino mass $M_\nu > m_\nu$, where M_ν is a fictitious mass and m_ν the true mass so:

$$M_\nu^2 - m_\nu^2 \cong 2\,\varepsilon\, R \qquad (9)$$

If e.g. $m_\nu = 0$ strictly we will have a fictitious mass:

$$M_\nu \cong \sqrt{2\varepsilon R} \qquad (10)$$

And this shows that if $R > m_\nu$ the accuracy increases not only improving the resolution but also increasing the knowledge of the resolution function which depends from each different detector.

At the end of this section we would like to recall another possible way to determine the neutrino mass in a three-body β decay using the internal bremsstrahlung electron capture and measuring the photon spectrum shape[13]. For several nuclei e.g. Pt^{193} the Δ value is so low that K (or L) capture is forbidden. This leads to an enhacement of the photon rate near the end point.

The atomic final state effects discussed above are also essentially absent; nevertheless, the pratical utilization of this method is very difficult on account of low efficiency of the resolution (few eV) photon spectrometer required.

3. RECENT EXPERIMENT ON H^3 BETA DECAY

Starting from the conclusions of the previous section we would like to recall the main requirements on the apparatus: good energy resolution, good luminosity, since only a very small fraction $\sim 10^{-6}$ of the spectrum is of interest, and low background, furthermore at the level of accuracy achieved in recent experiments 20 ÷ 50 eV resolution, some additional difficulties arise.

People have to take account that the He^3 atom may be left in an excited state with a probability around 30 - 40%, depending on the chemical composition of the source; the excitation energies are tipically 13-20 eV and the calculation of the spectrum of final states is straightforward for free atoms, but more complicated for molecules. As we said the shape of the energy response function giving the spread of a monoenergetic beam through the spectrometer must be known with great accuracy: overstimating the width can minic a neutrino mass; underestimating it can hide a mass.

An important contribution to the response function comes from the energy losses in the source which must be known with good accuracy.

The difficulties stemming from final-state effects and energy losses were solved in an elegant way by Simpson(12),(14), (15). He implanted H^3 atoms in a Si(Li) detector which was functioning as an active source. Despite of Monte Carlo treatament the data were consistent with a zero mass, and Simpson's result was $m_{\nu e} < 65$ eV (95% CL).

Unfortunately this metod is limited by the resolution of the Si(Li) counter: 300 eV FWHM at 18 KeV, which cannot be significatly improved.

The best instrument to study H^3 beta decay seems to be the magnetic spectrometer, here to minimise energy very thin sources are used.

Bergkvist[11] performed a very careful experiment using a $\pi \sqrt{2}$ spectrometer with central circle radius of 50 cm.

The energy resolution was 40 eV FWHM, and the source consisted of a thin foil of Al with tritium ions implanted near the surface.

Bergkvist found no evidence for a finite non zero mass and reported the limit $m_{\nu e} < 55$ eV (90% CL).

At ITEP in Moscow a toroidal spectrometer with total deflection angle 720° was built and a source of H^3 enriched Valine $C_5 H_{11} NO_2$ was used in all the measumeremts[16], [17], [18].

In the last upgraded version of the spectrometer the scans are performed eletrostatically at a constant magnetic field so that all the electrons hitting the detector have the same energy, reducing sistematic uncertainties. The eletrons emerging from the source are accelerated to an energy above the end-point energy; the threshold of the detector is high enough so that only these accelerated electrons are seen.

This drastically reduces the background due to H^3 contamination of the entire apparatus.

They used an extended source giving a high activity mounted on a resistive plate with an electric fied gradient in order to make it look like a point source. A multiwire proportional chamber was used as detector. The last analysis, incorporating additional data and the most recent calculation of the final-state spectrum of Valine[19], presented by Boris et al.[20] gives a good agreement between the experimental points and the fitted curve for $m_{\nu e} = 34.8 \pm 1.9$ eV. Just to show that the sensitivity to the final states is not too bad let us mention that taking the distribution for free H^3 atoms one obtains

$$m_{\nu e} = 30.9 \quad \text{eV}.$$

Taking, quite unrealistically, the distribution for for H^3 nuclei one obtains $m_{\nu e} = 13.8$ eV.

Trying to take into account the uncertainty from the final states, Boris et al. conclude that a realist estimate is

$$20 < m_{\nu e} < 40 \text{ eV}$$

The most advanced of the competing experiments is being performed by a group at the University of Zurich(21), a toroidal spectrometer, similar to the ITEP design, has been built. Here, however, the electrons are decelerated after the source which allows them to achieve a good energy resolution with a comparatively large luminosity; the source consists of tritium implanted in carbon.

The response function was determined by using X-ray source and an electron gun, the background was low and they refer a resolution of 25 eV but the data are again consistent with a vanishing mass. Taking into account statistical errors only the limit $m_{\nu e} < 10$ eV (95% CL) is derived, but an analysis making allowance for possible systematic errors, including the uncertainty on the final states, gives the firm upper limit

$$m_{\nu e} < 18 \text{ eV}$$

in disagreement with the ITEP result.

Another interesting programme is being pursued at Los Alamos [22] also with a toroidal spectrometer, but the most salient feature is the source, which will consist of free tritium atoms. In this case the final-state spectrum is particularly simple.

4. CONCLUSION AND OUTLOOK

Looking back at all the preceding sections one concludes that the theoretical situation is clear enough but in spite of a vast experimental effort, there is presently no strong evidence for ν_e mass. The result obtained from the study of tritium β decay by Moscow group $20 < m_{\nu e} < 40$ eV is contradicted by the Zurich limit $m_{\nu e} < 18$ eV.

Other competing experiments are in progress, but at present level of accuracy the experimental problems are quite formidable, and the results are coming out slowly. Using electron Spectrometer with tunnel junction in superconducting regime made by material with low energy gap (0.2 ÷ 20 meV) we will expect for the energy resolution[23]:

$$\Delta E/E = 2.35 \sqrt{FW/E} \tag{11}$$

where F is the Fano factor ($0<F\leq1$) and W is twice the gap energy of the material. We shall consider the case of the tin having $E_{gap} \cong 0.5$ meV and the worse situation with F = 1 so we have:

$$\Delta E/E = 7.4 \cdot 10^{-2}\ E^{-1/2} \tag{12}$$

In FIG. 4 we plot ΔE versus E to show the expected behaviour of this energy resolution in the 0 ÷ 20 KeV region (where one shoud have to do the measurements for β spectrum from H^3).

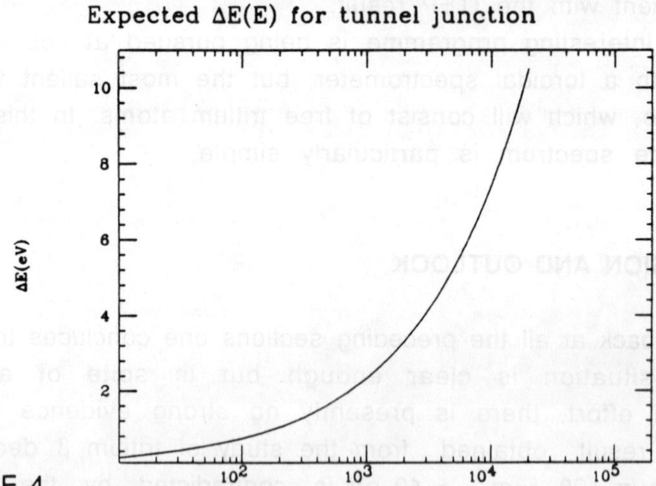

FIGURE 4

In the same previous approximation we have for the energy resolution function:

$$R(E) \propto E^{1/2} \tag{13}$$

while for the tritium end point $\Delta \cong 18$ KeV it will have an energy resolution $R(\Delta) \cong 10$ eV about one order of magnitute below the present apparatuses.

Doing the convolution between this resolution function (13) and the theoretical spectrum given by (7)

$N(E) = C \quad F(E,Z)\, p_e E \quad (\Delta - E) \quad [(\Delta - E)^2 - m_\nu{}^2]^{1/2}$, we calculate the expected experimental spectrum N_s:

$$N_s = \int N(E')\, R(E' - E)\, dE' \tag{14}$$

So in our case we have, assuming $F(E,Z)\, p_e\, E$ constant:

$$N_s \propto \int_E^\Delta (\Delta - E') \, [(\Delta - E')^2 - m_\nu]^{1/2} \, (E' - E)^{1/2} \, dE' \tag{15}$$

where $0 < E < \Delta$.

In the semplified case assuming $m_\nu = 0$ we have

$$N_s \propto \int_E^\Delta (\Delta - E')^2 \, (E' - E)^{1/2} \, dE' \tag{16}$$

$$N_s \propto (\Delta - E)^{7/2} \tag{17}$$

So for the Kurie plot we expect for the variable K a dependence from the energy E of the kind:

$$K \propto (\Delta - E)^{7/4} \qquad (18)$$

as shown in Fig. 5.

Actually this is the more reasonable prediction we can do about what we expect to see of a tritium β spectrum using this kind of devices.

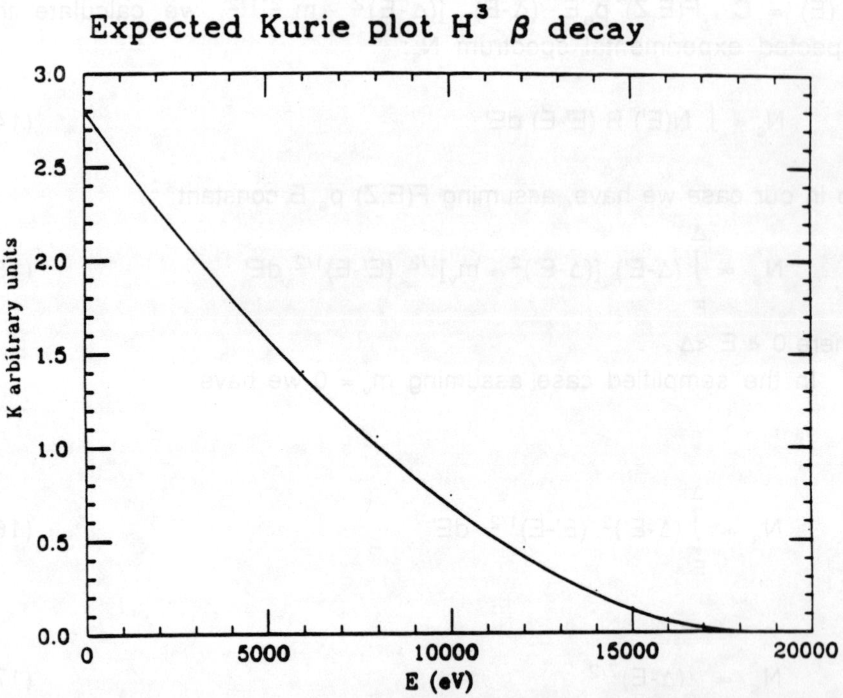

FIGURE 5

REFERENCES

(1) XII International Conference on Neutrino Physics and Astrophysics, Sendai, Japan, June 1986.

(2) M. Fritschi et al., Phys. Lett. B173, 485 (1986).

(3) H. Kawakami et al., Phys. Lett. B187, 198 (1987).

(4) E. Bellotti et al., Phys. Lett. B146, 450 (1984).

(5) D.O. Caldwell et al., Phys. Rev. Lett. 54, 281 (1985).

(6) Workshop on Neutrino Oscillations, BNL, February 1987, ed. M. Murtagh.

(7) K. Hiraka et al., Submitted to Phys. Rev. Lett. (1987).
R.M. Bionta et al., Submitted to Phys. Rev. Lett. (1987).

(8) R. Abela et al., Phys. Lett. B146, 431 (1984).

(9) H. Albrecht et al., Phys. Lett. B163, 404 (1985).

(10) K.E. Bergkvist, Phys. Scr. 4, 23 (1971).

(11) K.E. Bergkvist, Nucl. Phys. B39, 317 (1972).

(12) J.J. Simpson, Phys. Rev. D23, 649 (1981).

(13) A. De Rujula, CERN TH 3045 (1981), Nucl. Phys. B 188 (1981) 414.

(14) J.J. Simpson, Phys. Rev. D24, 2971 (1981).

(15) J.J. Simpson, Phys. Rev. Lett. 54, 1891 (1985).

(16) E. Tretjakov, et al. Proc. Int. Neutrino Conference (1976) Aachen, ed. H. Faissner, H. Reithler and P.Zerwins (Brauschweig : Vieweg), 663.

(17) V. Lubimov, et al., Phys. Lett. 94B, 266 (1980).

(18) S. Boris, et al., Proc. Int. Europhys. Conf. on High Energy Physics (1983) Brighton, ed J. Guy and C. Costain (Chilton : Rutherford Appleton Laboratory), 386.

(19) I. Kaplan, V. Smutnyi and Smelov, Sov. Phys. - JETP 57, 483 (1983), Dokl. Akad. Nauk. 279, 1110 (1984).

(20) S. Boris, et al., Phys. Lett. 159B, 217 (1985).

(21) M. Fritschi, et al., Phys. Lett. 173B, 485 (1986).

(22) R. Robertson, et al., Proc. Int. Conf. Neutrino 82. Balatonfuered, Hungary vol. 1 ed. A. Frenkel and L. Jenik, Budapest : Roland Eotvös Physical Society, 51 (1982) .

(23) S. Pace, F. Celani et al., Proc. ISEC 87, 349 (1987).

LIMITS TO THE ENERGETIC RESOLUTION OF IONIZING PARTICLE DETECTORS BASED ON SUPERCONDUCTING TUNNEL JUNCTIONS IN THE GIAEVER REGIME

S. PACE, R. SCAFURO
Dipartimento di Fisica, Università di Salerno,
84100 Salerno, Italy

F. CELANI, U. GAMBARDELLA, A. SAGGESE
INFN-Laboratori Nazionali di Frascati, C.P. 13,
00044 Frascati, Italy

1. - INTRODUCTION

In the past only few experiments on the performances of superconducting tunnel junctions (STJ) used in the Giavier regime as proportional detectors of nuclear radiation have been performed [1,2,3,4] . Moreover, the experimental energetic

resolution was worse than the expected one. We believe that in order to increase the resolution it is necessary both to point out all the limiting factors, and to analyse which and how such unwanted factors can be reduced.

An ionizing particle detection system basically follows the simple scheme, made of a bias circuit for the detector, which is connected to a suitable amplifier, eventually with a filter to improve the signal to noise (S/N) ratio [5] . If all the ionizing particle energy is lost within the detector, the limits to the energetic resolution can be intrinsically connected to the STJ or can be derived by extrinsic phenomena. The first are due to the junction noise, and to fluctuations both of the excess quasi-particles created by the ionizing particle and of the thermal quasi-particles. The extrinsic factors, which further reduce the resolution, are the amplification noise and spurious phenomena in the junction. In order to analyse some of these problems we previously summarize how the signal is generated.

When a particle bumps into the superconducting film, it creates primary excitations within the metal; the bumping process may go on producing a shower of secondary excitations. In this way the energy of the incoming particle is spread on a large number of excited states located nearly around the impact zone, where at the end of the process Cooper pairs are broken, so that a non-equilibrium quasi-particles state is generated. At low temperature the excess quasi-particle determine a reduction of the local gap Δ. These excited quasi-particles will diffuse within the film before they recombine in Cooper pairs with a recombination time τ_r. To keep τ_r exponentially increasing as the temperature T decreases, we have to assume a not far from equilibrium state. This request will be satisfied by a fast diffusion in a large area. At the end of the process a local heating of the film will be found.

The forementioned processes will produce variations in the STJ current-voltage (I-V) characteristic. Handling the STJ with the simple semi-conductor model, see Fig.1, if the tunnel process does not affect the equilibrium state of the electrodes, in the absence of external excitations we can write the net current

resolution was worse than the expected one. We believe that in order to increase the resolution it is necessary both to point out all the limiting factors, and to analyse which and how such unwanted factors can be reduced.

An ionizing particle detection system basically follows the simple scheme, made of a bias circuit for the detector, which is connected to a suitable amplifier, eventually with a filter to improve the signal to noise (S/N) ratio [5] . If all the ionizing particle energy is lost within the detector, the limits to the energetic resolution can be intrinsically connected to the STJ or can be derived by extrinsic phenomena. The first are due to the junction noise, and to fluctuations both of the excess quasi-particles created by the ionizing particle and of the thermal quasi-particles. The extrinsic factors, which further reduce the resolution, are the amplification noise and spurious phenomena in the junction. In order to analyse some of these problems we previously summarize how the signal is generated.

When a particle bumps into the superconducting film, it creates primary excitations within the metal; the bumping process may go on producing a shower of secondary excitations. In this way the energy of the incoming particle is spread on a large number of excited states located nearly around the impact zone, where at the end of the process Cooper pairs are broken, so that a non-equilibrium quasi-particles state is generated. At low temperature the excess quasi-particle determine a reduction of the local gap Δ. These excited quasi-particles will diffuse within the film before they recombine in Cooper pairs with a recombination time τ_r. To keep τ_r exponentially increasing as the temperature T decreases, we have to assume a not far from equilibrium state. This request will be satisfied by a fast diffusion in a large area. At the end of the process a local heating of the film will be found.

The forementioned processes will produce variations in the STJ current-voltage (I-V) characteristic. Handling the STJ with the simple semi-conductor model, see Fig.1, if the tunnel process does not affect the equilibrium state of the electrodes, in the absence of external excitations we can write the net current

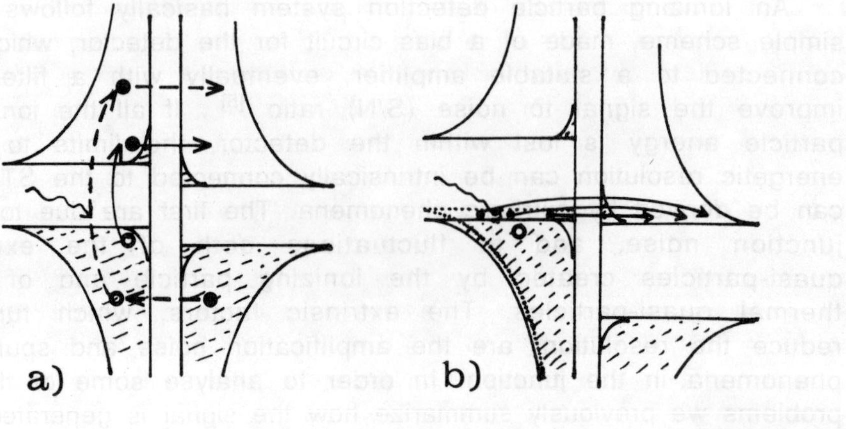

Fig.1. Semiconductor model of a tunnel junction in the perturbed state : a) excess current determined by excess quasi-particles, the dashed line shows the useless high energy excitations. b) multiplicative effects generated by the gap reduction.

flowing through the tunnel barrier:

$$I = \mathcal{F}_0(V) = \iiint P(x,y)\ N_1(E-eV)\ N_2(E)\ [f_0(E-eV)-f_0(E)]\ dE\ dx\ dy \qquad (1)$$

where the tunnel probability $P(x,y)$ is not constant due to the barrier dishomogeneities; $N_1(\Delta,T)$ and $N_2(\Delta,T)$ are the densities of states of the two superconducting films, which depend on the local value of the gap $\Delta_{1,2}(x,y)$, and on the temperature T; $f_0(E)$ is the Fermi-Dirac distribution function. The presence of excess quasi-particles, as shown in Fig.1a, increases the number of occupied states in the upper band, empting the lower band. In this way the distribution function $f(E,x,y)$ cannot follow the usual Fermi-Dirac laws, but Eq.1 is still valid if we consider the local temperature of the film higher than the neighbouring temperature, or if we assume that the distribution $f(E)$ depends on a chemical potential μ^* different for the quasi-particles: $f(E)=f_0(E,\mu^*)$. However, not regarding the particular description of the non-equilibrium state, low energy $(E<\Delta_1+\Delta_2)$ excess quasi-particles determine an increasing of the tunnel current. On the contrary, as shown in Fig.1a when high energy quasi-particles with corresponding empty states at energies lower than $\Delta_1+\Delta_2$ are created, the direct tunneling from the lower band of the superconductor on the right to the empty states in the lower band of the superconductor on the left will be very probable, leading to a vanishing tunnel current. Moreover since the excess number of quasi-particles will decrease the amplitude of the energy gap, we expect a sudden increase of the tunnel current if $(\Delta_1+\Delta_2)/e$ becomes lower than the bias voltage V. This phenomenon, as shown in Fig.1b, will lead to a multiplicative effect of the excess current. For both these two contributes to the excess current, due to the diffusion process of the quasi-particles, the observed perturbed I-V characteristic of the whole junction $I=\mathcal{F}(V)$ is determined by the excess current generated in the portion of the STJ common with the diffusion area. In this way large diffusion lengths will contribute to the reduction of the signal, if the impact of the ionizing particle occurs near the junction border.

Experimental values on the diffusion length seem to be one order of magnitute larger than the expected theoretical value.

Recenly [5,6], taking into account only the noise generated by the junction itself in the absence of leakage currents, and by supposing the junction uniformely perturbed by the ionizing particle, the qualitative behaviour of the signal to noise (S/N) ratio has been analysed as a function of temperature and tunnel probability.

In this paper, in Sec.2, we modify the previous model taking into account the gap variations. By means of an equivalent model, we analyse the S/N ratio of the STJ coupled to a voltage or to a charge amplifier in different bias conditions. In Sec.3 we discuss the limits to the energetic resolution generated by: true leakage currents, tunnel probability dishomogeneities, gap dishomogeneities, excitation of only a limited portion of the junction area. The detailed analysis of the excess quasi-particles diffusion process and its influence on the signal generation is not performed here, though, in our opinion, these phenomena may originate effects which seriously limit the energetic resolution. In Sec.4 some conclusions on the STJ design, on the amplifier choice, and on the bias and temperature conditions are suggested.

2.- S/N RATIO IN IDEAL CONDITIONS

The tunnel current between two electrodes in a first approximation can be given by: I=epN, where p is the tunnel probability for unit time and N are the electrons which can contribute to the tunnel process. For normal metal this approach gives the tunnel probability τ_t = 1/p as a function of the normal tunnel resistance:

$$\tau_t = e^2 N(0) A d R_{nn} \quad (2)$$

where A and d are respectively the surface of the junction and the thickness of the electrode; N(0) is the density of states per unit energy and volume at the Fermi level.

In the same approximation in a STJ the excess current under the excitation generated by the ionizing particle is:

$$I_e = epn'(t) \quad (3)$$

where n'(t) is the number of excess quasiparticles. Since n'(t) decreases both for the recombination and for the tunnel process, we have:

$$n'(t) = n'(0)\exp(-t/\tau_R) \quad (4)$$

$$\text{where: } 1/\tau_R = 1/\tau_t + 1/\tau_r \quad (5)$$

By substituting the Eq.4 in the Eq.3 we get:

$$I_e = I_e(0)\exp(-t/\tau_R) \quad (6)$$

with $I_e(0) = epn'(0)$.

In a better approximation, following the approach of the Eq.1, a voltage dependence of the excess current could be derived; in any case, for voltages $V<(\Delta_1+\Delta_2)/e$, the excess current is slowly dependent on the voltage.

The Eq.6 is valid only for a voltage bias. For bias conditions given by a current generator I^* with a parallel resistor R_B, the bias current $I_B=I^*-V/R_B$ must be equal to the current flowing in the junction in the excited state: $I_B=\mathfrak{F}(V)+C_D dV/dt$, where C_D is the STJ capacitance. A first power expansion of $\mathfrak{F}(V)$ gives: $\mathfrak{F}(V)=\mathfrak{F}(V_0)+G_D(V-V_0)$, where $G_D=R_D^{-1}=(\partial\mathfrak{F}/\partial V)_{V=Vo}$ is the dynamical conductance and V_0 is the voltage of the static bias point. Since the excess current $\Delta I(V_0)$ is related to the perturbed and the unperturbed I-V characteristic $(\Delta I(V_0)=\mathfrak{F}(V_0)-\mathfrak{F}_0(V_0))$, we get:

$$-\Delta V(1/R_D + 1/R_B) = \Delta I(V_0) + C_D\, d(\Delta V)/dt \qquad (7)$$

where $\Delta V = V-V_0$.

The capacitor C_D determines the detector time $\tau_D = C_D R_D$. Of course near the gap voltage the linear approximation of $\mathfrak{F}(V)$ is no longer valid. In the perturbed and unperturbed I-V characteristics two regions are separately examined. In the first $(eV<\Delta_1+\Delta_2)$ the variations are mainly sensitive to the excess quasiparticles: $\Delta I=I_e$, so that, neglecting C_D, the voltage signal in optimum conditions (i.e. $R_B=\infty$) is $\Delta V=I_e R_D$, that is p indipendent.

In the region two $(eV \approx \Delta_1+\Delta_2)$ at low temperature $R_D\rightarrow 0$, so that the gap variation $\delta\Delta$ determines the voltage signal:

$$\Delta V \approx 1/e\ \{\Delta[n'(t)]\ -\Delta^*\} = \delta\Delta(t)/e \qquad (8)$$

where Δ^* is the gap in the unperturbed state and $\delta\Delta(t)$ is p indipendent and can be approximated by $\delta\Delta(t) = \delta\Delta_0\exp(-t/\tau_R)$.

At constant bias voltage, the current variations shall be given by:

$$\Delta I(t) = 1/e\ \delta\Delta(t)/R_D \qquad (9)$$

In the analysis of the detector performances we must consider the STJ noise, whose spectral density is given by:

$$I_n^2(\omega) = F(V,T)/\pi R_s \tag{10}$$

where F(V,T), in the limit $eV/kT >> 1$, is given by eV, and R_s is the static resistance at the bias point.

The Eqs.7 and 10 lead to the equivalent circuit of the junction in dynamical conditions given by Fig.2.

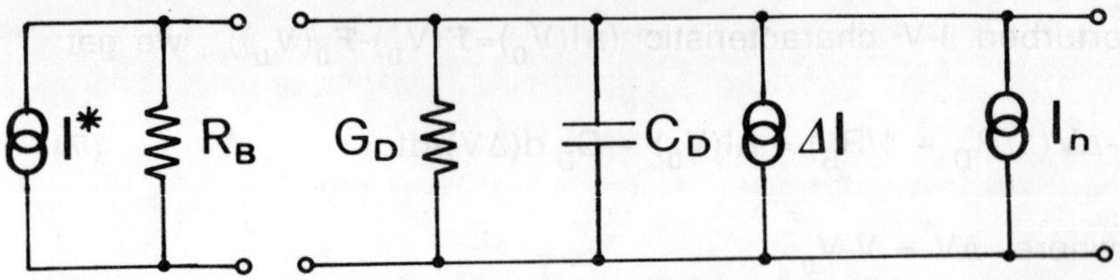

Fig.2. Equivalent circuit of the detector; the current signal generator ΔI is given by Eq.6 or by Eq.9 respectively for $eV<\Delta_1+\Delta_2$ and $eV\approx\Delta_1+\Delta_2$.

In order to compute the signal to noise ratio in the case of an ideal voltage amplifier coupled to the detector, the maximum value of the voltage signal must be derived.

In the best conditions $R_B = \infty$, and $Z_i = \infty$, where Z_i is the input impedance of the amplifier; from Eq.7 with $\Delta I = \Delta I_0 \exp(-t/\tau_R)$ we have:

$$\Delta V = R_D \Delta I_0 [\exp(-t/\tau_R) - \exp(-t/\tau_D)]/(1-\tau_D/\tau_R) \tag{11}$$

which reaches its maximum value at t=t*

$$\Delta V(t^*) = R_D \Delta I_0 \, (\tau_R/\tau_D)^{\tau_D/(\tau_D-\tau_R)} \tag{12}$$

in this way the S/N is:

$$S/N= \frac{R_D \Delta I_0 \left(\tau_R/\tau_D\right)^{\tau_D/(\tau_D-\tau_R)}}{(eV/\pi R_s)^{1/2} R_D\ (BW)^{1/2}} \tag{13}$$

where in a first approximation the filter determines only a high frequency cut off with a bandwidth BW equal to the lower value between $1/\tau_R$ and $1/\tau_D$ .Since both R_s and R_D are proportional to 1/p , as discussed by S.Pace et al. [5], the maximum of S/N accours in the high p limit, that is in the low τ_D limit.

As the temperature decreases the two voltage regions must be separately analysed. In the first region ($eV<\Delta_1+\Delta_2$) both the dynamic resistance R_D and the static one increase exponentially. On the contrary in the second region ($eV \approx \Delta_1+\Delta_2$) R_D decreases almost exponentially, while the static one R_s is nearly constant. As the temperature decreases in the first region we get high τ_D limit:

$$S/N = \frac{\Delta I_0^{ex}\ \ \tau_R}{(eV/\pi R_S)^{1/2}\ \ \tau_D^{-1/2}} \tag{14}$$

which decrease as the temperature decreases.

On the contrary in the second region we get $\tau_D \to 0$ and

$$S/N = \frac{\delta\Delta_0/e\ \ \tau_R^{1/2}}{(eI/\pi)^{1/2}\ \ R_D} \tag{15}$$

which increases with the R_D decreasing, so that the voltage amplifier can really operate in these conditions.

Let us now examine a charge amplifier connected to the STJ as shown in Fig.3. In this configuration a current bias in static conditions and a voltage bias in dynamical conditions is

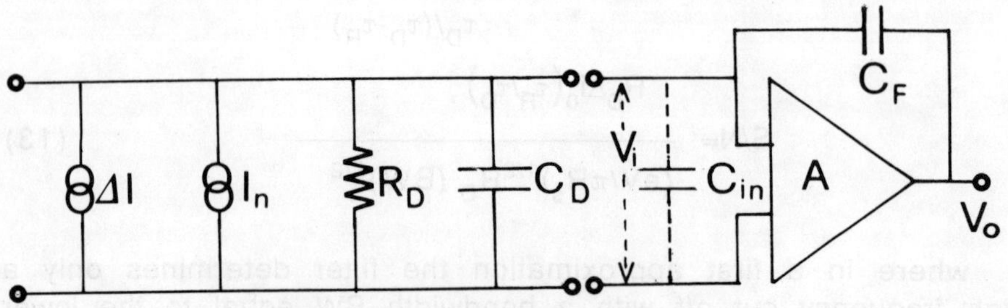

Fig.3. Ideal charge amplifier used with a STJ.

necessary, where the dynamical load line must be determined by the low input impedence of the amplifier. In the frequency range where the amplifier gain $A(\omega)$ can be considered constant (i.e. $A(\omega)=A$), the equivalent input capacitance C_{in} is:

$$C_{in} = (1+ |A|)C_F$$

The total capacitance at the amplifier input is: $C' = C_{in} + C_D$, so that only a fraction $\beta = C_{in}/C'$ of the excess charge flowing in the junction can be really collected by the amplifier. The collected charge Q is equal to $Q(t) = V_i(t)C_{in}$, where the input voltage $V_i(t)$ is given by Eq.11 with the substitution $\tau_D \to \tau'_D = R_D C'$. The maximum input voltage signal amplitude corresponds to the maximum collected charge Q_{MAX}.

$$Q_{MAX} = \Delta V(t^*)\, C_{in} \tag{16}$$

By Eqs.16 and 12 the total collected charge goes to zero if the dynamical resistance of the detector goes to zero. The optimum collection process is obtained if the current integration time necessary for the charge collection, τ_R is lower than the detector time τ'_D. In the limit of $\tau'_D \to \infty$ we get for the collected charge $Q_{MAX} = \Delta I_0 \tau_R \beta$, where, for $eV<\Delta_1+\Delta_2$ in the limit $\tau_R \to \infty$, $\Delta I_0 \tau_R$ is equal to the total number of the excited quasiparticles $n'(0)$. In the limit $\tau_D' \to \infty$ the maximum output voltage $V_{oMAX} = A\,\Delta V(t^*)$:

$$V_{0MAX} = I_e(0)\tau_R/C'_F \tag{17}$$

where $C'_F = C_F + (C_D + C_F)/A$.

The spectral density of the input voltage noise $V_n{}^2(\omega)$ will be equal to: $V_{in}{}^2(\omega) = I_n{}^2(\omega)\ |Z_{in}(\omega)|^2$, where Z_{in} is the total impedance at the amplifier input. If the condition $\tau_D > \tau_R$ is satisfied, we can consider the frequency ω high enough to have $|Z_{in}| = 1/\omega C'$. The spectral density of the output noise is:

$$V_{on}{}^2(\omega) = A^2 V_{in}{}^2(\omega) = I_n{}^2(\omega)/\omega^2 C'_F{}^2 \tag{18}$$

As usual for a charge amplifier a filter with a transfer function $S(\omega)$ is necessary, which limits both low and high frequency:

$$V_{on}{}^2 = \int_o^\infty (I_n{}^2(\omega)/\omega^2 C'_F{}^2)\ |S(\omega)|^2\, d\omega \tag{19}$$

The choise of the optimum filter, depending on the particolar shape of the signal is out of purpose of this paper. The very simple choise: $S(\omega) = 1$ for $\omega_1 < \omega < 2\omega_1$ and $S(\omega) = 0$ elsewere, gives $V_{on}{}^2 = (I_n/C'_F)^2/(2\ BW)$, with a bandwidth $BW = \omega_1$, so that the signal to noise ratio is :

$$S/N = \frac{(I_e(0)\ \tau_R)\ (\pi R_S)^{1/2}\ (2BW)^{1/2}}{(eV)^{1/2}} \tag{20}$$

With $BW = 1/2\ \tau_R$ the Eq.20 is formally identical to the expression of the S/N of the voltage amplifier in the low τ_D limit, but in the first case the signal is strictly related to the peak voltage, while in the second one the signal is a function of the total collected charge.

As discussed by S.Pace et al. [5] the S/N is an increasing

function of the total tunnel probability. As the temperature decreases, R_D increases exponentially in the first region of the I-V characteristic, so that the high $\tau_D{}'$ limit is reached and the performed analysis is valid. On the contrary in the second one the R_D goes to zero and by Eq.16 the collected charge should go to zero. Some compensation effect could arise from moltiplication phenomena but in general a charge amplifier cannot well operate in the low R_D case.

3 - COMMENTS ON ENERGETIC RESOLUTION LIMITING PHENOMENA

3.1 - Leakage Currents

Often the measured currents at voltages below $(\Delta_1+\Delta_2)/e$ are larger than the values predicted by the tunnel theory for homogeneous BCS superconductors at the thermal equilibrium. Among the several causes which produce such effect we have:

a) true leakage (metallic shorts);

b) non BCS density of states , due to a weakening of the superconducting state at the superconductor-barrier interface;

c) multiplication effects, like in the superconducting transistor proposed by K.E.Gray.

We analyse the true leakage current only as a purely ohmic resistance R_L added to the scheme of Fig.2 . In the following we put $1/R_L=G_L=\alpha(T)G_{NN}$, $R_S=r_S(T)R_{NN}$, and $R_D=r_D(T)R_{NN}$, where $\alpha(T)<<1$ in case of good junctions. At low temperatures, $\alpha(T)$ is proportional to $1/T^{\gamma}$, being $\gamma\geq1$. The presence of G_L will change G_D in $G'_D=G_D+G_L$ and τ_D in τ'_D. The S/N ratio for the voltage amplifier will be:

$$S/N=\frac{R_{NN}{}^{1/2}\ \Delta I_0\left(\tau_R/\tau'_D\right)^{\tau'_D/(\tau'_D-\tau_R)}}{\{\,(\alpha(T)\ 2k_BT)/\pi + eV/\pi r_S\,\}^{1/2}\ (BW)^{1/2}} \tag{21}$$

For $V<(\Delta_1+\Delta_2)/e$, as T decreases the S/N ratio gets lower both for the reasons discussed in the previous section, and for the increasing of the extra noise generated by G_L. On the contrary for $V\approx(\Delta_1+\Delta_2)/e$ the equivalent current signal generator increases exponentially when T decreases, while the current noise will be mainly determined by the leakage resistance ($\approx T^{(1-\gamma)/2}$). This will result in an increasing S/N.

The additional noise generated by the leakage currents must be added also for the S/N ratio computed for a charge amplifier coupled to the detector. In this case the Eq.20 shows that the S/N ratio increase as the temperature decreases, that is in the high τ_D limit for a voltage bias $V<(\Delta_1+\Delta_2)/e$. In these conditions the additional noise leads to :

$$S/N= \frac{pn'(0)\ R_{NN}^{1/2}\ \tau_R^{1/2}}{\{ (\alpha(T)\ 2k_BT)/\pi + eV/\pi r_S \}^{1/2}} \tag{22}$$

due to the presence of leakage currents the Eq.22 shows that serious problems arise. As T goes to zero the static resistance exponentially encreases, so that the term $(\alpha(T)\ 2k_BT)/\pi$ become the predominant noise contribution. In this way, assuming $\alpha(T)\approx T$, the S/N will saturate to a limit value.

3.2 - Dishomogeneities

Several dishomogeneities may be found in a STJ: gap inhomogeneities, barrier dishomogeneities, dishomogeneities and anisotropy of diffusion processes. In most STJ applications the presence of small local dishomogeneities is not relevant, because these applications only ask to get many junctions whose I-V characteristics are identical within 1%. This requirement implies that the tunnel probability, averaged on the single junction surface, does not change for each junction. On the contrary using a

STJ as a detector, local dishomogeneities, averaged on the diffusion area, play an important role.

The locally induced gap amplitude variations depend on the used materials in the STJ. They generate variations in the excited quasi-particle number, which is directly related with the current signal $\Delta I(t)$, and their effect on the energetic resolution cannot be reduced.

Barrier dishomogeneities can be due both to the oxide thickness disuniformity, and to the dielectric property variations of the insulating oxide, which affect the barrier height. These dishomogeneities cause local tunnel probability fluctuations which seems to be one of the largest limiting factor. In order to analyse this effect we consider first signal variations induced by the variation of p in a uniformly perturbed junction, then we consider the signal generation due to the perturbation of a portion of a STJ having a constant p, and finally we combine both effects.

For a uniformly perturbed STJ in the best condition the maximum collected charge is $Q_{max}=n'(0)p\tau_R$, so that p fluctuations (dp) are related to Q_{max} fluctuations (dQ_{max}) by :

$$dQ_{max}/Q_{max} = [1/(1+ \tau_r/\tau_t)]\ (dp/p) \qquad (23)$$

that respectively for $\tau_r << \tau_t$ and $\tau_r >> \tau_t$ leads to :

$$dQ_{max}/Q_{max} = dp/p \qquad \text{and} \qquad dQ_{max}/Q_{max} = (\tau_t/\tau_r)\ (dp/p).$$

The second case can be obtained decreasing the temperature, while it is not convenient to decrease τ_t by decreasing the barrier thickness. Indeed in this case larger barrier dishomogeneities are produced. For what concerning the voltage signal, as remarked in Sec.2, there is no effect on the signal coming from tunnel probability variations.

If we consider that only a fraction σ of the STJ area, determined by the diffusion process, will be uniformly perturbed, the detection system can be described by the equivalent circuit of Fig.4a, where the STJ conductance has been splitted in two parts:

(1-σ)G represents the unperturbed and constant STJ conductance, while σG is the varying conductance of the perturbed junction area. The equivalent model is shown in Fig.4b . The load

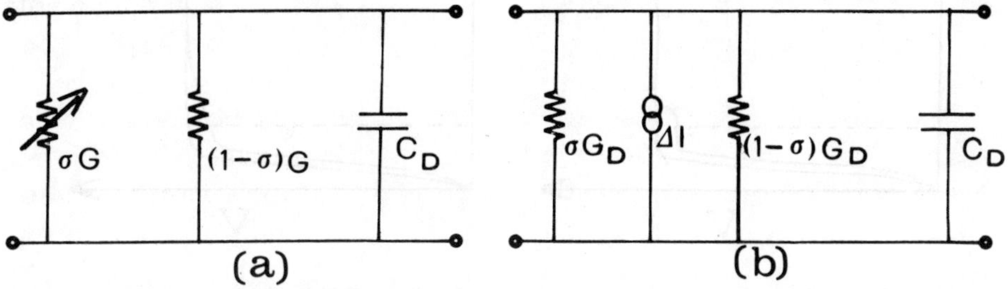

Fig.4. a) equivalent circuit of a STJ with only a portion of the area perturbed by the ionizing particle b) equivalent model in dynamical conditions.

line of the perturbed part will be mainly determined by the unperturbed part, and this corresponds to an almost voltage bias for the perturbed STJ fraction. Biasing a STJ near $(\Delta_1+\Delta_2)/e$, even for an external current bias, multiplicative effects occur. Because these effects depend on the value of the perturbed STJ area, also ΔI(t) will depend directly on this value. On the contrary below $(\Delta_1+\Delta_2)/e$, the equivalent signal generator ΔI(t) becomes independent of σ, so that, from the equivalent model of Fig.4b, both voltage and current signal will be unaffected by σ.

A locally perturbated STJ with barrier dishomogeneities will show fluctuations in the generated signal ΔI(t), which depends on the local value of p. At bias voltages below the gap, for charge amplifier the Eq.23 is still valid, were in this case p will be the mean value and dp its fluctuation. As far as the voltage signal is concerned, the amplitude is given by the low τ_D limit of the Eq.11: $\Delta V = \Delta I(p(x))/G_D$, where G_D is the whole junction conductance, which is pratically unaffected by the p local fluctuations. Thus the voltage fluctuations will be linearly dependent on the p fluctuations.

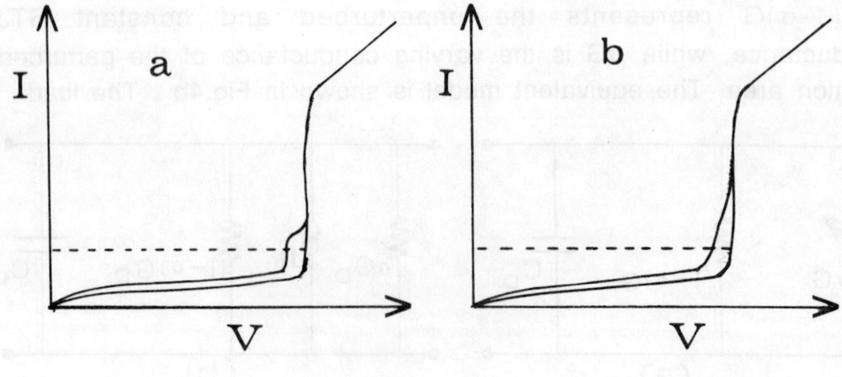

Fig.5. Sketch of the variations of the STJ I-V characteristic determined by a perturbation of the portion of the area in the junction : a) uniform perturbation b) not uniform perturbation

The analysis near the gap voltage is quite difficult, but a qualitative discussion can be performed . At any voltage the total current will be the sum of the two currents flowing through the perturbed and the unperturbed area of the STJ; in the case of uniformly perturbed fraction of the STJ the total I-V characteristic will look like in Fig.5a, where the knee just below the gap is generated by the gap reduction in the perturbed area. The voltage signal at the current bias indicated in Fig.5a should be the same as the unperturbed portion of the STJ did not exist. Because the gap reduction is not uniform within the perturbed area, the knees will be smeared as in Fig.5b.

4. - CONCLUSION

Our analysis shows that two different voltage ranges should be considered in the analysis of the I-V characteristic variations induced by an ionizing particle. In the first range, below the gap voltage, the excess current is directly connected with the excess quasi-particles. In the second range, near the gap voltage, the voltage variations comes from the gap variations induced by the excess quasi-particles. If we assume a uniformly perturbed junction, without leakage currents, and dishomogeneities, the current signal increases with the tunnel probability while, neglecting τ_D effects, the voltage signal is p independent. Because of the different behavior of R_D in the 1st or in the 2nd region as the temperature decreases, for a charge amplifier the S/N ratio increases only in the 1st region, while for a voltage amplifier it increases only in the 2nd one.

The leakage currents can strongly affect charge amplifier performances, but it should not seriously decrease the S/N ratio of the voltage amplifier operating in region 2.

Taking into account both the barrier dishomogeneities and the local excitation of the junction area, the tunnel probability fluctuations gives neglecting effects on the charge signal if the temperature is enough low to get $\tau_r >> \tau_t$. On the contrary the voltage signal, for voltages below $(\Delta_1+\Delta_2)/e$, is directly affected by the p fluctuations. At voltages near $(\Delta_1+\Delta_2)/e$ the voltage signal should be nearly the same produced by a small junction of area equal to the perturbed surface. The effect of a continuously varying gap in the perturbed zone should be deeply analysed.

In addition we believe that this analysis have to be completed taking into account also the noise generated by the voltage or charge amplifiers, and some care should be kept both in the filter design and eventually in the use of a transformer. Following this qualitative analysis a numerical evaluation of the S/N ratio is in progress.

REFERENCES

1) G.H. Wood and B.L. White, Can. J. Phys. , 51 (1973) 2032 .
2) M. Kurakado, J. Appl. Phys. , 55 (1984) 3185 .
3) A. Barone et al. , Nucl. Instr. and Meth. in Phys. Res. , A235 (1985) 254 .
4) D.Twerenbold , Phys. Rev. , 34B (1986) 7748 .

5) S. Pace, F. Celani et al. " II Soviet-Italian Symposium on Weak Superconductivity " , Naples (Italy) , May 1987
to be pubblished on " Advances in the Physics of Matter Condensed " 1987 , World Scientific .
Internal Report INFN-LNF 87/25(P),1987.

6) S. Pace, F. Celani et al. Contributed Paper at "1987 International Superconductive Electronics Conference " Tokyo (Japan) , Aug. 1987 . Internal Report INFN-LNF 87/73(P),1987.

STATUS OF STANFORD SUPERCONDUCTIVE DETECTORS FOR MONOPOLES AND WEAKLY INTERACTING PARTICLES

B. CABRERA

Physics Department, Stanford University,
Stanford, California 94305

In our laboratories at Stanford, we have undertaken several research programs to search for dark matter candidates in our galaxy with laboratory based detectors. The first effort is a search for magnetic monopoles in the cosmic rays. These would be supermassive (10^{16}-10^{19} GeV/c^2) and a density of only one per 10-10,000 km^3 would be sufficient to account for the local dark matter around our galaxy. We are now operating a 1.3 m^2 times 4π sr detector utilizing eight SQUIDs. At present it is the largest superconductive monopole detector. The second effort involves the development of large mass (~1 kg) elementary particle detectors capable of sensing weakly interacting particles. These utilize silicon crystals at temperatures below 1 K, have spatial resolution in three dimensions and would measure the total energy deposition. Such detectors will be used for direct dark matter searches and for neutrino experiments capable of setting better limits on the neutrino mass. Both of these efforts are summarize in this paper.

PART A: SUPERCONDUCTIVE MAGNETIC MONOPOLE DETECTORS

Introduction

In 1931, P.A.M. Dirac proposed the existence of magnetically charged particles to explain the observed quantization of electric charge. He showed that only integer multiples of a fundamental magnetic charge g (Dirac charge) = $hc/4\pi e$ are consistent with quantum mechanics. Many years of experimental searches produced no convincing candidates. In 1974 't Hooft and independently Polyakov showed that in all true unification theories (those based on simple or semi-simple compact groups) magnetically charged particles are necessarily present. The modern theory predicts the same long-range field and thus the same charge g as the Dirac solution; now, however, the near field is also specified leading to a calculable mass. The standard SU(5) model predicts a monopole mass of 10^{16} GeV/c^2, much heavier than had been considered in previous searches. More recent unification theories based on supersymmetry or Kaluza-Klein models yield even higher mass values up to the Planck mass of 10^{19} GeV/c^2.

Such supermassive magnetically charged particles would possess qualitatively different properties from those assumed in earlier searches. These include necessarily nonrelativistic velocities from which follow weak ionization and extreme penetration through matter. Thus such particles may very well have escaped detection in earlier searches based on heavy ionization of relativistic monopoles.

Although GUTs theories are very clear in their prediction of the existence of monopoles, cosmological theories based on GUTs lead to impossibly high or unobservably low predictions for monopole particle flux limits with the latter results being exponentially model dependent. Thus, only astrophysical arguments provide guidance for the relevant detector sensing areas for experiments. An upper bound of 10^{-15} cm^{-2} sr^{-1} s^{-1} is obtained assuming an isotropic flux from arguments based on the existence of the 3 microgauss galactic magnetic field (Parker bound). The Parker bound becomes less severe linearly with monopole mass for masses above 10^{17} GeV/c^2. In addition, several authors have demonstrated that models incorporating monopole plasma oscillations would allow a much larger particle flux, approaching in some cases the local galactic dark mass limit. All of these bounds assume particle velocities in gravitational virial equilibrium, i.e., very near 10^{-3} c.

Much more severe astrophysical limits can be obtained based on proton decay catalysis. It has been shown theoretically that the supermassive monopoles arising from many grand unification theories would catalyze nucleon decay processes. If the cross section for such events is of order the hadron cross section, as has been suggested, then all attempts at direct detection of the monopoles from such theories may be doomed to failure. Arguments based on x-ray flux limits from galactic neutron stars, which assume a strong interaction cross section for proton decay catalysis, lead to an upper bound for magnetic particle flux of about 10^{-21} cm^{-2} sr^{-1} s^{-1}. There remain unanswered questions concerning the detailed theoretical understanding of the catalysis cross section and concerning the astrophysical arguments based on our incomplete understanding of neutron stars. In addition, proton decay has not yet been observed. Thus, from an experimental point of view more weight should be given to the less model dependent astrophysical limits.

Superconductive technologies, many developed at Stanford University over the last decade, have led naturally to very sensitive detectors for magnetically charged particles. These superconductive detectors directly measure the magnetic charge independently of particle velocity, mass, electric charge and magnetic dipole moment[1]. In addition, the detector response is based on simple and fundamental theoretical arguments which are extremely convincing. Because of their velocity-independent response, these detectors are a natural choice in searches for a particle flux of supermassive (and therefore slow) magnetically charged particles. By far, the most definitive positive identification of a magnetic charge would be made with a superconductive detector. Other detectors, though larger, are less satisfactory for positive identification of monopoles since the interaction of monopoles with matter through ionization processes is less well understood [2].

Recent detailed reviews of the theoretical, the experimental and the astrophysical work can be found in **Magnetic Monopoles** [R.A. Carrigan and W.P. Trower, eds., Plenum 1983] and **Monopole '83** [J. Stone, ed., Plenum 1984].

Operation of Three Loop Detector

Between February, 1983 and March, 1986, we operated a three loop coincident superconducting magnetic monopole detector [3]. The rms current noise levels in an effective noise bandwidth of ≈ 0.16 Hz were $0.02\ \phi_o/L$ in all three loops, less than 1% of the signal expected from the passage of a Dirac magnetic charge through one of the two turn loops ($4\ \phi_o/L$). The sensing area based on this low noise operation was 476 cm^2 (71 cm^2 loop area and 405 cm^2 near miss area) for events greater than 0.1 ϕ_o/L in at least two of the three loops. The data were extremely clean and no candidate events were seen. Based on the 1008 days of accumulated active running time, these data set an upper limit of 4.4×10^{-12} cm^{-2} sr^{-1} s^{-1} (90% C.L.) on any uniform flux of magnetic monopoles passing through the earth's surface at any velocity [3].

New Large Area Octagonal Detector

During this past year, our group at Stanford, consisting of B. Cabrera, M. Huber, M. Taber and R. Gardner, has designed [4], constructed, and brought into continuous operation an eight loop detector with a cross section averaged over 4π sr of 1.3 m^2 for double coincident events and a signal-to-noise ratio of 30 for the passage of a single Dirac charge. This detector shown schematically in Fig 1, is composed of eight planar superconducting detection coils arranged around a cylinder with an octagonal cross section. Each coil is a gradiometer 16 cm wide and 6 m long and is connected to a high sensitivity rf SQUID current sensor. The entire assembly is surrounded by a superconducting lead shield and housed in a dewar which is enclosed in a μ-metal shield. This detector has a sensing area 30 times larger than that of the three loop detector.

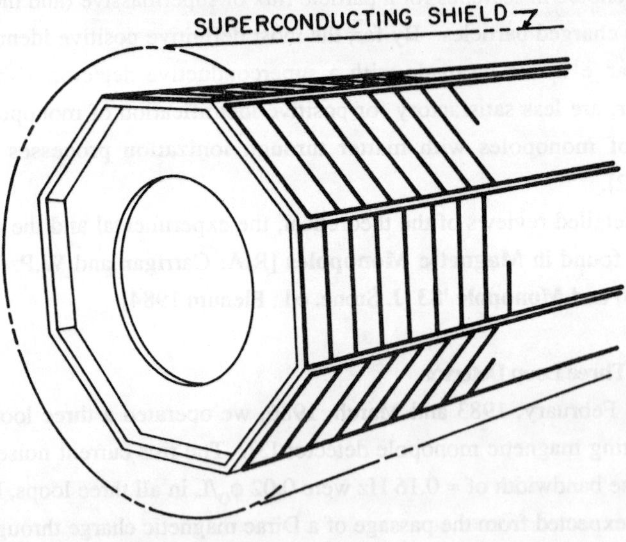

Fig 1. Schematic diagram of octagonal detector.

A very important design feature for the large scale use of superconducting coils as monopole detectors is the use of a gradiometer winding pattern [5,6]. The sensitivity of a gradiometer to external magnetic field changes is substantially reduced over that from a simple coil whereas the sensitivity to the passage of a magnetic charge remains high since the particle passes through only one element of the gradiometer pattern. We have utilized computer calculations to optimize our coil design under constraints from dewar size and minimum acceptable signal to noise ratio. The optimum pattern which we have used for our design is shown in Fig 2. The conducting elements are NbTi ribbon, 2mm wide and 50 μm thick. It is a repeating rectangular pattern with an aspect ratio of 0.6 to 1.0. A further improvement was achieved by breaking the loop up into a number of parallel elements which are connected together to one SQUID. This technique reduces the coupling losses to the SQUID from a 1/L proportionality to 1/√L, where L is the inductance of the single series loop [7].

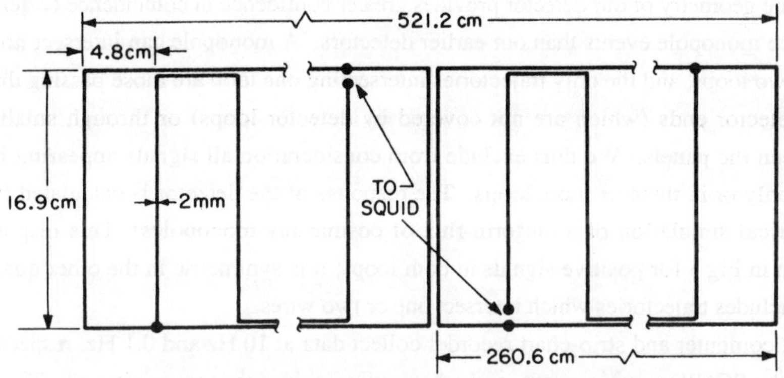

Fig 2. Loop parameters for octagonal detector.

The new 1.3 m^2 monopole detector requires eight low-noise SQUID systems, one for each panel of the detector array. We have chosen to use rf SQUIDs manufactured by Biomagnetic Technologies Inc. [B.T.i.], (formerly S.H.E. Corp.) together with rf electronics from Quantum Design. A shield of high magnetic permeability metal (μ-metal) has been designed and constructed to shield the detector from the earth's magnetic field during the initial cooldown, as the inner superconducting lead shield goes through its transition temperature. The external shield consists of 0.038 inch-thick sheets of μ-metal mounted on an aluminum frame. The shield provides an absolute field below 10 milligauss throughout most of its interior. We are utilizing a Model 1200 closed-cycle liquid helium refrigeration system from Koch with the new octagonal detector. We have modified the operation of the helium liquifier to maintain a constant dewar pressure relative to atmosphere. This modification eliminates SQUID signals resulting from the pressure variations of the original cooling-capacity control system. A very slow drift remains, caused by atmospheric pressure changes, but it is well below the bandwidth of our detection system and poses no degradation of our signal to noise ratio for monopoles. There are two significant advantages derived from using this system: lower long-term operating costs, and the virtual elimination of disturbances that are produced by liquid cryogen transfers. Now that the new system is cooled down and stabilized, we achieve essentially 100% live time.

The geometry of our detector provides greater confidence in coincidence correlations for true monopole events than our earlier detectors. A monopole can intersect no more than two loops, and the only trajectories intersecting one loop are those passing through the detector ends (which are not covered by detector loops) or through small gaps between the panels. We thus exclude from consideration all signals appearing in one loop only or in three or more loops. The response of the detector is calculated from a numerical simulation of a uniform flux of cosmic ray monopoles. This response is shown in Fig 3 for positive signals in both loops; it is symmetric in the other quadrants and includes trajectories which intersect one or two wires.

A computer and strip-chart recorder collect data at 10 Hz and 0.1 Hz, respectively, from the SQUIDs, strain gauge, and other anti-coincidence instrumentation. The strain gauge senses mechanical disturbances which sometimes induce spurious offsets. Also recorded are a line voltage monitor, ultrasonic motion detector, and flux-gate magnetometer in the anti-coincidence instrumentation. We will add a wide-band radio frequency voltmeter if EMI causes noticeable signals.

We began obtaining low noise data on May 5, 1987 [8] and as of September 17, 1987, we have obtained 103 days of live time. A preliminary analysis of these data contain no candidate events and set a limit of 1.6×10^{-12} cm^{-2} sr^{-1} s^{-1} (90% C.L.) for any particle flux of magnetically charged particles passing through the surface of the earth. We intend to run the detector continuously for at least three years.

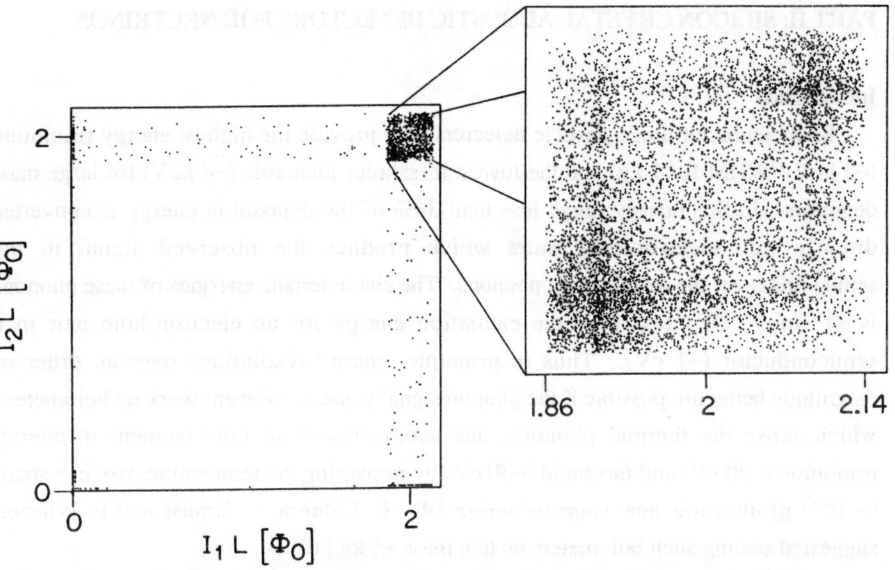

Fig 3. Calculated response of the detector to a uniform flux of cosmic ray monopoles. The signal sizes for the two loops intersectted by the monopole trajectory are plotted for each Monte Carlo event.

Conclusions

The Parker bound as modified to include supermassive monopoles has a mass independent floor at ~10^{-15} cm^{-2} sr^{-1} s^{-1} and rises linearly with mass above ~10^{17} GeV/c^2 until it intersects the local dark matter bound around the Planck mass (10^{19} GeV/c^2). Since more recent unification theories suggest a monopole mass approaching the Planck mass, searches at a flux level of ~10^{-13} cm^{-2} sr^{-1} s^{-1} are already important.

The current particle flux limit obtained from our eight loop superconductive detector of 1.6×10^{-12} cm^{-2} sr^{-1} s^{-1} at 90% C. L. is within a factor of 16 of the peak of the Parker bound. This limit is ~887 times lower than the original 151 day run [1] with the prototype single loop detector and ~1600 above the mass-independent Parker bound. A similar limit has been set by the group at IBM, Yorktown Heights with their 1 m^2 detector.

This research is funded in part by DOE Contract DE-AM03-76-SF00-326.

PART II: SILICON CRYSTAL ACOUSTIC DETECTORS FOR NEUTRINOS

Introduction

Semiconductor diode particle detectors now provide the highest energy resolution (~3 keV FWHM for 1 kg) and the lowest thresholds available (~4 keV) for large mass detectors. In this energy range less than 30% of the deposition energy is converted directly into electron-hole pairs which produce the observed signal in the semiconductors, the rest forming phonons. The characteristic energies of these phonons is ~1 meV, 10^3 less than the excitation energy for an electron-hole pair in a semiconductor (~1 eV). Thus in principle, energy resolutions over an order of magnitude better are possible if the phonon signal is used. Recent work on bolometers, which sense the thermal phonons, has demonstrated an improvement in energy resolution (~30 eV) and threshold (~30 eV) by measuring the temperature rise in a small (~ 10^{-5} g) ultra-low heat capacity sensor [9]. B. Cabrera, L. Krauss and F. Wilczek suggested scaling such bolometers up to a mass ~1 kg [10].

Motivated by this suggestion, our group, now composed of B. Cabrera, J. Martoff, B. Young and A. Lee at Stanford and B. Neuhauser at San Francisco State University, has proposed direct sensing of the ballistic phonons produced by an event in a large insulating single crystal [11]. Since this wavefront carries information on the event total energy and location within these silicon crystal acoustic detectors (SiCADs), substantial improvements in background rejection are possible over detection schemes which measure total energy deposition only. A threshold of 1 keV or better is important for our primary interest in using SiCADs as neutrino detectors [10]. For example, at power reactors all nuclear recoil signals from elastic neutrino scattering are below 10 keV with 60% above 1 keV, and a 1 kg SiCAD would register an event rate of ~100 events/day.

Ballistic Phonons from Point Events in Crystal Cubes

For a deposition of energy in a well localized volume within a single crystal of silicon, a thermal-like spectral phonon distribution is generated with a characteristic temperature of 10-20 K. The reason for this universal spectral distribution is that the first excitations are plasmons or very short wavelength phonons, both of which quickly relax to longer wavelength phonons within less than ~10 nsec. Since the decay rates are very strongly dependent on phonon energy and crystal temperature as soon as the distribution has reached wavelengths of even 10 lattice spacings further decays are negligible. These longer wavelength phonons propagate throughout the crystal with no dispersion and little isotopic scattering. This mode has come to be called the the ballistic phonon mode. Since the discovery of ballistic phonons in 1964 an enormous body of research has established that phonons with wavelengths greater than a few lattice spacings propagate freely through single crystals such as Si, Ge, CaF_2, SiO_2, Al_2O_3 and other insulating crystals [12]. This work has led to an improved understanding of phonon physics.

An interesting and important aspect of ballistic phonon propagation is that although the propagation in dispersionless, strong focussing effects occur within the crystal. These focussing patterns projected onto each face of a silicon cube cut with its faces orthogonal to the principal axes of the crystal permits three dimensional reconstruction of event locations and can resolve tracks or multiple scattering events. We have performed detailed Monte Carlo calculations [11]. Fig 4 shows the result of such a calculation for a point energy deposition in a crystal cube. The number of points used in the calculation

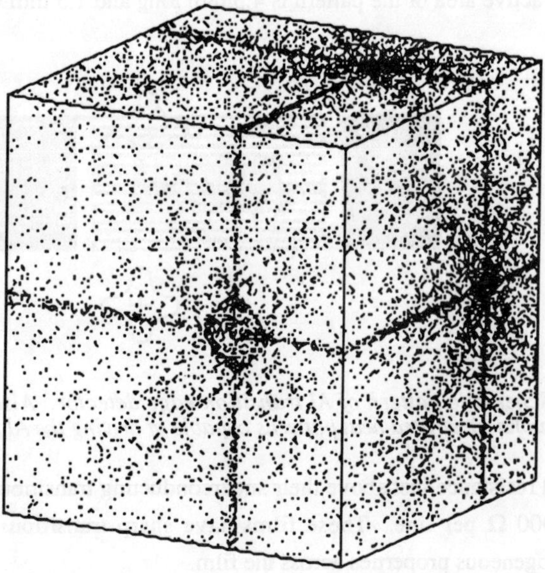

Fig 4. Monte Carlo calculation of phonon energy density arriving at the surfaces of a silicon cube from a point deposition of energy.

corresponds to the number of phonons expected from a 1 keV energy deposition at a point within the crystal. It is clear that high resolution information on the event location as well as tracking information is available.

Experiments on Aluminum Transition Edge Sensors on Silicon Wafers

In our laboratory effort, we have investigated aluminum transition edge sensors as phonon detectors deposited on the surfaces of a silicon crystal [13]. A thin superconducting line biased with a current near the critical current, exhibits self-terminating voltage pulses when bombarded with alpha particles. These pulses are caused by the heating of line segments above the superconducting transition temperature.

The sensors were made by depositing 40 nm of aluminum on crystalline silicon wafers which are 275 μm thick. The polished wafer faces are along the [100] axis. The meander patterns, as shown schematically in Fig 5, were produced using conventional photolithographic techniques and consist of 299 parallel lines each 2 μm wide with 3 μm space between lines. The pattern is aligned with the crystalline axes of the wafer. The active area of the pattern is 4.5 mm long and 1.5 mm wide.

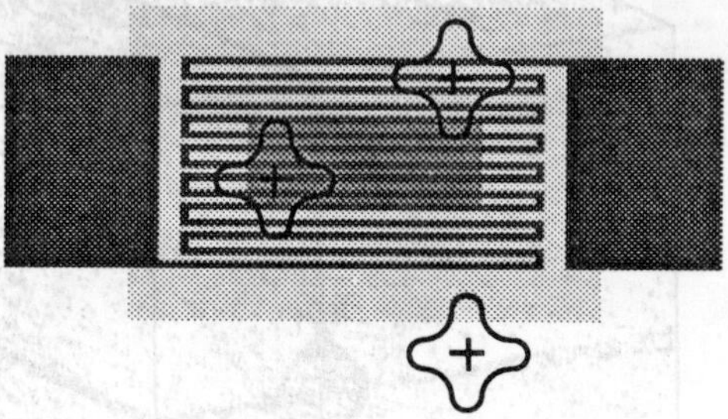

Fig 5. Schematic pattern of Al transition edge detector. Also shown are alpha-generated heat patterns capable of driving the Al film normal.

The normal resistance just above their superconducting transition temperature (T_c = 1.35 K) is ~1000 Ω per line. These films have sharp transition widths of <3 mK indicating homogeneous properties across the film.

Experiments were performed with a 0.1 μCi source (I_o = 3,700 disintegrations per sec, ±15%) of ^{241}Am. The half-life is 432 yrs. and the spectrum is dominated by two alpha particle energies at 5.486 MeV (85%) and at 5.443 MeV (13%). A thin coat of nickel (~20 nm) was evaporated over the ^{241}Am to prevent removal, but this film does not detectably degrade the source energy spectrum. The alphas are distributed uniformly over solid angle, so that half of them are absorbed in the platinum foil. We calibrated the source intensity and energy spectrum using a semiconductor detector at room temperature. The results were in agreement with the nominal source strength (±10%) and set a limit on the intrinsic linewidth of <1% FWHM (full width half max).

For the experiments with alpha particles incident on the front of the wafer, the total number of alphas incident on the pattern was ~115 counts/sec. Fig 6 shows the pulse height spectrum from a 1200 sec run. The bias current was 1.2 μA and the temperature 1.329 K. Returning to Fig 5, the peak in the spectrum corresponds to alpha hits within the small central rectangle where the entire heated area sufficient to drive the film normal is contained within the pattern, and the continuum between the peak and zero corresponds to those alphas which hit within the large rectangle but outside of the small one. The shape and magnitude of this continuum distribution is used to determine the size and some aspects of the shape for the heat spot produced by each hit. From Fig 6, the diameter of the heat spot is determined to be ~100 μm with the qualitative shape

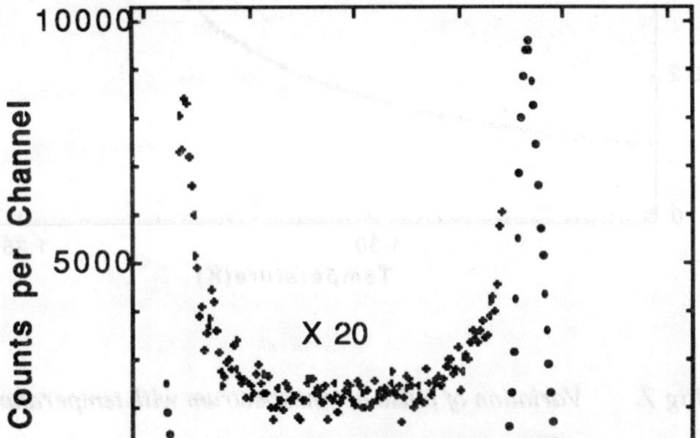

Fig 6. Pulse height spectrum for ^{241}Am from Al transition edge detector with I_{bias}= 1.2 μA and T = 1.329 K.

shown in Fig 5, as determined by Monte Carlo calculations. In fact, the area is also in agreement with the signal size ~.3 mV which indicates that ~ 1.5 mm of line was driven normal.

In a second series of experiments we carefully varied the temperature between runs and mapped out the variation of pulse peak position with temperature. For each point shown in Fig 7, the temperature was regulated to better than ~ 0.2 mK. The transition temperature of these films is $T_c \sim 1.35$ K. As the temperature approaches T_c, the signal size increases sharply, and within ~10 mK of T_c the pulses are lengthened and become erratic, greatly broadening the pulse height spectrum (not shown). Empirically, the pulse height is proportional to bias current and varies as $1/[1-(T/T_c)^2]$ with temperature. Limited data on the variation with alpha energy indicate a roughly linear dependence.

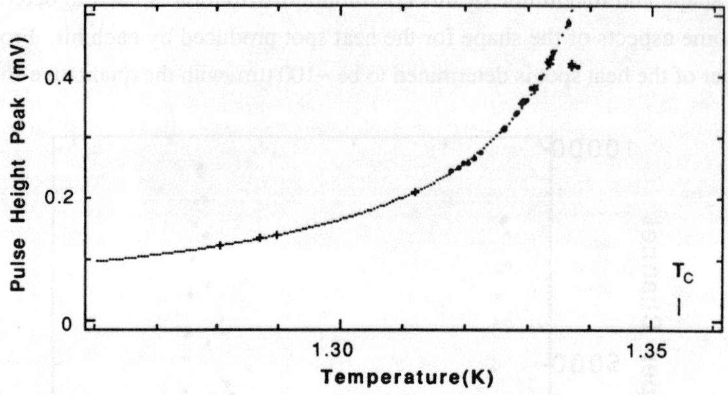

Fig 7. Variation of pulse height spectrum with temperature.

Back Alpha Experiments

The most significant step to date in our program to develop silicon crystal acoustic detectors has been the detection of alpha particles incident on the *back* of the substrate. The range of a 5.5 MeV alpha particle in silicon is only ~25 μm. Therefore the detector response must be due to phonons generated in the stopping process. Fig 8 shows a typical pulse height spectrum using our 299 line sensor. The FWHM is nearly identical with that shown in Fig 6 for the front alpha data and the pulse height is a factor of ten smaller. The size of the area driven normal (~30 μm square) is consistent with the size of the central ballistic phonon focussing spot for silicon (~0.15 radians) across the 275 μm thick wafer (see Fig 4).

To our knowledge, the pulse height spectra shown in Fig 6 for the front alpha signals represent the best alpha spectra taken with a superconducting device. Improvements in pulse height signal-to-noise may be possible using other superconducting materials such as niobium for the patterns.

The ultimate phonon sensitivity of these devices is not likely to surpass that already demonstrated for superconducting tunnel junctions. Nevertheless, these devices may prove very useful for phonon studies because of their ease of manufacture and their durability.

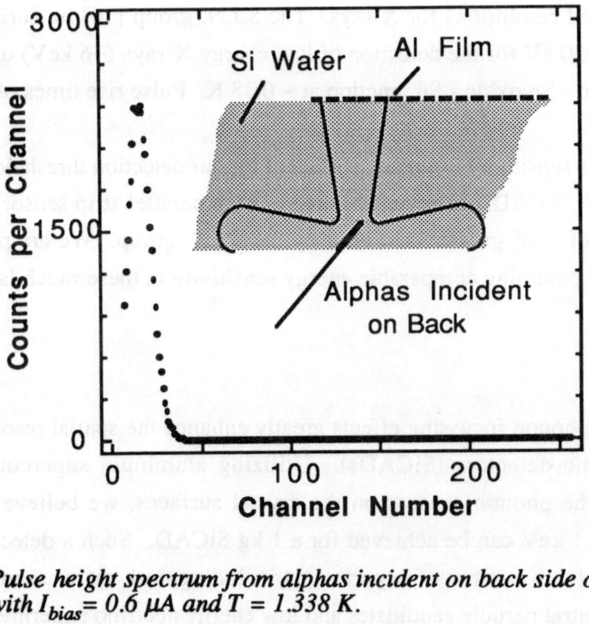

Fig 8. Pulse height spectrum from alphas incident on back side of wafer with I_{bias}= 0.6 μA and T = 1.338 K.

Superconducting Tunnel Junctions as Phonon Detectors

The most promising sensors for high sensitivity phonon detection are superconducting tunnel junctions. For our proposed application as a SiCAD readout , we are beginning the development of aluminum junctions. These are formed by a ~200 nm thick aluminum film deposited directly onto the surface of the silicon crystal, followed by the formation of a thin oxide tunnel barrier (~1-2 nm thick) and then deposition of a second aluminum upper film. Phonons from an event within the crystal reach the surface and enter the aluminum, which has an excellent acoustic match with silicon. Once inside the superconducting aluminum, most phonons have energies greater than 2Δ (where Δ is the superconducting energy gap) and are strongly absorbed by breaking Cooper pairs and forming electron-like excitations called quasiparticles. In fact for phonon energies well above threshold, several Cooper pairs will be dissociated. We are operating junction tests down to a temperature of 250 mK utilizing a sealed ^{3}He system with a charcoal cryopump. Ultimately we intend to operate at $T < T_c/10$ (~100 mK for aluminum), where quasiparticle recombination with thermal quasiparticles is negligible.

Recently several European groups [14,15] have reported excellent energy sensitivities and resolutions for X-rays. The S.I.N. group [14] in Zurich, Switzerland has reported ~30 eV for the detection of low energy X-rays (~6 keV) using a 100μm × 100μm area Sn - Sn oxide - Sn junction at ~ 0.38 K. Pulse rise times of ~15 μsec were observed.

In fact, this sensitivity is already sufficient for our detection threshold requirement of 1 keV in a 1 kg SiCAD. However, the area of each parallel strip sensor needs to be ~ 1 cm^2, a factor of ~10^4 greater than used by the S.I.N. group. We are pursuing several techniques for obtaining comparable energy sensitivity in these much larger area tunnel junctions.

Conclusions

Ballistic phonon focussing effects greatly enhance the spatial resolution in silicon crystal acoustic detectors (SiCADs). Utilizing aluminum superconducting tunnel junctions as the phonon sensors on the crystal surfaces, we believe that an energy threshold of ~1 keV can be achieved for a 1 kg SiCAD. Such a detector would be of great interest for a number of experiments including dark mater searches for weakly interacting neutral particle candidates and low energy neutrino experiments to set better limits on the neutrino mass.

This work has been funded in part by a Research Corporation Grant, a Lockheed Research Grant and DOE Contract DE-AM03-76-SF00-326.

References

[1] B. Cabrera, *Phys. Rev. Lett.* **48**, 1378 (1982).

[2] See the review article by D. E. Groom, *Phys. Rep.* **140**, 323 (1986).

[3] B. Cabrera, M. Taber, R. Gardner and J. Bourg, *Phys Rev Lett.* **51**, 1933 (1983); and R. Gardner, B. Cabrera, M. Taber and M. Huber, to be submitted to *Phys. Rev.* **D**.

[4] R. D. Gardner, PhD. Thesis, Stanford University (1987).

[5] S. Bermon, P. Chaudhari, C. C. Chi, C. D. Tesche, and C. C. Tsuei, *Phys. Rev. Lett.* **55**, 1850 (1985).

[6] J. Incandela, M. Campbell, H. Frisch, S. Somalwar, M. Kuchnir, and H. R. Gustafson, *Phys. Rev. Lett.* **53**, 2067 (1984).

[7] S. Somalwar, H. Frisch, J. Incandela, and M. Kuchnir, *Nucl. Inst.* **A226**, 341 (1984).

[8] M.E. Huber, B. Cabrera, M.A. Taber and R.D. Gardner, *Jap. J. of Appl. Phys.* **26**, 1687 (1987).

[9] S.H. Moseley, J.C. Mather and D. McCammon, J. Appl. Phys. **56**, 1257 (1984); D. McCammon, S.H. Moseley, J.C. Mather and R.F. Mushotzky, J.Appl. Phys. **56**, 1263 (1984).

[10] B. Cabrera, L. Krauss and F. Wilzcek, Phys. Rev. Lett. **55**, 25 (1985).

[11] B. Cabrera, J. Martoff and B. Neuhauser, Stanford Preprint No. BC44-86.

[12] See for example: **Nonequilibrium Phonon Dynamics**, ed. W.E. Bron, NATO ASI Series **B124**, Plenum Press, N.Y., 1985. An excellent review of recent developments in phonon physics.

[13] B. Neuhauser, B. Cabrera, C.J. Martoff and B.A. Young, *Jap. J. of Appl. Phys.* **26**, 1671 (1987).

[14] D. Twerenbold, Europhys. Lett. **1**, 209 (1986).

[15] H. Krauss, T. Peterreins, F. Probst, F.v. Feilitzsch, R.L. Mossbauer and E. Umlauf, Europhys. Lett. **1**, 161 (1986).

EXPLOITATION OF PHONON FOCUSING TO OBTAIN SPATIAL RESOLUTION IN THE DETECTION OF α-PARTICLES

Th. PETERREINS, F. PROBST, F.V. FEILITZSCH, H. KRAUS

Physik-Department,
Technische Universitat Munchen,
D-8046 Garching, Germany

1. INTRODUCTION

The detection of ballistic phonons created by the absorption of nuclear radiation in a crystal may provide a new interesting concept of cryogenic particle detection [1]. These phonons are the final product of the relaxation of the excitations caused by the incident radiation. In pure single crystals of dielectric material kept at low temperature ($T < 10$ K) phonons travel ballistically, that means at the speed of sound without scattering, over macroscopic distances [2]. The wave front carries information on the deposited energy and on the location of the event. Position information may be derived from the arrival times of the phonon pulse at the crystal surfaces and the characteristic spatial pattern of phonon flux arriving at the surfaces. This pattern results from the phonon focusing effect [2], which is induced by an anisotropy of the elastic constants of the crystal. Its main consequence is the concentration of the energy flux from a source of phonons with isotropically distributed $\vec{k}$-vectors into certain crystal directions. It is theoretically well understood [3,4,5] and has been investigated experimentally in great detail [6]. For detection of the phonons arriving at the crystal surface, one may use superconducting tunnel junctions, which have already demonstrated their sensitivity as X-ray detectors [7,8].

An advantageous feature of this device is its sensitivity to non-ionizing interactions. Furthermore, it may offer substantial improvement in background rejection compared to detection methods relying on the measurement of an overall temperature rise alone [9,10].

In this article we report on the detection of ballistic phonons originating from α-particles incident on the back side of a single crystalline Si substrate [11,12]. The experiment is designed to investigate the possibility of using phonon focusing patterns and the timing information to obtain spatial resolution.

2. EXPERIMENT

Three $Al/Al_2O_3/Al$ tunnel junctions were vacuum deposited through suitable masks onto the top surface of a 3mm thick silicon wafer, which was parallel to the (001) planes of the crystal. The dopant concentration of the sample was about 10^{13} cm^{-3}. The natural oxide was removed with hydrofluoric acid prior to evaporation. Both top and bottom surfaces had optical quality. Geometrical details are shown in fig.1.

An ^{241}Am source emitting α-particles with energies of 5.486 MeV (85%) and 5.443 MeV (15%) was mounted facing the opposite (bottom) side of the wafer. The active area of the 38 μCi source had a diameter of about 5 mm. Using a semiconductor detector an intrinsic line width of the source of 15% (FWHM) was determined. The set-up is shown schematically in fig.2. Two distinct regions of α-absorption were defined by holes of 0.2 mm and 0.3 mm diameter in a 0.1 mm thick copper sheet. The distance between the centers of the two holes was 0.75 mm. Since the range of 5.5 MeV α-particles in Si is only about 25μm the phonons are produced very close to the surface opposite to the tunnel junctions. The position of the holes was chosen such that junction No.1 is in the [001]-direction of the crystal seen from the center of the left hole, and junction No.2 is in [111]-direction seen from the larger right hole (Fig.2). In the [001]-direction of the crystal very strong phonon focusing for the dominant [3] slow transverse phonons occurs, whereas a much weaker phonon flux is expected from the larger hole to junction No.1. The focusing in [111]-direction is rather complicated

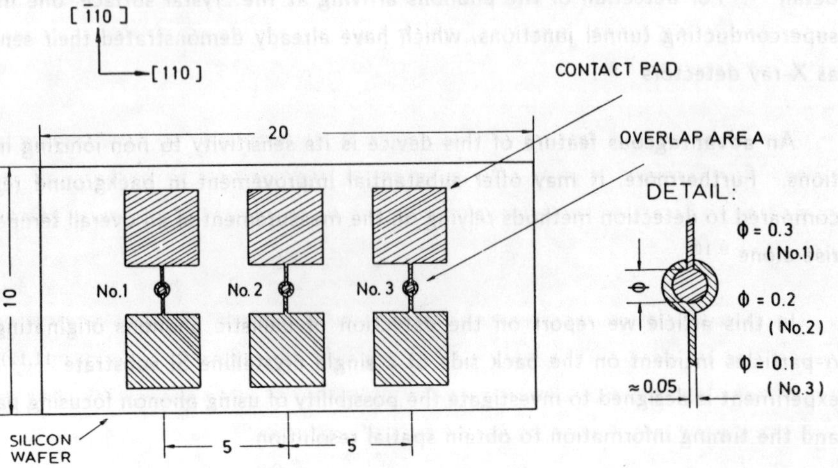

Fig.1: Geometrical arrangement of the Al tunnel junctions on the Si wafer. The aluminum films evaporated onto the substrate are shown shaded. The junctions are confined to the overlap area of two films, which is shown enlarged in the insert. Dimensions are given in mm. The junctions themselves are not drawn to scale.

and much less pronounced. Longitudinal phonons are focused with a broad maximum lying in the [111]-direction. The contour of maximal energy flux of the slow transverse mode lies a few degrees off the [111]-direction [4,5].

The whole assembly was placed into the vacuum chamber of a ^{3}He-cryostat and cooled down to T=0.38 K. Signals from the junctions were amplified by charge sensitive preamplifiers and main amplifiers with independently selectable time constants for integration and differentiation. To study correlations in time between signals in different junctions, signals from two junctions were recorded in coincidence. One channel was fed into a transient recorder. From the other channel a trigger was derived to initiate the post trigger sampling of the transient recorder. The digitized pulses from the transient recorder were stored on tape for later processing on a computer. Time correlated signals measured this way arise from phonons produced in the stopping process of a single α-particle with subsequent detection of the phonon wave front in two different tunnel junctions.

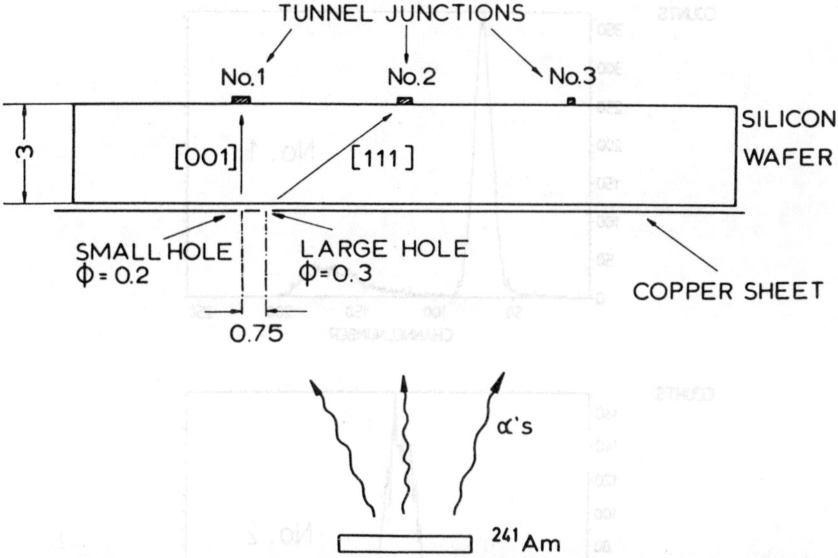

Fig.2: Schematic cross-sectional view of the experimental set-up. Dimensions are given in mm.

3. RESULTS AND DISCUSSION

Pulse height spectra from the three tunnel junctions are shown in fig.3. The integrated number of counts in the peaks of the spectrum of junction No.1 allows to attribute the narrow peak at lower pulse height to α-particles passing through the larger right hole. The wider peak at larger pulse height corresponds to the small left hole. Junction No.1 lies in [001] direction as seen from the small hole, where strong phonon focusing occurs. A much weaker phonon flux is expected in the direction from the large hole. The pulse height ratio of the two peaks of about 2.5 : 1 is much larger than one would expect in the hypothetical case of isotropic phonon propagation (1.1 : 1) but falls somewhat below the value roughly estimated from tabulated focusing patterns (about 10 : 1) [5]. This discrepancy may originate from an onsetting nonlinear response of the tunnel junction due to the very large phonon flux. A possible slight misalignment of the holes relative to the junctions may also contribute. Nevertheless the large separation of the two peaks clearly demonstrates that phonon focusing patterns can be exploited to obtain spatial resolution.

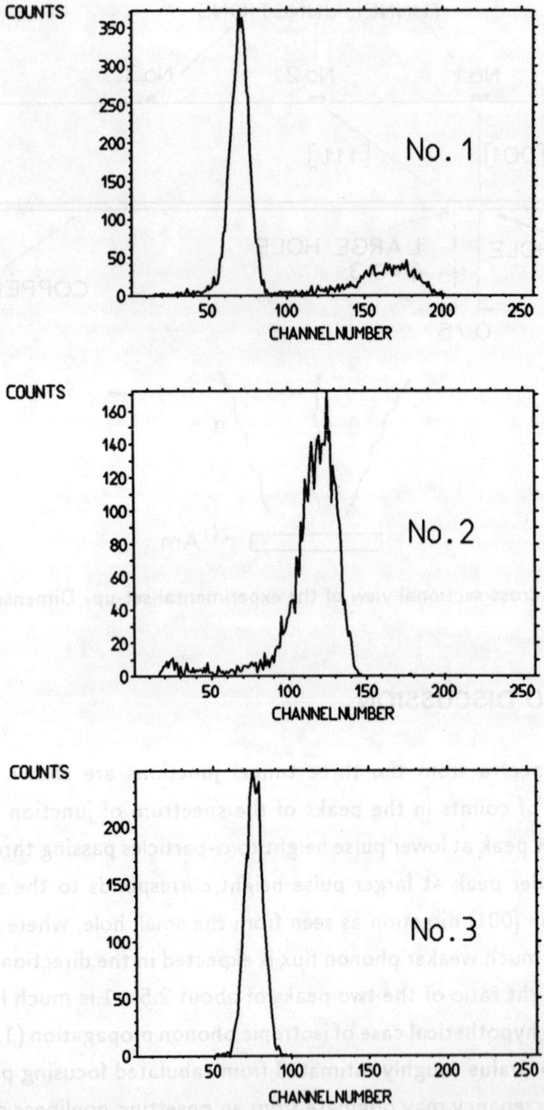

Fig.3: **Pulse height spectra from the three tunnel junctions. Low pulse heights have been cut off. Pulse heights from different junctions are not directly comparable because of differences in junction properties and amplifier gain settings.**

The pulse height spectrum of junction No.2 (fig.3) shows only a single peak with a slight broadening on the low energy side. Assuming isotropic phonon propagation the amplitude of pulses from α-particles passing the left hole amounts to 70% of the amplitude from those passing the right hole, which would result in a pulse height spectrum of well resolved double peak structure. Inspection of time correlated appearance of pulses from junctions No.1 and No.2 revealed that the ones with the lower pulse height from No.1 correlate with pulses on the high energy side of the peak in the spectrum of No.2, and also that larger pulses of No.1 correlate with pulses on the low energy side of the peak in the spectrum of No.2. The larger distance of the left hole may have been partly outbalanced by a focusing of slow transverse phonons from the left hole onto junction No.2. The maximal flux enhancement for this focusing pattern lies a few degrees off the [111]-direction, favoring signals from the left hole (Fig.2). It has a stronger effect than the focusing of the longitudinal mode with its broad maximum in [111]-direction.

Junction No.3 is more than 10 mm away from the points of absorption and lies out of any focusing direction. Nevertheless it yields signals with a signal-to-noise ratio of about 10:1 at the output of an amplifier with a bandwidth from 100 Hz to 1.5 MHz. The width of the single peak in the pulse height spectrum of junction No.3 is 17% (FWHM), which only slightly exceeds the intrinsic width of the α-source (15% FWHM). Isotropic phonon propagation predicts a pulse height ratio of 0.8 : 1 for the signals from the two holes so that a broadened peak would be expected. A qualitative inspection of focusing patterns for Si [4,5] does not provide an obvious explanation of the narrow experimental peak width. Detailed calculations of the focusing in the region of interest have certainly to be done, before speculations about an alternative origin of this effect appear justified, like for example significant detection of phonons that have undergone multiple reflections.

Finally the relative timing of pulses from the junctions was evaluated by determining the time difference between pulses in two junctions induced by the same event of α-absorption. First, the pulses from junction No.1 were recorded on tape with a trigger signal derived from pulses of junction No.2. After that, the pulses from junction No.2 were passed through the same processing and recorded on tape, leaving the trigger condition unchanged. Thus the finite trigger level and delays caused by the electronics could be compensated. The time when the first noticeable phonon intensity reaches a junction was determined by fitting a model pulse to the digitized experimental pulses within a fixed time span around the onset of the pulse. This was done for pulses both of junctions No.1 and No.2. The parameterized pulse $F(t)$ was described by

$$F(t) = \begin{cases} C + A \cdot [1 - exp(-(t - t_0)/\tau)], & \text{if } t > t_0; \\ C, & \text{otherwise.} \end{cases}$$

Here the offset C, the amplitude parameter A, the rise time τ and the time t_0 of the onset of the pulse were adjustable parameters. The method is illustrated in fig.4 where an experimental pulse (amplifier bandwidth from 100 Hz to 1.5 MHz) from junction No.1 and the fitted curve are shown. The experiment was not optimized for good time resolution. To derive a trigger signal, the preamplifier output pulse of junction No.2 was suitably shaped and a fixed threshold was used to initiate the trigger. This simple trigger condition together with the relatively long rise time of the pulses (20-30 μs at the preamplifier output) determines the limited time resolution of our measurement, as the trigger serves as a common time reference. Using a charge sensitive preamplifier the rise time of the preamplifier output pulse is essentially given by the lifetime of the excess quasiparticles in the junction material (Al) at a temperature of T=0.38 K. We used charge sensitive preamplifiers for signals from both junctions to ensure that different pulse shapes do not systematically falsify the difference in times of pulse onset determined by the fit procedure.

In fig.5 the distribution of times of pulse onset (as determined by the fit procedure) for pulses from junctions No.1 and No.2 is shown. The time scale is given relative to the mean time of pulse onset in junction No.2. The different pulse heights in No.1 were used to discriminate between signals from both holes. The onset of pulses in junction No.2, caused by α-particles hitting the wafer through the small left hole, was found to be delayed by 3.4 μs relative to corresponding pulses from junction No.1. A delay of 1.9 μs was determined for events from the large right hole.

The interpretation of these time differences is complicated by the following facts:

1. The sound velocity of a given phonon branch in Si shows considerable anisotropy, which amounts to up to 20%. In certain directions, it can even be multi-valued.

2. The relative contribution of the three phonon branches, the longitudinal, fast

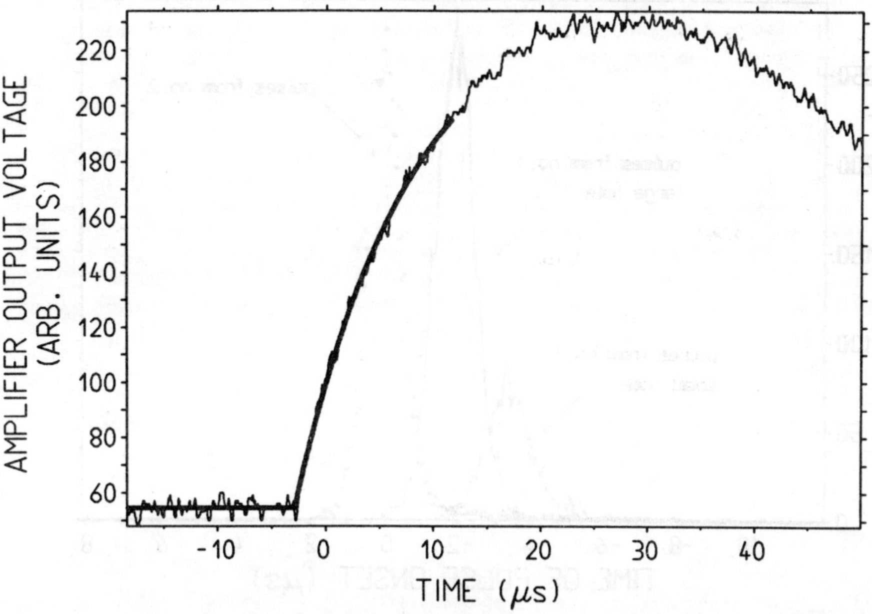

Fig.4: Experimental pulse of tunnel junction No.1 with a function fitted to determine the onset of the pulse. The time scale refers to the mean time of onset of a corresponding pulse in junction No.2.

transverse, and slow transverse branch, depends on the direction of propagation. In Si, longitudinal phonons may play a significant role only in the [111]-direction. Sound velocities for different branches range from about 5000 m/s to more than 9000 m/s, but we cannot resolve them due to the long time constant of the Al tunnel junction. The onset of a pulse, of course, is determined by the fastest phonon wave front which arrives at the diode with a flux exceeding the detection threshold.

The measured time differences are much larger than expected even if these effects are taken into account. This statement is further supported by measurements of time differences between signals of junctions No.1 and No.3 as well as between signals of junctions No.2 and No.3. The origin of these long times is presently not understood. One speculation is that phonons emitted from the absorption region might still carry information about the direction of the incoming α-particle, which is not perpendicular to the wafer surface (see fig.2). Several reflections on the wafer surface may then be necessary until a detectable phonon flux reaches the more distant junctions. Another

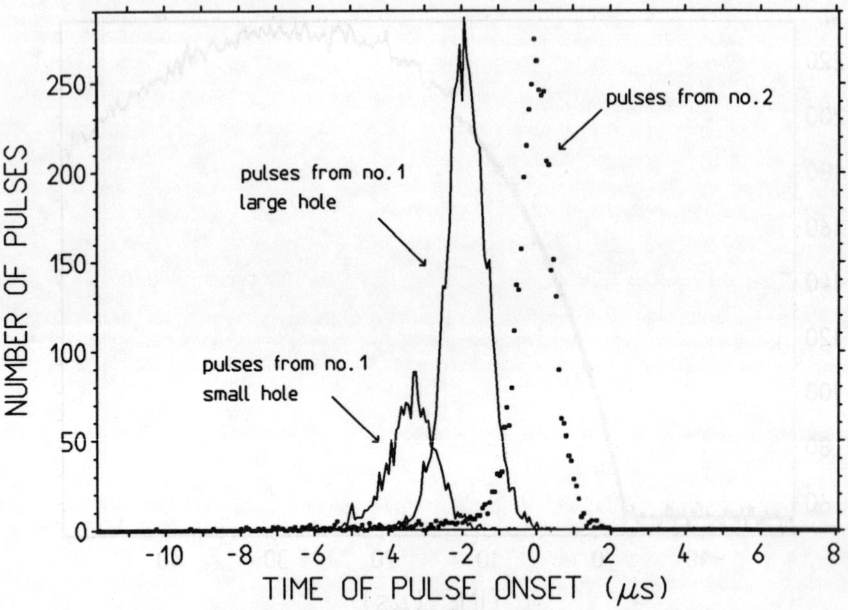

Fig.5: Time of pulse onset versus corresponding number of pulses. The time scale is given relative to the mean time of pulse onset in junction No.2.

possible explanation may be given by detection of significant amounts of phonon intensity in the contact leads. Analogous to the case of X-ray detection [7], diffusion phenomena might then contribute to the time differences observed.

4. CONCLUSIONS

For the first time, the benefit of phonon focusing patterns for obtaining spatial resolution by pulse height analysis in the detection of nuclear radiation has been experimentally demonstrated. In Si the strong focusing in [100]-direction seems especially favorable in this respect. Additional information can be obtained from time resolved measurements of the correlations of pulses in different junctions, but a final understanding of the processes involved requires further experimental work. In particular, the spatial distribution and the spectrum of phonons emitted in the absorption process of nuclear radiation have to be carefully investigated, as well as potential effects of the detailed detector geometry.

This work was supported by the Bundesministerium für Forschung und Technologie.

References

1) Neuhauser,B., Cabrera,B., Martoff,C.J., and Young,B.A., "Phonon-Mediated Detection of Alpha-Particles with Aluminum Transition Edge Sensors," In: Proc. 18th Int. Conf. on Low Temperature Physics, Kyoto, 1987, Japanese Journal of Applied Physics, 26, Supplement 26-3, 1671 (1987).

2) For a review, see for example : Wolfe,J.P., "Ballistic heat pulses in crystals", Physics Today , December 1980, p.44

3) Rösch,F., and Weis,O., Zeitschrift f. Physik B 25, 101 (1976).

4) Rösch,F., and Weis,O., Zeitschrift f. Physik B 25, 115 (1976).

5) Elbaum,C., "Anisotropic heat conduction in cubic crystals", In: Proceedings of the International Conference on Phonon Scattering in Solids, Paris, ed. by Albany,H.J., Comptes Rendus, Saclay, 1972, pp. 1-17.

6) Northrop,G.A., and Wolfe,J.P., Phys. Rev. B 22, 6196 (1980).

7) Kraus,H., Peterreins,Th., Pröbst,F., v.Feilitzsch,F., Mössbauer,R.L., Zacek,V., and Umlauf,E., Europhys. Lett., 1, 161 (1986).

8) Twerenbold,D., and Zehnder,A., J. of Appl. Phys. 61, 1 (1987).

9) Moseley,S.H., Mather,J.C., and McCammon,D., J. of Appl.Phys., 56, 1257 (1984); also McCammon,D., Moseley,S.H., Mather,J.C., and Mushotzky, R.F., J. of Appl. Phys., 56, 1263 (1984).

10) Coron,N., Dambier,G., Focker,G.J., Hansen,P.G., Jegoudez,G., Jonson, B., Leblanc, J., Moalic, J.P., Ravn, H.L., Stroke,H.H., and Testard,O., Nature 314, 75 (1985).

11) Peterreins,Th., Pröbst,F., v.Feilitzsch,F., and Kraus,H., to appear in IEEE Trans. Nucl. Science

12) v.Feilitzsch,F., Hertrich,T., Kraus,H., Peterreins,Th., Pröbst,F., and Seidel,W., "Low Temperature Calorimeters", In: Proceedings of the International Conference on Solar Neutrino Detection with ^{205}Tl and Related Topics, Dubrovnik 1986, to appear in Nucl. Instr. and Methods

ELECTRON-BEAM IRRADIATION OF JOSEPHSON TUNNEL JUNCTIONS

R. GROSS, R.P. HUEBENER, and U. KLAβ

Physikalisches Institut, Lehrstuhl II,
Universitat Tubingen
D-7400 Tubingen, FRG

1. INTRODUCTION

Recently superconductors have been discussed for applications in a possible new family of solid state particle detectors.[1)-4)] Because of the low energy gap in a superconductor of typically about 1 meV, such a material appears highly favorable as a solid state particle detector compared to a semiconductor with a typical energy gap of about 1 eV. The low energy gap in a superconductor results in a correspondingly larger number of charge carriers generated by the particle irradiation and promises a higher energy resolution compared to a semi-conducting detector. An important advantage of a superconducting particle detector lies also in the fact that the superconducting energy gap is smaller than the Debye energy. Therefore, phonons generated by the particle irradiation can contribute to the detector signal (by breaking of Cooper pairs), in con-trast to the situation for a semiconducting detector.

The number δN of charge carriers generated in the detector by the particle irradiation is given by

$$\delta N = E_{dep}/\varepsilon \quad , \qquad (1)$$

where E_{dep} is the energy deposited by the irradiation and ε the minimal energy required for the generation of one free charge carrier. Due to competing processes such as phonon or photon generation, the value of ε is always larger than the superconducting energy gap Δ of the detector material.

Counting the excess number of quasiparticles generated by the particle irradiation represents another essential function which must be performed by the detector. Here the configuration of a superconducting tunnel junction can be utilized by measuring the excess quasiparticle tunneling current. However, it is important to note that in many cases only a fraction of the generated excess number of quasiparticles is tunneling within their lifetime and is detected by this counting process. Hence, in an evaluation of superconducting tunnel junctions as particle detectors the short quasiparticle recombination time (and phonon trapping effects) must be taken into account carefully. A detailed discussion of this point including model calculations has been reported recently.[5)]

The quasiparticle current-voltage characteristic of a superconducting tunnel junction shows highly nonlinear behavior resulting in a sensitive dependence of the magnitude of the detector signal upon the bias point of the junction. In the thermal tunneling regime the current is determined by the quasiparticles which are thermally excited at the operating temperature of the junction. Here the detector signal directly results from the additional quasiparticles generated by the particle irradiation. In the gap regime the situation is more complicated. Here, in addition to the excess number of quasiparticles, the reduction of the energy gap due to the particle irradiation contributes significantly to the detector signal. This reduction of the energy gap is most important for materials having a small quasiparticle diffusion length and results in a distinct maximum of the detector signal for a bias point at the gap sum voltage.[5] For materials having a large quasiparticle diffusion length this effect is less important due to the fast reduction of the excess quasiparticle density due to quasiparticle diffusion.

Experimentally the electronic response of superconducting tunnel junctions to particle irradiation can be investigated using a scanning electron microscope equipped with a liquid-He stage. We have utilized such a system extensively for imaging spatial structures in tunnel junctions by means of low temperature scanning electron microscopy (LTSEM)[6)-9)]. In these experiments the junction is scanned with the electron beam and a proper junction response signal is recorded as a function of the coordinates of the beam focus, while the back side of the substrate carrying the junction is in direct contact with liquid He. In addition, the principle of LTSEM can serve for quantitatively evaluating the application of tunnel junctions as cryogenic particle detectors.[5)] In the present paper we report on recent experiments of this second category performed with Pb-alloy junctions.

In a discussion of superconducting tunnel junctions as possible cryogenic particle detectors beyond operational problems (suitable signal amplification, temperature stability, etc.) three major problem areas arise which have to be clarified. First, the magnitude of the detector signal and the detector sensitivity must be studied. Second, the stability of the tunneling barrier and possible barrier changes occuring during the particle irradiation must be known. Third, the degree of spatial homogeneity of the tunneling barrier and of the thermal coupling to the substrate must be evaluated, because of its influence on the energy resolution which can be obtained. In the following we address ourselves to all three subject areas.

2. SAMPLES AND EXPERIMENTAL PROCEDURES

Our experiments were performed using thin-film superconducting tunnel junctions fabricated by standard vacuum deposition techniques. Both the base and the top electrode consisted of Pb-In alloy containing 10 weight%In (deposition by flash evaporation). The substrates were single-crystalline sapphire disks of 20 mm diameter and 1 mm thickness. During the deposition of the junction electrodes the substrates were in good thermal contact with a metallic plate cooled with liquid N_2. The natural oxide of the base electrode served as the tunneling barrier (oxidation at room temperature at O_2-pressure of about 30 mbar; oxydation time 12 - 18 h). The tunneling area was defined by a square window of 28.3 μm side length in a SiO film thermally evaporated upon the base electrode. Microfabrication was performed using stand photolithographic techniques.

The geometry of the tunnel junctions is shown schematically in Fig. 1. A barrier-parallel magnetic field could be applied by passing an electric current along the base electrode. On each substrate two electrically separated tunnel junctions were prepared. A total of 10 junctions have been investigated. The characteristic parameters of these junctions are listed in Table 1.

The electron beam irradiation experiments at low temperatures were performed using our standard configuration for low temperature scanning electron microscopy (LTSEM).[6)-9)] Here the substrate carrying the tunnel junction detector at the top is mounted on the low-temperature stage of a scanning electron microscope

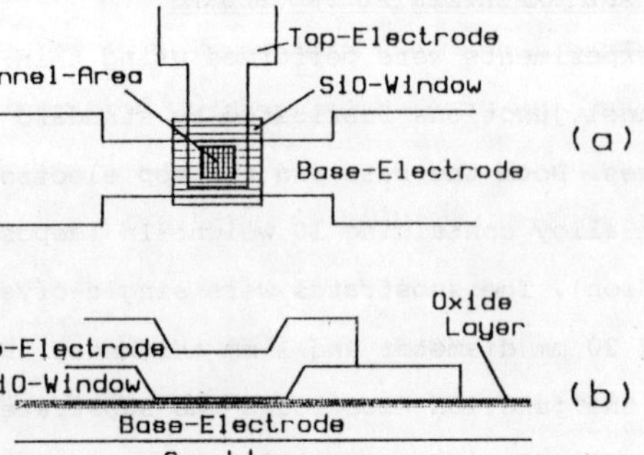

Fig. 1: Tunnel junction geometry. (a) Top view.
(b) Cross section.

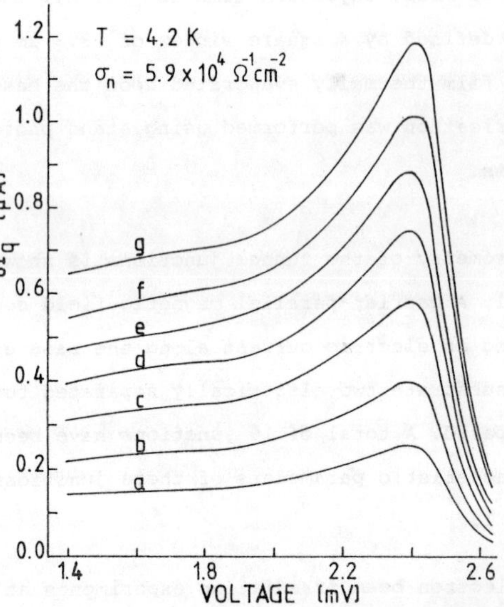

Fig. 2: Beam-induced signal δI_q versus the junction bias voltage for the following values of the electron beam power absorbed in the junction: 3.9 (a), 4.3 (b), 4.8 (c), 5.6 (d), 6.2 (e), 6.8 (f), and 7.8×10^{-7} W (g). Beam voltage = 25 kV, beam modulation frequency = 25 kHz.

Sample No.	Thickness [nm] Base Electrode	Thickness [nm] Top Electrode	Oxidation Time [h]	$\sigma_n[\frac{1}{\Omega\cdot cm^2}]$	$\frac{\sigma_n}{\sigma_{SG}}$
1/I	280	320	16 (at 670 HPa)	$6.3\cdot10^4$	8.0
1/II				$1.0\cdot10^4$	7.5
2/I	212	511	18	$5.7\cdot10^4$	2
2/II				$2.9\cdot10^4$	7.4
4/I	200	668	14	$7.2\cdot10^4$	8.0
4/II				$1.1\cdot10^5$	8.6
1/6/I	190	590	15	$3.2\cdot10^4$	6.4
1/6/II				$1.0\cdot10^5$	4.4
2/6/I				$1.6\cdot10^4$	6.4
2/6/II				$2.2\cdot10^4$	6.2

Table 1: Sample characteristics. σ_n is the normal tunneling conductivity and σ_{SG} the sub-gap tunneling conductivity at T = 4.2 K.

(SEM) in such a way, that the detector can be irradiated directly with the electron beam, whereas the bottom of the substrate is in direct contact with the liquid helium bath. The electron beam of the SEM can be periodically modulated up to frequencies of about 5 GHz. Further, short beam pulses can be obtained. The coordinate point of the beam focus can be chosen arbitrarily. Typical beam parameters are the following: voltage = 25 kV, power = 10^{-5} - 10^{-8} W, diameter < 1000 Å, pulse width $\geq$ 10 ns. During most experiments the temperature of the liquid-He bath was 4.2 K.

The beam-induced response signal generated in a tunnel junction depends upon the junction bias condition. If the junction is voltage biased on the quasiparticle voltage-current characteristic, the beam irradiation causes a current signal δI_q. Correspondingly, a voltage signal δV_q results for current-biased operation. For increasing the sensitivity it is convenient to modulate the beam and to measure the modulated response signal using a lock-in amplifier. A more detailed description of the electronic principles for signal detection in LTSEM of superconducting tunnel junctions can be found elsewhere.[10]

3. ELECTRON BEAM INDUCED DETECTOR SIGNAL

In our experiments the total thickness of the electrodes of the tunnel junction detectors was chosen to be larger than the penetration depth of the electron beam, the latter being about 1 μm for 25 keV electrons and the electrode material of our samples. Therefore, the total beam energy is deposited in the junction electrodes. Typical experimental results are shown in Fig. 2. Here the increment of the tunneling current δI_q induced by the beam irradiation is plotted versus the detector bias voltage for different values of the electron beam power absorbed in the sample. The curves in Fig. 2 show a distinct maximum at the gap sum voltage. This maximum can easily be understood in terms of the usual tunneling theory if the reduction of the energy gap in the irradiated region and the strong increase of the quasiparticle tunneling current at the gap sum voltage is taken into account. The temperature dependence of the signal δI_q for constant electron beam power and bias voltage was found to be very weak in the range investigated (1.5 - 4.2 K). A detailed discussion of the signal δI_q and its variation with the bias voltage and the electron beam power will be given elsewhere.[11]

It is interesting to estimate from the measured quantitiy δI_q the deposited beam energy needed to produce a single additional charge carrier, i.e., the energy ε following an extension of the definition in Eq. (1). Writing

$$\delta I_q = \frac{\delta N}{t} \cdot e_o \ , \qquad (2)$$

we obtain

$$\varepsilon = \frac{E_{dep}}{\delta N} = \frac{P_{dep}}{\delta I_q} \cdot e_o \ . \qquad (3)$$

Here t, e_o, and P_{dep} denote time, the elementary charge, and the deposited beam power, respectively. We note that in Eq. (3) we have extended the original definition of ε in Eq. (1), referring only to the <u>generated</u> number of charge carriers, to the number of carriers generated and detected. The quantity ε calculated from Eq. (3) is shown in Fig. 3 and 4 for T = 4.2 K and T = 1.7 K, respectively as a function of the beam current (at constant beam voltage of 26 kV) and for different values of the junction bias voltage. For obtaining the data of Fig. 3 and 4 we have assumed that 66 % of the beam power is absorbed in the junction electrodes. We see that the energy ε is much larger than the gap energy of about 1.2 meV. Of course, this is a consequence of the quasiparticle recombination process reducing the fraction of the quasiparticles which is tunneling following the generation process. At the lower temperature the energy ε is seen to be distinctly smaller. This can be understood qualitatively in terms of the temperature dependent quasiparticle recombination rate which is proportional to $\sqrt{T}\cdot\exp(-\Delta(0)/k_B T)$, decreasing with decreasing temperature.

So far we have described only measurements performed by modulating the beam power such that high sensitivity can be achieved by utilizing lock-in techniques for signal detection. However, in their application as cryogenic particle detectors superconducting tunnel junctions must record statistical events, and phase-sensitive detection methods are ruled out in general. Again, a scanning electron microscope is well suited for evaluating experimentally the junction sensitivity to uncorrelated short pulses of particle irradiation. For determining experimentally the energy resolution one can simply vary the accelleration voltage of the electron beam. Of course, such experiments require state-of-the-art fast electronics for signal pulse detection. Whereas electron beam irradiation experiments of this type have not yet been

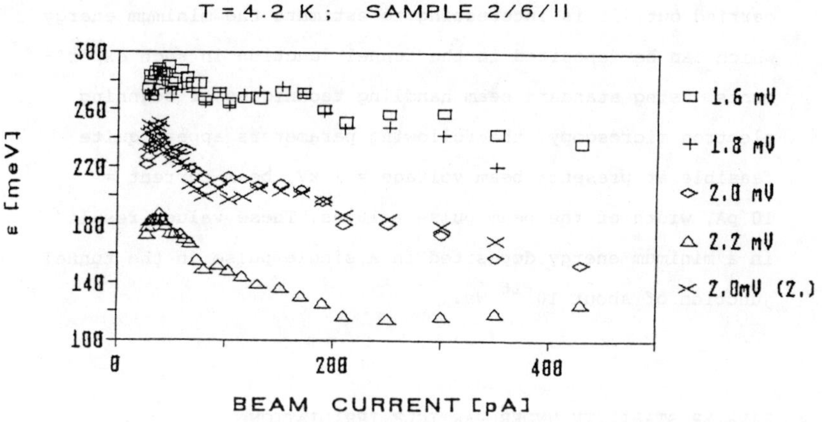

Fig. 3: Quantity ε from Eq. (3) versus the beam current for different values of the junction bias voltage. (The crosses for 2.0 mV bias voltage refer to a second run). T = 4.2 K, beam voltage = 26 kV, beam modulation frequency = 20 kHz, $\sigma_n = 72 \cdot 10^3\ \Omega^{-1}\ cm^{-2}$.

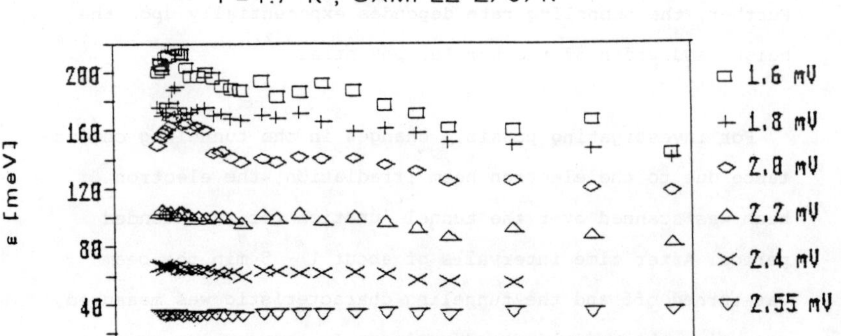

fig. 4: Quantity ε from Eq. (3) versus the beam current for different values of the junction bias votlage. T = 1.7 K, beam voltage = 26 kV, beam modulation frequency = 20 kHz, $\sigma_n = 127 \cdot 10^3\ \Omega^{-1}\ cm^{-2}$

carried out, it is interesting to estimate the minimum energy which can be deposited in the tunnel junction in such experiments. Using standard beam handling techniques of scanning electron microscopy, the following parameters appear quite feasible at present: beam voltage = 1 kV, beam current = 10 pA, width of the beam pulse = 10 ns. These values result in a minimum energy deposited in a single pulse in the tunnel junction of about 10^{-16} Ws.

4. BARRIER STABILITY UNDER PARTICLE IRRADIATION

In addition to experimentally evaluating the sensitivity of superconducting tunnel junctions as particle detectors, it is important to address also the question of the barrier stability and possible changes of the tunneling barrier under particle irradiation. After all, the barrier is a highly delicate object with a thickness of only a few atomic layers. Further, the tunneling rate dependes exponentially upon the height and width of the barrier potential.

For investigating possible changes in the tunneling conductance due to the electron beam irradiation, the electron beam was scanned over the tunnel junction for an extended period. After time intervalls of about 1 - 5 min the beam was turned off and the tunneling characteristic was measured, repeating this procedure many times. In these experiments the total scanned area was about 9 times larger than the tunneling window, the latter being in the center of the scanned area. The He-bath temperature was 4.2 K. Holding the beam voltage constant at 25 or 26 kV, the beam current was kept small in the beginning (typically 30 - 100 pA) and subsequently was raised gradually (up to about 1 nA) until no appreciable changes in the tunneling conductance could be observed any more.

Typical results are presented in Figs. 5 - 7, showing the tunneling conductivity in the normal state versus the number of beam electrons which have struck the tunneling window. (This number was calculated from the irradiation time, the beam current,and the fractional portion of the total scanned area taken up by the tunneling window of 28.3 x 28.3 μm^2 area. At the top of Fig. 5 and 6 the different values of the beam current and the corresponding irradiation times are indicated). In Figs. 5 - 7 the tunneling conductivity is normalized to the conductivity at the beginning of the particular irradiation run. The latter conductivity values are indicated in the figure legends.

Figure 5 refers to the first irradiation of tunnel junction 2/6/I. The normal conductivity is seen to increase during the irradiation initially, to pass through a maximum and then to approach pradually a conductivity value smaller than that at the beginning. A similar result has been observed for a second irradiation run with the same tunnel junction, if the junction had been warmed up to room temperature in the meantime (see Fig. 6. Between the first and second irradiation run the junction was held at room temperature for 3 - 4 days). If the junction was always kept below 77 K between two subsequent irradiation runs, the second irradiation generally effected a monotonic increase of the tunneling conductivity.

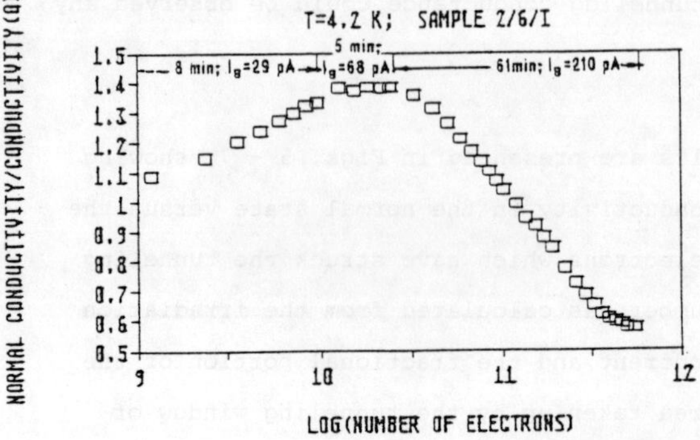

Fig. 5: Tunneling conductivity in the normal state versus the number of beam electrons which have struck the area of the tunneling window. The conductivity is normalized to the value $\sigma_n(0) = 29 \cdot 10^3\ \Omega^{-1}\ cm^{-2}$ at the beginning of the run. Beam voltage = 26 kV. First irradiation of this sample.

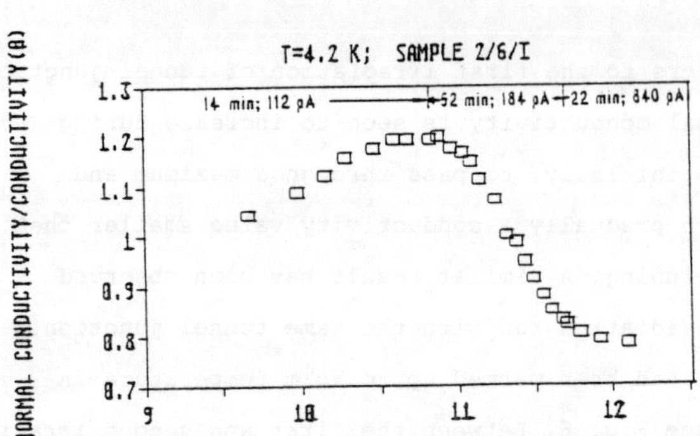

Fig. 6: Tunneling conductivity in the normal state versus the number of beam electrons which have struck the area of the tunneling window. The conductivity is normalized to the value $\sigma_n(0) = 168 \cdot 10^3\ \Omega^{-1}\ cm^{-2}$ at the beginning of the run. Beam voltage = 26 kV. Second irradiation of this sample after it was held at room temperature for 3 - 4 days.

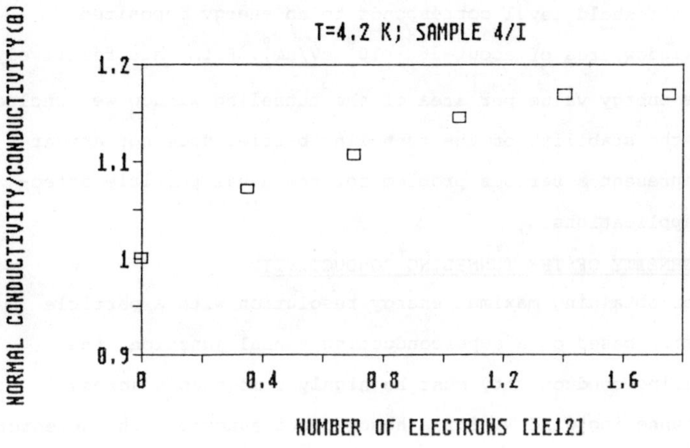

Fig. 7: Tunneling conductivity in the normal state versus the number of beam electrons which have struck the area of the tunneling window. The conductivity is normalized to the value $\sigma_n(0) = 75 \cdot 10^3\ \Omega^{-1}\ cm^{-2}$ at the beginning of the run. Beam voltage = 25 kV. Second irradiation of this sample after it was held below 77 K.

Typical results are shown in Fig. 7 for such a case.

An explanation of the change in the tunneling conductivity effected by the electron beam irradiation and shown in Figs. 5 - 7 appears difficult at present. One might speculate about additional electronic states generated by the irradiation within the barrier or in its immediate viciuity. However, a satisfactory clarification of these effects requires further

The fact that appreciable changes of the tunneling conductivity do not occur below the irradiation level of about 10^9 electrons striking the area of the tunneling window represents an important result of our measurements. With the window area of 812 μm^2 and the beam energy of 26 keV this threshold level corresponds to an energy deposited per window area of about $26 \cdot 10^9$ eV/μm^2. From this relatively large energy value per area of the tunneling window we conclude that the stability of the tunneling barrier does not appear to represent a serious problem for the usual particle detector applications.

5. HOMOGENEITY OF THE TUNNELING CONDUCTIVITY

For obtaining maximum energy resolution with a particle detector based on a superconducting tunnel junction, the tunneling conductivity must be highly homogeneous across the tunneling window. For a homogeneous junction, the detector signal only depends upon the energy of the particle irradiation and not upon the coordinate point at which the particles impinge upon the tunneling window. Hence, an accurate evaluation of possible spatial variations of the tunnelign conductivity becomes highly important. Such an evaluation can be performed conveniently by means of LTSEM.[6)-9)] Simply one has to record the beam-induced voltage or current signal as a function of the coordinate point of the beam focus, while the junction is biased properly on its quasiparticle voltage current characteristic.

Inhomogeneities in the tunneling conductivity can originate in the tunneling barrier as well as in the junction electrodes, the former case being more likely than the latter. Further details regarding this question of the junction homogeneity can be found in a recent report.[5)]

ACKNOWLEDGMENT

This work originated from stimulating discussions between A. Barone and one of the authors (R.P.H.).

References

1) A. Barone, S. De Stefano,K. E. Gray, Nucl. Instr. and Meth. A 235, 254 (1985)

2) A. Barone, G. Darbo, S. De Stefano, G. Gallinaro, A. Siri, R. Vaglio, S. Vitale, Nucl. Instr. and Meth. A 234, 61 (1985)

3) D. Twerenbold, Phys. Rev. B34, 7748 (1986)

4) D. Twerenbold and A. Zehnder, J. Appl. Phys. 61, 1 (1986)

5) R. Gross, U. Klaß, and H. Bender, Extended Abstracts of 1987 International Superconductivity Electronics Conference (ISEC '87), Tokyo, Japan, p. 345 (S 10-4)

6) R. P. Huebener, Rep. Prog. Phys. 47, 175 (1984)

7) R. P. Huebener, Advances in Electronics and Electron Physics, P. W. Hawkes, ed., Vol. 70, 1987

8) J. Bosch, R. Gross, and R. P. Huebener, Josephson Effect-Achievements and Trends, A. Barone, ed., World Scientific, 1986, p. 174

9) R.P. Huebener, J. Bosch, R. Gross, and J.Mannhart, Extended Abstracts of 1987 International Superconductivity Electronics Conference (ISEC '87), Tokyo, Japan, p. 353 (S 11-1)

10) R. Gross, Thesis, University of Tübingen (1987)

11) U. Klaß, R. Gross, and R. P. Huebener, to be published

SUPERCONDUCTIVE DEVICES FOR ULTRA-FAST DISCRIMINATION

A. BARONE, R. CRISTIANO, and P. SILVESTRINI

Istituto di Cibernetica del CNR,
Arco Felice, Napoli, Italy
Dipartimento di Scienze Fisiche, Università di Napoli,
Napoli, Italy
INFN - Sezione di Napoli, Italy

1. INTRODUCTION

The unique properties of superconducting tunnel junctions offer the possibility of using such devices as radiation detectors for three different purposes: high energy resolution (1), fast discrimination (2) and high spatial resolution (3).

In principle each one of these issues requires a specific device to satisfy rather stringent prescriptions although the possibility of a single structure performing two or even all three operations cannot be ruled out.

The present work deals with the search for the optimization of some junction parameters to realize a fast particle (radiation) discriminator. Such a kind of detector is based on the ultra-high speed characteristic of the operation of a Josephson junction.

In this context at least three aspects should be investigated:

1) Actual switching time
2) Effects of noise
3) Material requirements.

Moreover other aspects would certainly deserve attention such as device reliability, concrete areas of applications where superconductive detectors can offer performances significantly competitive over existing devices, system considerations such as the required cryogenic environment etc. We shall confine our attention to general aspects of the switching operation of the device in connection with the problem of intrinsic thermal noise.

2. DEVICE OPERATION

Quite schematically the device would operate in the following way. The radiation interacting with the tunneling junction (Josephson type) will break a number of Cooper pairs reducing ultimately the Josephson supercurrent. If the junction is properly biased, i.e. as close as possible to its critical current value, then the zero voltage state becomes unstable and consequently a switch occurs to a finite voltage state.

The actual switching time would result from the combination of two characteristic times. The former, t* say, which corresponds to the time interval between the radiation interaction with the sensitive volume of the detector and the actual starting of the switching process. The latter is given by the intrinsic switching time of the Josephson junction.

2.1 Considerations on t*

The time t^* involves various steps. The first depends upon the specific interaction through which the energy of the impinging particle is released to the detector. There are reasons to believe that the primary interaction occurring in the order of 10^{-13} sec can result in quite energetic particles which in turn will transfer their energy directly to the condensate in a time of the order of 10^{-12} sec.

The whole non-equilibrium picture of the three-fluids system (quasiparticles, electron pairs and phonons) is quite complicated to describe (see the paper by K.E. Gray in this Volume). Pulses of the order of the phonon pair-breaking time (picoseconds) should be obtained when strong coupling superconductor such as Nb are employed. It is worth considering that in such a picture the following longer time (10^{-9} sec) process of the energy transfer from the condensate to the lattice should not affect the switching which indeed occurs as stated on a much shorter time scale. High energy phonons created in the early stages of the interaction would likely cross the thin film structure of the junction. The optimal choice of the substrate material also represents a delicate issue.

It appears obvious that the process will develop into a switching event provided that "enough" pairs are broken in the junction films so that the Josephson current can be appreciably lowered due to such a reduction of the superelectrons in the "reservoir". Macroscopically this implies that a suitable volume and in turn a large enough region A_p of the junction area has to be "activated".

2.2 Junction instrinsic switching time.

A concrete evaluation of the intrinsic switching time directly follows from the experiments performed in the context of the projects for a new computer based on the Josephson junction technology. Indeed measurements demonstrating picoseconds switching times have been performed since 15 years ago [4,5]. In the case of computer elements the switching is induced by a suitable control current which flowing in a film overlaying the junction produces a magnetic field which modulates (reduces) the Josephson current, leading thereby to the switching. In our case all this is produced by the radiation.

It is interesting to observe that, historically, digital logic components and particle detectors travelled already on a parallel track when both types of devices were studied by using a single superconducting film. In that case the two states were identified by the normal and superconducting state of the film. The transition was obtained in one case by magnetic field or local heating and the other by the impinging radiation. In the switching from the V=0 to the V≠0 state of a Josephson junction no phase transition is involved (which implies relatively slow phase boundary motion through the sample). Rather ultimate limits of switching time in a Josephson junction appear to be related to quanto-mechanical considerations.

An important aspect of the switching dynamics of the Josephson junction is that of the switching from the finite back to the zero voltage state . Indeed a loading indicated by one of the resistances in Fig. 1 leads to three different possibilities. In the first case the junction after the switching from the V=0 to the V≠0

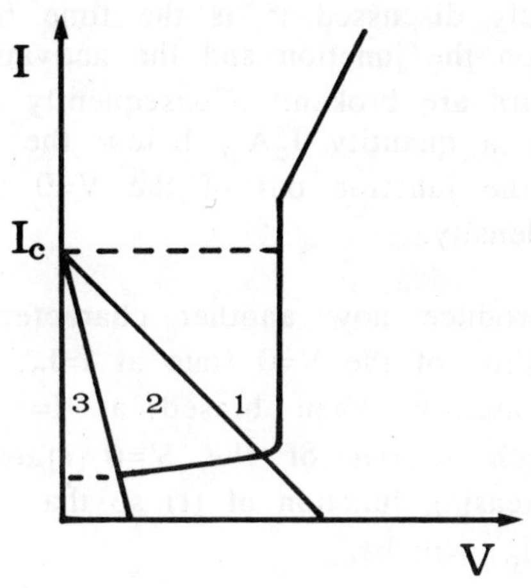

Fig. 1 - Typical I-V characteristic of a Josephson tunnel junction with the three different loading regions: latching mode (1), unlatching mode (2) and self-resetting mode (3).

state "remains" in the finite voltage state (latching mode operation) In the second case as soon as the cause is removed, the zero voltage state is recovered (unlatching mode). In the last case no matter how long the radiation effect holds the junction automatically switches back to the V=0 state (self-resetting mode). Back switching can be delayed by the presence of resonant structures (related to a.c. Josephson effect) within the gap voltages. It is worth noting that the detector will work as long as the dead time, i.e. the sum of t*, the switching time and the resetting time, is less than t_o, i.e. the minimum time between two events.

3. EFFECT OF INTRINSIC THERMAL NOISE.

As previously discussed t* is the time between the arrival of the radiation on the junction and the activation of an area A_p where Cooper pairs are broken. Consequently the critical current I_c is reduced by a quantity $J_c A_p$ below the bias current I_p in order to switch the junction out of the V=0 state, J_c being the critical current density.

Let us introduce now another characteristic time $\tau(I_p)$ which is the lifetime of the V=0 state at $I=I_p$, i.e. the mean time spent by the junction when biased at $I=I_p$ before thermal fluctuations switch it out of the V=0 state. $\tau(I_p)$ is an exponentially decreasing function of (I) so that the closer I_p is to I_c the smaller $\tau(I_p)$ will be.

Since the intrinsic switching time of a Josephson junction is of the order of 10^{-12} sec, the time response is dominated by t* which is related to A_p through a complicate diffusion process dependent on the specific superconductor material, on the radiation-junction interaction etc. To skip these problems which are extensively discussed in other chapters of this volume and which still need experimental support , we refer in the next to $A_p = A_p(t^*)$ instead of t*.

With respect to thermal fluctuations, working conditions obviously require that during t_o no switching out of V=0 occurs due to thermal noise, i.e.

$$\tau(I_p) >> t_o \qquad (1)$$

In fact the number of noise induced switching within t_o is

$$N(t_o) = 1 - \exp(- t_o/ \tau(I_p)) \qquad (2)$$

which under the condition (1) becomes:

$$N(t_o) \approx t_o / \tau\,(I_p)$$

Using the last relationship, condition (1) can be cast in the equivalent forms

$$\tau(I_p)/t_o = \varnothing\, A\, \tau(I_p) >> 1$$

or $$\varnothing/\varnothing_{eff} >> 1 \qquad (3)$$

where Ø is the radiation flux, A is the junction area and $\varnothing_{eff} = 1/A\ \tau(I_p)$. The physical meaning of $\varnothing_{eff}$ is that of an effective radiation flux which produces a number of switching equal to that produced by thermal fluctuations. Since Ø is fixed by the particular experimental frame, we shall replace condition (3) by $\varnothing_{eff} << 1$. ($\varnothing_{eff}$ very small). So that from the investigation of $\varnothing_{eff}$ as function of A, A_p, J_c, T and using the relation $A_p(t^*) = A(1-I_p/I_c)$ we can obtain a useful picture in order to optimize the design of the device with respect to the intrinsic noise and given radiation flux.

First we can consider how to realize the junction to obtain the desired response time (i. e. at a fixed A_p). In Fig. 2, $\varnothing_{eff}$ vs. A for various J_c and at T = 1.3 K is plotted in order to show which values of the junction parameters, A and J_c, are consistent with a given prescription for eq. (3). This figure refers to Nb-Nb_yO_x-Pb Josephson junction and it is based on the measured lifetime τ in junctions with low J_c ($J_c \sim$ 1-10 A/cm^2)[6] and on an extrapolation at higher J_c.

As expected the device sensitivity improves (small $\varnothing_{eff}$) as A decreases and J_c increases.

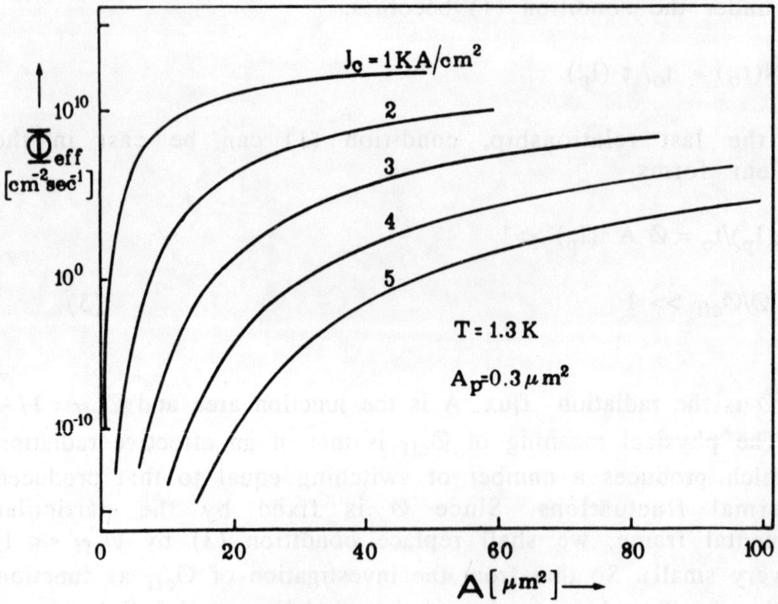

Fig. 2 - $\emptyset_{eff}$ vs. A at T = 1.3 K, A_p = 0.3 μm² and for different values of Jc. The curves are based on experiments on Nb-Nb_2O_5-Pb junctions (6).

4. SYSTEM DESIGN AND MATERIAL CONSIDERATIONS.

Let us summarize here what is the best response time we could expect for a given junction, using the presently available fabrication technology and which kind of superconducting material should be preferred.

It is possible to realize junction areas down to A = 1 μm² and J_c up to 10^4 A/cm². The lack of geometrical efficiency could be solved by suitable junction arrays of a proper design. In Fig. 3a $\emptyset_{eff}$ vs. A_p is shown at various J_c. The range of A_p shown in this figure would very likely lead to very short response times (20 : 100 ps) (7). The values of T is chosen equal to 1.3 K which is the minimum temperature available with simple LHe^4

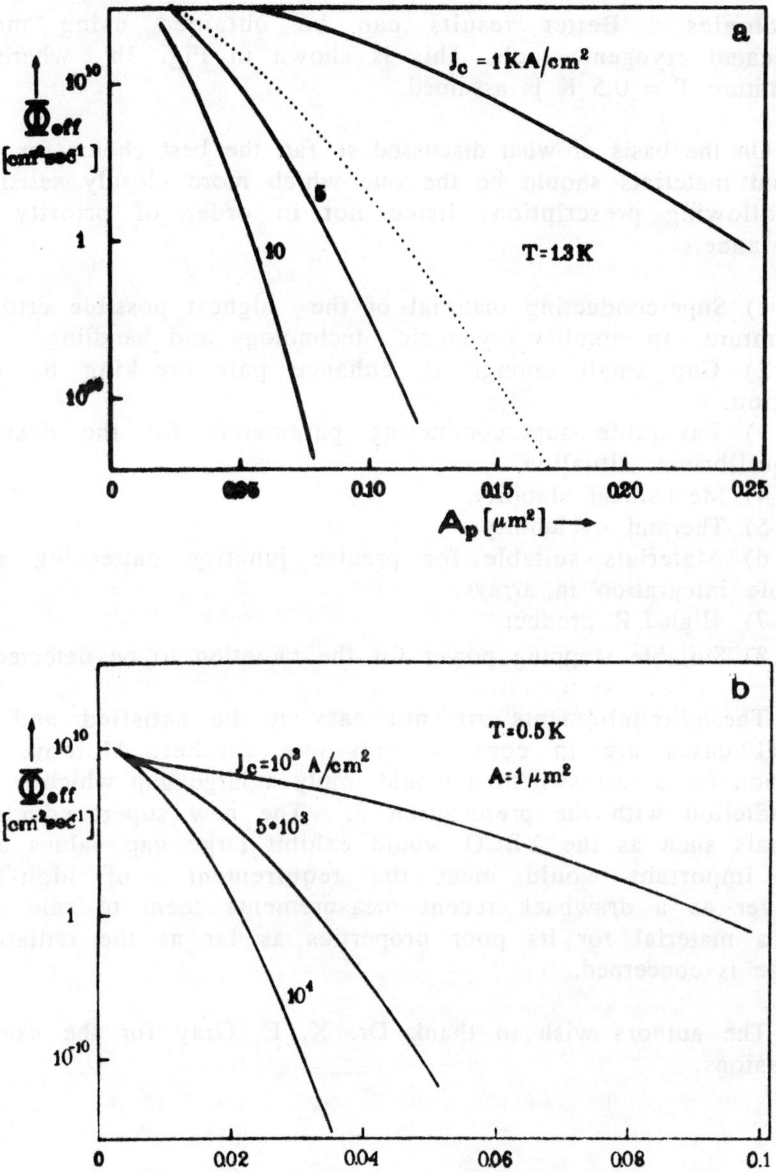

Fig. 3 - $\emptyset_{eff}$ vs. A_p for A =1 μm^2 and different values of Jc (solid line); A = 10 μm^2 and Jc = 10 KA/cm^2 (dotted line); a) T = 1.3 K; b) T = 0.5 K.

technologies. Better results can be obtained using more sofisticated cryogenic tools. This is shown in Fig. 3b where a temperature T = 0.5 K is assumed.

On the basis of what discussed so far, the best choice for the junction materials should be the one which more closely satisfies the following prescriptions listed not in order of priority of importance.

1) Superconducting material of the highest possible critical temperature to simplify cryogenic technology and handling.
2) Gap small enough to enhance pair breaking by the radiation.
3) Favorable superconducting parameters for the desired nonequilibrium situation.
4) Mechanical stability.
5) Thermal cyclability.
6) Materials suitable for precise junction patterning and possible integration in arrays.
7) High I_cR product
8) Suitable stopping power for the radiation to be detected.

These requirements are not easy to be satisfied and in several cases are in conflict with one another. Moreover a condition for a fast switching would imply a large gap which is in contradiction with the prescription 2. The new superconducting materials such as the YBCO would exhibit large gap values and more important would meet the requirement of high-Tc. However as a drawback recent measurements seem to rule out such a material for its poor properties as far as the radiation damage is concerned.

The authors wish to thank Dr. K. E. Gray for the useful discussions.

REFERENCES

1) G. H. Wood and B.C. White Can. J. Phys. 51, 2032 (1973)
M. Kurakado and H. Mazaki Phys. Rev. B22, 168 (1980)

A. Barone, G. Darbo, S. De Stefano, G. Gallinaro, A. Siri, R. Vaglio, S. Vitale - Nucl. Instrum. Methods A234, 61 (1985)

A. Barone, S. De Stefano and K.E. Gray, Nucl. Instrum. Methods A235, 254 (1985)

D. Twerenbold and A. Zehnder, J. Appl. Phys. 61, 1 (1987) and references therein.

2) A.Barone and S. De Stefano, Nucl. Instrum. Methods 202, 513 (1982)
Pu Huan Shun, Zhong Yuanyuan, Jiang Jiauhna, and Zhuang Jiejia,Acta Phys. Sin. 34, 941 (1985)

3) See for instance, O. Liengme in this Volume.

4) W. Jutzi, T.O. Mohr, M. Gasser and H.P. Gschwind, Electron. Lett. 8, 589 (1972)

5) H. H. Zappe and K. R. Greebe, J. Appl. Phys. A4, 865 (1973)

6) P. Silvestrini, O. Liengme, K. E. Gray, Phys. Rev. B37 (Feb. 1988)

7) K. E. Gray, Private Communication

PARTICLE INDUCED SWITCHING OF JOSEPHSON TUNNEL JUNCTIONS

A. CAMPBELL, M. NISENOFF, R. MAGNO, R. SHELBY[1], and J. KIDD

US Naval Research Laboratory
Washington, DC 20375 USA

INTRODUCTION

The interaction of energetic charged particles (ions) and neutrons with superconducting tunnel devices can result in at least three types of effects: (1) upsets or temporary changes in state that are reversible by cycling the bias current through zero bias conditions,[1,2,3] (2) changes in the operating state of the devices and/or device characteristics that can be removed by raising the temperature and then recooling to the superconducting state,[1,2,3] and (3) irreversible permanent damage characterized by changes in the device characteristics which persist even when the device is heated above room temperature.[4] The research reported here deals with upset events of the first type in Josephson tunnel devices. These transient switching events can be used as detectors of energetic particles and the statistics of the upsets as a function of bias current can be used to infer particle characteristics such as stopping power. Specifically, we report here on the effects of MeV alpha particles on niobium-amorphous silicon-niobium tunnel junctions.

EXPERIMENT

The effect of energetic particles on Josephson devices was studied by measuring the induced upset or switching rate as a function of bias current. The niobium-amorphous silicon-niobium Josephson devices,[5] with areas ranging from 8 to 110 square micrometers, and the Americium-241 radioactive alpha source were mounted in a chamber containing a helium exchange gas and immersed in a 60 liter liquid helium dewar. The alpha source was mounted on a rod so that it could be positioned near the device or withdrawn in order to obtain the upset rates due to background noise. The strength of the source was measured to be 130 microcuries. The source was a square plate approximately 9 mm on edge and during the irradiations its center was about 4 mm from the junction. Therefore, the alpha particles entered the junction with angles ranging from 0° to 60° with respect to the normal to the source. Some experiments were performed with masks between the source and junction to limit the maximum angle of incidence. The alphas from the bare source had an energy spectrum peaked near 4.4 MeV. This spectrum could be shifted to lower energies by placing mylar sheets over the source for some measurements. The spectra from the source was measured using a silicon surface-barrier detector. The measurements were made with an automated data acquisition system consisting of a Tektronix Model 4051 smart terminal, a Hewlett-Packard Model HP 2240A Measurements and Control Processor, low noise amplifiers, and a motor drive for positioning the alpha source with respect to the device. Since an automated data acquisition system and a long hold time dewar were used, it was possible to collect data, unattended, for several days. This feature was quite desirable as some of the times for upset events were longer than 1,000 seconds and the measurements had to be repeated a number of times in order to obtain good statistics.

The first step in an experiment was to determine the critical current with the alpha source removed from the locality of the device. The critical current was measured by slowly increasing the bias current until the junction switched from the Josephson state to the normal state. The measurement was repeated several times and if the values obtained reproduced within several tenths of one percent, comparable with the resolution of the A/D converters in the HP 2240A, the average value was taken to be I_C, the critical current. The alpha source was then moved close to the device, and the bias current was increased to some predetermined value, I_B. The elapsed time between reaching I_B and switching to the normal state while exposed to the alpha flux was measured. This process was repeated a number of times, at each bias current, in order to prepare histographs. The alpha induced upset rate was then obtained by an exponential fit to the data.[3] The upset rate due to background noise was also measured at each bias to insure that the observed upset rate was not due to extraneous sources. The critical current was also checked after each upset event and the fluctuations in it were of the same size as the uncertainty in the digitizing. At bias currents near $0.98*I_C$, it was possible to wait more than 100 times the observed alpha induced upset time without detecting noise induced event. At bias levels near $0.70*I_C$, noise induced events were not observed after waiting for times as long as 15 hours. These limits on the noise upset times are sufficiently long that all the upset events observed under alpha particle irradiation are considered to have been produced by the alpha particles.

During all of the exposures reported here, no permanent damage effects were reported, i.e. critical currents remained constant. Earlier work[1] showed the importance of shielding since magnetic flux trapping can result in an apparent change in critical current.

MODEL

When an ionizing particle passes through a target, it loses energy at a rate depending on its energy and the atomic number and density of the target.[6] For a 4.4 MeV alpha particle (the average energy of alphas emitted by our Americium-241 source at normal incidence) passing through a niobium target, the loss rate is about 370 keV per micrometer of track length. For the devices used here, the junction was approximately 0.5 micrometer deep so about 185 keV of energy was deposited in an ion track which can be approximated as a circular cylinder with a radius of 0.1 micrometer. Initially, the energy goes into the electronic system as the particle produces ionization along its path. At some later time the electrons thermalize and share their energy with the lattice. One possibility for a loss of superconductivity is the breaking of Cooper pairs. If one assumes that an energy of several times the superconducting energy gap (E_g, about 2 meV for niobium) is required for pair breaking, then approximately $9x10^7$ Cooper pairs should be affected, which would correspond to only 0.1 to 1.0 percent concentration of quasiparticles in the junction electrodes because of the small area of the ion track compared with a 100 square micrometer junction area. This is unlikely to produce a switching event.

Another possibility, and the one that was adapted for explaining the observed data, is that the energy loss was coupled into the lattice and electron gas and these, either separately or collectively, were heated to some temperature in excess of T_c . It is assumed that when a portion of the junction of area A is normalized, the bias current I_B flowing through the device is redirected to flow through the remaining portion of the device, A_0-A, where A_0 is the geometrical area of the device. If the current density through A_0-A exceeds the critical current density of the device, an upset event will occur. Thus an upset will occur at I_B whenever an area A given by

$$I_B / (A_0-A) > J_C = I_C / A_0 \tag{1}$$

or,

$$I_B/I_C > 1 - A / A_0 \tag{2}$$

is heated above the critical temperature. A similar "heating model" was used by Crittenden and Spiel[7] to explain the reduced critical current in superconducting ribbons under alpha particle irradiation.

Because of the finite thickness of the Americium-241 source, the alpha particles are not emitted monoenergetically or isotropically. Thus it is necessary to know the energy spectrum of the source for the range of angles of incidence of alphas striking the junction in order to calculate the number of alpha particles which can normalize an area A_N. If dE/dx is the rate of energy loss along the particle track or stopping power for a given particle energy, and θ is the angle of incidence of the particle on the junction, then the energy loss per unit thickness of the junction will be $(dE/dx)/\cos(\theta)$ and the resulting normalized area will be given by

$$A_N = K(dE/dx)/\cos(\theta) \quad (3)$$

Here K is a proportionality constant which is a function of specific heat, thermal diffusivity within each layer as well as between layers and thermal diffusivity between the device and the environment. With dE/dx in the units of keV/micrometer and area in the units of square micrometers, the units of K are square micrometers/keV/micrometer or cubic micrometers/keV.

A calculation of the size of the normalized region is very complex as the Josephson device is a multilayered structure (four metallizations, one barrier layer and at least two insulating layers), and the heat flow behavior between layers and from the device to the helium exchange gas is not known. However, a simple calculation still can be made to obtain a rough estimate of the size of this normalized region. It is assumed that all of the energy deposited into the niobium electrode material stays within that layer and uniformly heats a right circular cylindrical region (the axis of which is the particle track) from the bath temperature of 4.2°K to a temperature above T_c (i.e. 9.2°K). If the energy loss is coupled only to the superconducting electrons, the normalized region heated above 9.2°K is about 3.5 square micrometers/MeV/micrometer of device thickness (value of the constant K). However, if the energy can couple to both the lattice and the superconducting electrons, then the normalized region is about 2.4 square micrometers/MeV/micrometer of device thickness. This calculation is only rough estimate of the order of magnitude of the extent of the region normalized by an alpha particle.

Finally, the alpha induced upset rate is found by counting all the alphas which normalize an area A_N from Equation (3), and pass through or near enough to the junction to heat an area A from Equation (2). The counting is done by using the source energy spectrum to calculate the flux $F(E,\theta)$ of alpha particles entering near the device with an energy E and an angle θ. E and θ are also used in Equation (3) to calculate the appropriate A_N for these alpha particles. Equation (2) is then used to find the threshold area A at the bias, I_B, being studied. If $A_N>A$ alpha particles striking near the device will also cause upsets so the effective area of the device is assumed to be $A_{eff}=(L_0+\sqrt{(A_N/\pi)})^2$ where $A_0=L_0^2$ is the junction's area with L_0 being the length of the side of an equivalent square. The upset rate for alpha particles entering at E and θ is then $F(E,\theta)*A_{eff}$, and the total upset rate is found by summing over all values of E and θ. Those alpha particles which produce a value of $A_N<A$ will not cause upsets and those with $A_N=A$ will have an effective area $(L_0-\sqrt{(A_N/\pi)})^2$ which is less than the geometrical area of the junction. A schematic diagram of the variation of the upset rate with bias current from this model for a monoenergetic alpha source for two different junction areas is shown in Figure 1. Since the actual alpha

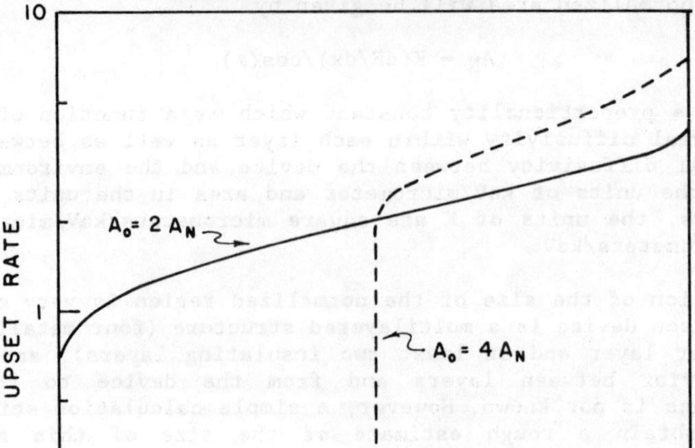

Figure 1 - Model prediction of the variation in upset rate with reduced bias current as a function of junction area. One curve is for the case of a junction area twice that normalized by an ion and one for the case of a junction four-times the normalized area.

source emits particles with a range of energies and angles of incidence, there will be a range in values of dE/dx deposited in the junction and there would be additional rounding off this upset rate curve.

RESULTS

The particle flux from the Americium-241 source was about 0.02

particles/square micrometer/second for the unmasked source. Spectra from the source taken at angles from 0° to 60° show average energies from 4.4 MeV at 0° to 3.0 MeV at 60° with corresponding stopping powers in niobium of 370 keV/micrometer to 455 keV/micrometer.[6] Note that as the angle of incidence is increased, the particles must traverse increasing thicknesses of source material and so the average energy decreases and the stopping power at the device increases. To determine the dependencies of the upset rate on particle energy and angle of incidence, masks were placed in front of the source to limit the maximum angle of incidence and mylar sheets were placed over the source to shift the spectrum to lower energies.

The only unknown in the model is the parameter K in Equation (3) and in this experiment it was adjusted to obtain the best fit to the data shown in Figure 2. This figure plots measured upset rate vs reduced

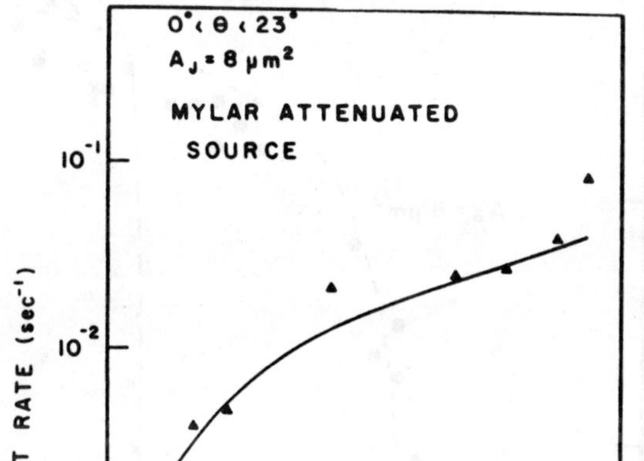

Figure 2 - Experimental data for upset rate vs reduced bias current with a fit of the data based on the heating model.

bias current, I_B/I_C with: a) a 8 square micrometer area device, b) a mylar sheet in front of the source to shift the energy spectrum to an energy with a stopping power of about 550 keV/micrometer, and c) a mask to limit the angles of incidence to less than 23°. Since most of the particles strike the device at normal incidence and the cosine(23°)=0.92, the factor in Equation (3) was considered constant. The solid line in Figure 2 is the fit obtained using the model described above with K=3.7 cubic micrometers/MeV. This value agrees well with the estimates of 3.5 to 2.4 cubic micrometers/MeV made in the preceding Model section.

As a further test of the model, the mylar sheets were removed and the data shown by the solid circles in Figure 3 were measured for the same

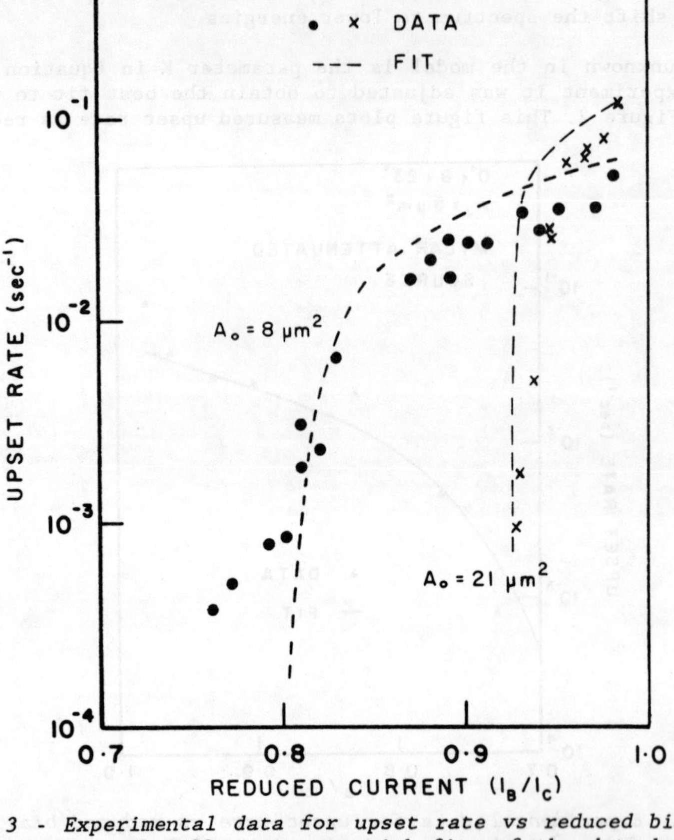

Figure 3 - Experimental data for upset rate vs reduced bias current for two devices of differing areas with fits of the data based on the heating model.

8 square micrometer junctions. Removing the mylar shifted the peak of the spectra upward to a maximum value of dE/dx of about 370 KeV/micrometer. A junction with an area of 21 square micrometers was also measured with the same masked bare source and the results for it are also shown in Figure 3. The dashed lines shown in the figure were obtained by applying the model described above along with the same K =

3.7 cubic micrometers/MeV which was found previously by fitting the data in Figure 2. These represent a good fit to the data, both in terms of the magnitude of the observed upset rate and also to the variation of the upset rate as a function of bias current. Thus, by taking into account the energy spectrum of the incident alphas, one value of K can provide good fits to upset rates obtained for two devices with different energy spectra. These results support the assumption that the heating model may be a suitable description of the effects of ionizing particles on the switching of Josephson devices.

Note also that from an estimate of the flux from the masked source used for the data of Figure 3, for values of I_B/I_C above about 0.9, every particle that interacts with the junction is causing an upset or switching event and thus being sensed or detected.

Because of the relatively low flux of particles and the extended source, it was not possible to study the angular dependence of the upset rate directly. It was investigated by using the unmasked source and series of masks with apertures of varying dimensions. Thus, data could be taken not with specific angles of incidence but for angles ranging to some maximum value θ_{max}. The upset rate as function of reduced bias current for the 8 square micrometer device for four different aperture sizes is shown in Figure 4. The data points for the 35° and 45° curves have been omitted for clarity and solid lines, which are not theoretical fits, have been drawn to show the observed trend for the data. The variation in the magnitude of the upset rates at high reduced bias current above about 0.9 reflects the differences in the particle flux as the source is masked to smaller areas. The increase in upset rate as a function of θ_{max} at reduced bias currents below about 0.8 reflects the shift in energy spectrum, to lower energy and higher dE/dx values as the angle of incidence moves away from the normal. In addition, at larger angles of incidence, the factor $1/\cos(\theta)$ can no longer be taken to be unity, as was done early. If one assumes that the energy spectra are the same, which is a fair approximation for angles less than 45°, then the shift of the knee in the curve should vary inversely with the $\cos(\theta)$ in Equation (3). With these assumptions, the values of K equal to 3.9, 3.7, and 3.8 cubic micrometers/MeV were obtained for the 23°, 35°, and 45° curves, respectively. These are in fairly good agreement with a K of 3.7 cubic micrometers/MeV determined from the data in Figure 2.

Attempting to explain the 65° curve of Figure 4 in light of the proposed model is not easy. For large angles, the energy spectrum is not only shifted toward lower energy, but there is a low energy tail with fairly large magnitude relative to that of the peak of the curve. However, it can be assumed that the apparent reduced upset rate at low bias reflects the small number of alphas that have high enough $(dE/dx)/\cos(\theta)$ to produce an upset at these low values of reduced bias current. Note however that for reduced bias currents greater than about 0.9, estimates from the source strength and geometry show that even for the 65° case, every incident particle does cause an upset or is detected.

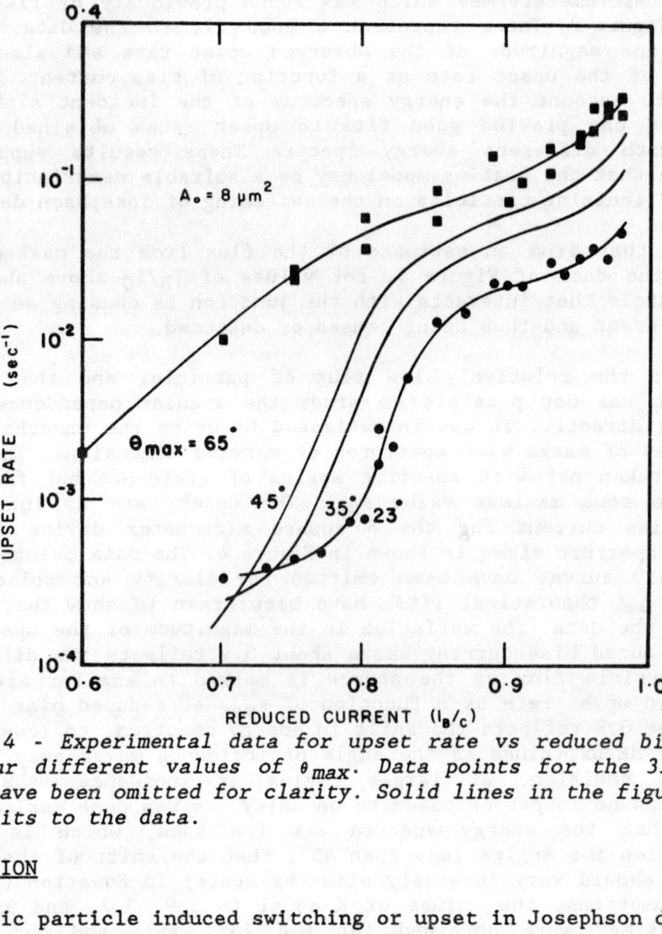

Figure 4 - Experimental data for upset rate vs reduced bias current for four different values of θ_{max}. Data points for the 35° and 45° cases have been omitted for clarity. Solid lines in the figure are <u>*not*</u> *model fits to the data.*

<u>DISCUSSION</u>

Energetic particle induced switching or upset in Josephson devices for the case of MeV alpha particles has been studied in some detail and several of the factors associated with this phenomena have been identified. A model has been developed that says that if an energetic particle deposits enough energy to cause a portion of the junction to undergo a transition from the superconducting to the normal state, the current flowing through the device will be redistributed into the remaining superconducting portion. An upset event will occur if the redistributed current density exceeds the critical current density of the device.

Based on this model, the area of the normalized region is proportional to the rate of energy loss along the particle track and inversely proportional to the cosine of the angle of incidence of the particle. Thus to know the response of a Josephson junction to any energetic particle it is only necessary to know the response to one particle with known energy and angle of incidence since the stopping power can be calculated with a knowledge of the material parameters of the device. The only adjustable parameter in this model is the constant of proportionality, K, between the normalized region and the energy loss per unit thickness of the device. This constant is only a function of the material parameters of the device. The validity of this model is supported by the observation that a single value of K yielded equally good fits to a variety of data taken with different device areas and energy spectra. This implies that for any particle or energy, the value of K obtained for one particle can be used to predict the response of the device.

The good fit of the model to the data also implies that for a given bias condition, every particle that can normalize an area larger in magnitude than required to satisfy Equation (2) will result in an upset event with unit probability. Thus these devices can be used as detectors of particles with 100% efficiency.

Although the proportionality constant K was treated as a "free parameter" in fitting the data reported here, it is not an arbitrary quantity. It is a quantity specific to one type of structure. According to the heating model, as the alpha particle penetrates through the device structure, it deposits energy along the track at a rate dE/dx which is a function of the energy of the particle. The energy will diffuse away from the location of the track raising the temperature of the device above ambient temperature. The temperature profile will be a function of time and will depend on the device specific heat and thermal diffusivity as well as the thermal diffusion across boundaries.

In regard to the particle radiation response of these Josephson devices, it has been shown that Nb-Si-Nb devices with areas of the order of 8 square micrometers when operated at biases as low as $0.6*I_C$ will upset or switch for each MeV alpha particle that enter with an angle of incidence of the order of 60 degrees. Assuming that there is no dramatic difference in the response of Josephson devices fabricated with different electrode materials, then all Josephson devices with areas of the order or less than 6 square micrometers may be extremely sensitive to energetic particles. To totally verify this heating model other particles such as protons and heavy ion must be used to see if the proposed model is still valid. Furthermore, higher irradiation fluxes, probably obtained from accelerators, will be needed to observe the onset of permanent damage. Finally, experiments will need to be formulated and carried out to identify the details of the mechanism responsible for the normalization of local regions of the device structure as an ionizing particle passes through the device.

In these experiments, the Josephson device was biased in the zero-voltage state and switched into the finite-voltage state whenever the energy deposited in the junction exceeded some value - the device functioned as a "threshold detector" of energetic particles. If the energy deposited was less than the threshold value, nothing happened but when the energy exceeded the threshold value, the device switched to the other state.

Another possible mode of operating a Josephson device as a radiation particle detector is to bias the device in the quasi-particle (finite resistance) branch of its current vs voltage characteristics. The voltage, V_g, at which there is an abrupt increase in current flowing through the device and the magnitude of the "leakage current" that will flow for bias voltages less than the gap voltage, V_g, will depend on the operating temperature relative to the critical temperature of the material. The ideal variation of sub-gap leakage current and the energy gap as a function of reduced temperature, that is, for T/T_c, is shown in Figure 5. As the energetic particles traverse the junction,

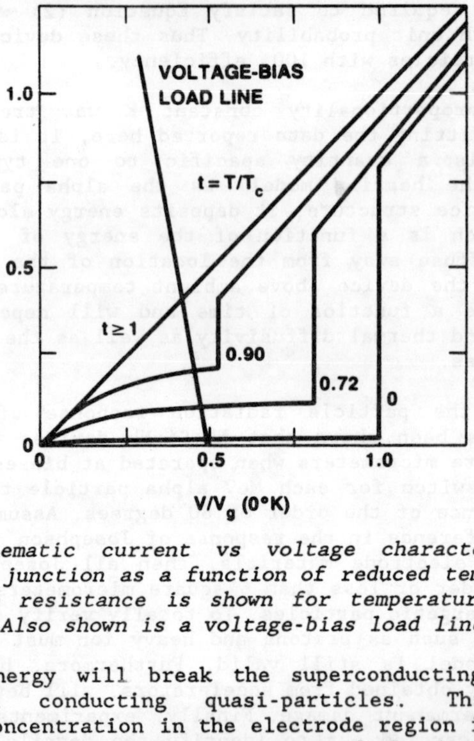

Figure 5 - Schematic current vs voltage characteristics for a Josephson tunnel junction as a function of reduced temperature, T/T_c. The normal state resistance is shown for temperatures equal to or greater than T_c. Also shown is a voltage-bias load line.

the deposited energy will break the superconducting Cooper pairs, creating normal conducting quasi-particles. This change in quasi-particle concentration in the electrode region is equivalent to an increase in the "effective" temperature of the device. This change in quasi-particle concentration (or effective temperature) can be monitored either by a voltage bias mode of operation, in which the load line crosses the leakage current below V_g (see Figure 5). As was inferred from the SEU experiments described above, the energy to switch a device into the normal state for an operating temperature of 4.2^oK (that is, a reduced temperature of 0.46) is about 400 keV/micrometer. Therefore, if one could monitor very small (effective) temperature increases by biasing the device into the quasi-particle branch, then much smaller energies deposited in the junction by a traversing particle could be detected. The limit to the energy resolution will depend on many experimental factors including the slopes of the various portions of the quasi-particle branch as functions of temperature, the signal bandwidth that must be used, and the noise in the electronic detection system. Modeling of the Josephson device in this mode of operation must be done and experiments carried out in order to make comparisons to other detector technologies to fully evaluate the merits of Josephson devices as energetic particle detectors.

ACKNOWLEDGEMENTS

We would like to thank Dr. Larry N. Smith formerly of the Sperry Research Center, Sudbury, MA, for providing the Josephson tunnel devices used in these experiments. We would also like to acknowledge discussions with M. Rosen and M.I. Haftel of NRL on various aspects of the heating model.

REFERENCES

1) R. Magno, R. Shelby, M. Nisenoff, A.B. Campbell, & J. Kidd, "Alpha Particle Induced Switching in Josephson Tunnel Junctions,", IEEE Transactions on Magnetics MAG-29, 1286 (1983).

2) R. Magno, M. Nisenoff, R. Shelby, J. Kidd, & A.B. Campbell, "Alpha Induced Upsets in Josephson Tunnel Junctions," IEEE Transactions on Nuclear Science NS-29, 2090 (1982).

3) R. Magno, M. Nisenoff, R. Shelby, A.B. Campbell, & J. Kidd, "Upset Events in Josephson Digital Devices Under Alpha Particle Irradiation," IEEE Transactions on Nuclear Science NS-28, 3994 (1981).

4. W. Dietsche and G. Wehr, "Tunnel Spectroscopy in Neutron Irradiated Pb," Sol. State Comm. 24, pp. 591-593, 1977; Y.H. Lee and P.R. Brosious, "Electron Irradiation Effects in Lead Alloy Josephson Junctions," Appl. Phys. Lett. 40, 347 (1982).

5. H. Kroger, L. N. Smith and D.W. Jillie, "Selective Niobium Anodization Process for Fabricating Josephson Tunnel Junctions," Appl. Phys. Lett. 39, 280 (1981).

6. J.F. Ziegler, Helium Stopping Powers and Ranges in All Elemental Matter, (Pergamon Press), Vol. 4.

7. E.C. Crittenden, Jr. and D.E. Spiel, "Superconducting Thin-Film Detectors of Nuclear Particles," J. Appl. Phys. 42, 3183 (1981).

THE ULTIMATE ENERGY RESOLUTION OF CRYOGENIC DETECTORS

C. BORAGNO, G. GALLINARO, A. SIRI, S. VITALE

Dipartimento di Fisica, Università di Genova and
Istituto Nazionale di Fisica Nucleare,
Sezione di Genova, Genova, Italy

In 1961 Martin and Bloor realized a superconducting bolometer able to detect an absorbed power of $10^{-13}W$ with a bandwidth of $1Hz$(1). In the same year similar results were obtained by Low(2) who used a doped semiconductor operating at low temperature. In a short time the semiconductor cryogenic bolometers were commonly used to detect infrared radiation, microwaves and, since 1967(3), atomic and molecular beams. In 1949 Andrews, Fowler and Williams(4) detected single α particles with a superconducting bolometer and in 1974 Niinikoski(5) proposed to use semiconductor cryogenic bolometers to detect ionizing particles, but only in the last years bolometers have been realized that, because of the good performances, are ideal candidates as detectors in various experiments where very high energy resolution (neutrino mass measurement(6)) or large sensitive volumes (double β decay(7), solar neutrino spectrum(8)) are required.

For instance a silicon bolometer of a mass of $10Kg$ operating at $20mK$ should detect a minimum energy of 500 eV FWHM. Remarkable results in the detection of X rays and α particles have been obtained. The smallest energy actually detected by a bolometer operating at $0.1K$ is $35eV$(9), the minimum exponential decay time achieved at $0.3K$ is $20\mu sec$(10).

In 1964 it has been proposed the use of tunnel superconducting junctions as ionizing particles detectors (11,12). In 1969 have been published the first positive results indipendently obtained by Wood and White (13). The best result so far obtained is of 90 eV even if in a very small sensitive volume(14).

In this paper we discuss the behaviour of bolometers and tunnel superconducting junctions as particle detectors and we compare their performances. The conclusion is that at operating temperatures lower than $0.3K$ for a given sensitive volume generally the bolometers have better performances.

BOLOMETER BEHAVIOUR ANALYSIS

A bolometer consists of an energy bulk absorber, in weak thermal link with a thermostat. The absorber itself can be the thermometer (for instance an homogeneously doped semiconductor), but often the bolometer is of composite type with the temperature sensitive element in good thermal link with the bulk absorber.

At low temperatures good absorbers are the crystalline pure silicon, sapphire or diamond because of the low thermal capacity and high diffusivity. The sensitive element can be: a) a semiconductor thermistor (for silicon absorber this can be obtained doping the surface[15]); b) a superconducting film at the transition temperature; c) a superconductor-insulator-normal metal (S-I-N), or a superconductor-insulator-superconductor (S-I-S) tunnel junction[16].

Many other suggestions can be made for the absorber (e.g. a superconductor) and for the thermometer.

When the temperature sensitive element behaves as a thermistor the analysis of the system can be made as follows. For temperature measurement a voltage source yields a current I in the bolometer resistance R through a load resistance R_L, the voltage V across the bolometer is then amplified. If the bolometer resistance is high, the high frequency limit is determined by the electrical capacity of the leads between the bolometer and the amplifier (scheme 1).

The bolometer can operate at constant voltage connecting one lead to the voltage source and the other lead to a current-voltage transducer made by an operational amplifier, with a feedback resistor R_F. Because of the virtual ground at the input of the transducer the leads capacity does not limit the bandwidth (scheme 2).

In both cases the total electronic chain gain is A (including the factor R_F/R). We suppose that at high frequency the gain decreases at 6 db/octave with an integrating time τ_F.

If the heat capacity of the link is neglected (in reference [17] this capacity is taken into account), the bolometer thermal balance equation can be written as:

$$C_b \frac{dT}{dt} + \dot{Q}(T, T_C) = V^2 \frac{R}{(R_L + R)^2} + W_0 + W(t) \tag{1}$$

Where C_b and T are the bolometer heat capacity and temperature, $Q(T, T_C)$ is the power flowing from the bolometer to the thermostat at the temperature T_C, W_0 and $W(t)$ are the continuous and time variable absorbed power. If $W(t) \ll W_0 + I^2 R$ we have:

$$C_b \frac{d\Delta T}{dt} + G_{eq} \Delta T = W(t) \tag{2}$$

with:

$$G_{eq} = G + G_P, G = \frac{d\dot{Q}(T, T_C)}{dT}, G_P = -\alpha P \frac{R_L - R}{R_L + R}, \alpha = \frac{1}{R}\frac{dR}{dT}, P = I^2 R$$

If $W(t) = E\delta(t), G_{eq} \geq 0$ and $\tau = C_b / G_{eq}$ the solution of the equation (2) is:

$$\Delta T = \frac{E}{C_b} e^{\frac{-t}{\tau}} \tag{3}$$

In the case of the scheme 1 the relevant signal is the voltage ΔV_B developed across the bolometer:

$$\Delta V_B = I \frac{R_L}{R_L + R} \frac{dR}{dT} \Delta T = S \frac{E}{\tau} e^{-\frac{t}{\tau}} \tag{4}$$

where $S = I \frac{R_L}{R_L + R} \frac{dR}{dT} \frac{1}{G_{eq}}$ is the zero frequency bolometer responsivity.

For the scheme 2 the current ΔI yielded in the bolometer is:

$$\Delta I = S \frac{E}{\tau R} e^{-\frac{t}{\tau}} \tag{5}$$

This current gives at the output of the operational amplifier the voltage:

$$\Delta V = S \frac{E}{\tau} \frac{R_F}{R} e^{-\frac{t}{\tau}} \tag{6}$$

The voltage output of the electronic chain is:

$$\Delta V_A = AS\frac{E}{\tau}\frac{e^{-\frac{t}{\tau}} - e^{-\frac{t}{\tau_F}}}{1 - \frac{\tau_F}{\tau}} \tag{7}$$

Neglecting the noise due to the load resistance and the blackbody radiation fluctuations the output noise voltage is:

$$\overline{V_N^2} = \frac{A^2}{4\tau_F}(4S^2kT^2G + 4kTR + \overline{e_N^2} + \overline{i_N^2}R^2) \tag{8}$$

k is the Boltzmann constant, the first two terms are due to the thermal link phonon noise and to the bolometer Johnson noise, e_N and i_N are the input voltage and current noise of the electronic chain.

We have the best energy resolution if the phonon noise dominates and $\tau_F = \tau$. In this case we can infer from (7) and (8) that the minimum detectable energy is given by (18):

$$E_{min\,FWHM} = 1.5 \cdot 10^8 T\left(C_b\frac{G}{G_{eq}}\right)^{\frac{1}{2}} eV \tag{9}$$

For an homogeneously doped silicon bolometer Mac Cammon et al. have as typical values at $0.3K$ (19):

$$S \simeq 3 \cdot 10^7\frac{V}{W},\ G \simeq 1.5 \cdot 10^{-8}\frac{W}{K},\ R \simeq 10M\Omega$$

In this case the Johnson noise dominates and the minimum detectable energy increases by about a factor 3 compared with what calculated in (9). Similar results are expected for a composite type bolometer with a superconducting film thermometer if suitable, low temperature operated matching transformer is used (16); this thermometer can be easily fabricated but operates at a fixed temperature.

The S-I-N and S-I-S tunnel junctions are good candidates as sensitive elements when it is necessary to have a wider operating temperature range (16). We discuss the behaviour of these devices as thermometers. For the S-I-N junctions if $\Delta \gg kT$ the current is given by:

$$I = \frac{(2\pi\Delta kT)^{\frac{1}{2}}}{eR_N}e^{-\frac{\Delta}{kT}}\sinh\frac{eV}{kT} \tag{10}$$

e is the electron charge, 2Δ the superconductor energy gap and R_N the normal state junction resistance [16]. We have:

$$\left(\frac{dI}{dV}\right)_T = \frac{1}{R_N}\left(\frac{2\pi\Delta}{kT}\right)^{\frac{1}{2}} e^{-\frac{\Delta}{kT}}\cosh\frac{eV}{kT} \tag{11}$$

$$\left(\frac{dI}{dT}\right)_V = \frac{(2\pi\Delta kT)^{\frac{1}{2}}}{eR_NT}e^{-\frac{\Delta}{kT}}\left((0.5+\frac{\Delta}{kT})\sinh\frac{eV}{kT} - \frac{eV}{kT}\cosh\frac{eV}{kT}\right) \tag{12}$$

$$\left(\frac{dV}{dT}\right)_I = -\frac{k}{e}\left(\frac{\Delta}{kT}\text{tgh}\frac{eV}{kT} - \frac{eV}{kT}\right) \tag{13}$$

If $eV \geq 2kT$ with an error less than 2% it results:

$$\left(\frac{dV}{dT}\right)_I = -\frac{k}{e}\frac{\Delta - eV}{kT} = 8.6\cdot 10^{-5}\frac{\Delta - eV}{kT}\frac{Volt}{Kelvin} \tag{14}$$

$$\left(\frac{dV}{dI}\right)_T = R_N\left(\frac{2kT}{\pi\Delta}\right)^{\frac{1}{2}} e^{\frac{\Delta - eV}{kT}} \tag{15}$$

For the S-I-S junctions, if the two superconductors have the same energy gap $2\Delta \gg kT$ the current can be well approximated by:

$$I = \frac{2}{eR_N}e^{-\frac{\Delta}{kT}}\left(\frac{2\Delta}{eV+2\Delta}\right)^{\frac{1}{2}}(eV+\Delta)\,f(x) \tag{16}$$

with $x = \frac{eV}{kT}$ and $f(x) = \sinh(x)K_0(x), K_0(x)$ is a modified Bessel function of the second kind [20]. $f(x)$ has a maximum value of about 0.5 for $x \simeq 0.8$. We have:

$$\left(\frac{dI}{dV}\right)_T = \frac{2}{R_N}e^{-\frac{\Delta}{kT}}\left(\frac{2\Delta}{eV+2\Delta}\right)^{\frac{1}{2}} f(x)\left(\frac{eV+3\Delta}{2eV+4\Delta} + \frac{eV+\Delta}{kT}\frac{df(x)}{dx}\frac{1}{f(x)}\right) \tag{17}$$

$$\left(\frac{dI}{dT}\right)_V = I\left(\frac{\Delta}{kT^2} - \frac{eV}{kT^2}\frac{df(x)}{dx}\frac{1}{f(x)}\right) \tag{18}$$

$$\left(\frac{dV}{dT}\right)_I = -\frac{k}{e}\frac{\frac{eV+\Delta}{k^2T^2}\left(\Delta - eV\frac{df(x)}{dx}\frac{1}{f(x)}\right)}{\frac{eV+3\Delta}{2eV+4\Delta} + \frac{eV+\Delta}{kT}\frac{df(x)}{dx}\frac{1}{f(x)}} \tag{19}$$

For a bias voltage corresponding to the maximum of $f(x)$ $eV \ll \Delta$ and we can write:

$$\left(\frac{dV}{dI}\right)_T = \frac{4R_N}{3}e^{\frac{\Delta}{kT}} \tag{20}$$

$$\left(\frac{dV}{dT}\right)_I = -\frac{4k}{3e}\left(\frac{\Delta}{kT}\right)^2 = 1.15 \cdot 10^{-4}\left(\frac{\Delta}{kT}\right)^2 \frac{Volt}{Kelvin} \tag{21}$$

An $Al - Al_2O_3 - Al$ junction could be used as thermometer for a composite bolometer operating at $0.3K$. In this case from (20) and (21) we have for the operating voltage, dynamic resistance and bolometer responsivity:
$V \simeq 20\mu V$, $R \simeq 10^3 R_N$, $S = \left(\frac{dV}{dT}\right)_I \frac{1}{G_{e_{eq}}} \simeq 5 \cdot 10^{-3} \frac{1}{G_{eq}} \frac{V}{W}$ with $G_{eq} = G - I\left(\frac{dV}{dT}\right)_I$
For typical values of $R_N \simeq 10^{-2}\Omega$ and $G_{eq} \simeq 2.5 \cdot 10^{-8} \frac{W}{K}$ we have $R \simeq 10\Omega$ and $S \simeq 3 \cdot 10^5 \frac{V}{W}$.

In this case the phonon noise dominates if a low temperature transformer with turns ratio 1 : 30 is used. So the ultimate energy resolution of the equation (9) can be achieved. The S-I-S thermometer heat capacity is negligible; if we have a pure silicon absorber of volume $V(in\, cm^3)$ which has a heat capacity $C = 6.7 \cdot 10^{-7} T^{-3} \frac{joule}{cm^3 K^3}$, assuming $G = G_{eq}$ we can write:

$$E_{min\, FWHM}(S_i) = 1.2 \cdot 10^5 T^2 (TV)^{\frac{1}{2}} eV \tag{22}$$

MODEL FOR S-I-S DETECTORS: COMPARISON WITH THE BOLOMETERS

It is known that tunnel junctions between superconductors can be used as ionizing particles detectors[11,12,13,14,21,22,23,24,25,26].

Together to the group of Naples we have obtained remarkable results in attempting to clarify the detection process of tunnel junctions $Nb - Nb_xO_y - Pb$ irradiated by α particles using a new geometry, an island type geometry, in which the area of the superconducting contacts is negligible respect to the junction area. In this way it is avoided the diffusion effect that spoils the resolution and we have obtained detectors with good response uniformity. An ionizing particle which loses an energy E in a superconductor with energy gap 2Δ excites a quasiparticles number $N(0) = \frac{hE}{\Delta}$ where h is a parameter introduced to take into account that the energy required to excite two quasiparticles is greater than the energy gap. The tunneling probability per unit time W_T is given by:

$$W_T = \frac{1}{e^2 N_m V R_N} \tag{23}$$

e is the electron charge, N_m is the electron density per cm^3 and per eV at the Fermi surface of the lead, V is the lead volume ($in\, cm^3$) and R_N is the normal state resistance of the junction.

For the island type geometry the quasiparticles diffusion out of the barrier region can be neglected. In this case we can write:

$$\frac{dN(t)}{dt} = -N(t)(W_t + W_R) \tag{24}$$

$N(t)$ is the number of quasiparticles in the lead at the time t after the absorption of the radiation and W_R is the excited quasiparticles recombination probability per unit time. We have $W_R = \frac{b}{V}(N_T + N(t)$ with b constant and $N_T \simeq 2N_m \Delta e^{-\frac{\Delta}{kT}}$ is the thermally excited quasiparticles number. Putting $a = W_T + \frac{b}{V}N_T$ we can deduce from(24):

$$\frac{dN(t)}{dt} = -aN(t) - \frac{b}{V}N^2(t) \tag{25}$$

The solution of differential equation (25) is:

$$N(t) = \frac{aN(0)e^{-at}}{a + \frac{b}{V}N(0)(1 - e^{-at})}$$

If we suppose that the operating voltage is within the linearity range of the $I(V)$ junction characteristic and we put $R = \frac{V}{I}$, the voltage pulse across the junction due to the energy loss in the lead is given by:

$$\Delta V(t) = eW_T R N(t) \tag{27}$$

If $N_T \ll N(0)$ we can write:

$$\Delta V(t) = \Delta V(0) e^{-\frac{t}{\tau}} \tag{28}$$

where $\frac{1}{\tau} = W_T + W_R$ and $\Delta V(0) = \frac{heEW_T R}{\Delta}$

If the voltage across the junction is amplified with a gain A decreasing at high frequency at 6 db/octave with an integrating time τ_F the voltage output of the amplifier is:

$$\Delta V_A(t) = \frac{AheEW_T R}{\Delta} \frac{e^{-\frac{t}{\tau}} - e^{-\frac{t}{\tau_F}}}{1 - \frac{\tau_F}{\tau}} \tag{29}$$

The output noise voltage is:

$$\overline{V}_N^2 = \frac{A^2}{4\tau_F}(4kTR + 2eVR + \overline{e}_N^2 + \overline{i}_N^2 R^2) \tag{30}$$

The first two terms are due to the junction Johnson noise and the current shot noise, e_N and i_N are the input voltage and current noise of the amplifier, k is the Boltzmann constant and T is the junction temperature; the statistical fluctuations of $N(0)$ are neglected. We can infer from (29) and (30) that we have the best energy resolution when the Johnson noise dominates, $\tau_F = \tau$ and $W_T \gg W_R$ so that $\frac{1}{\tau} = W_T$. In this case the minimum detectable energy is given by:

$$E_{min\,FWHM} = 6.4\Delta \left(kTN_m V \frac{R_N}{R}\right)^{\frac{1}{2}} \tag{31}$$

For the lead we have $\Delta = 1.34 meV$ and $N_m = 1.42 \cdot 10^{22}$ electrons per cm^3 and per eV. Therefore:

$$E_{min\,FWHM}(Pb) = 9.5 \cdot 10^6 \left(TV \frac{R_N}{R}\right)^{\frac{1}{2}} eV \tag{32}$$

Actually, for the $Nb - Nb_xO_y - Pb$ junctions, it is very difficult to have a $\frac{R_N}{R}$ ratio greater than 10^3 because of the junction defects, also if a temperature lower than $1K$ the theoretical value for an ideal junction is very high.

At $0.3K$, if the energy loss in the lead is low enough, $W_T \gg W_R$, because of the small number of thermally excited quasiparticles (a few unities for a volume of $10 cm^3$).

To compare the performances of the bolometer and tunnel junction detectors with the same sensitive volume and temperature we can calculate the ratio $\frac{E(S_i)}{E(P_b)}$ between the minimum detectable energies [18]. From (22) and (32) we deduce:

$$\frac{E(S_i)}{E(P_b)} = 1.3 \cdot 10^{-2} T^2 \left(\frac{R}{R_N}\right)^{\frac{1}{2}} \tag{33}$$

REFERENCES

(1) Martin D.H. and Bloor D.: *Cryogenics*, **1**, 159 (1961).

(2) Low F. J., *J. Opt. Soc. Am*, **51**, 1300 (1961).

(3) Cavallini M., Gallinaro G. and Scoles G.: *Z. Naturforsch. A.*, **22**, 413 (1967).

(4) Andrews D. H., Fowler R.D. and Williams M. C.: *Phys. Rev.*, **76**, 154 (1949).

(5) Niinikoski T. O.: *Cosmic-Ray disturbances in Thermometry and Refrigeration in Liquid and Solid Helium* (New york, Halsted, 1975).

(6) Blasi A., Gallinaro G., Osculati B., Siri A., Valbusa U. e Vitale S.: INFN Internal Report (1985).

(7) Fiorini E. and Niinikoski T. O.: *Nucl. Instrum. Methods A*, **224**, 83 (1984).

(8) Cabrera B., Krauss L. M. and Wilczek F.: *Phys. Rev. Lett.* **55**, 25 (1985).

(9) Mc Cammon D., Juda M., Zhang J., Kelley R. L., Moseley S. H. and Szymkowiak A. E. *I.E.E.E. Trans Nucl. Sci.* **NS-33**, 236 (1986).

(10) Niinikoski T. O. *et al.*: *Conference of the American Physical Society*, April 1985.

(11) Gallinaro G.: unpublished report INFN (1964).

(12) Pecorella A.: Tesi Università di Genova (1965-66).

(13) Wood G. H. and White B. L.: *Appl. Phys. Lett.*, **15** 237 (1969).

(14) Twerenbold D.: *Europhysics Letters*, **1**, 209 (1980).

(15) Downey P. M., Jeffries A. D., Meyer S. S., Weiss R., Bacher F. J., Donnelly J. P., Lindley W. T., Mountain R. W. and Silversmith D.J.: *Appl. Opt.*, **23**, 910 (1984).

(16) Clarke J., Hoffer G.I., Richards P.L. and Yeh N.H. *J. Appl. Phys.*, **48**, 4865 (1971).

(17) Gallinaro G., Terreni S. and Salvo C.: *Cryogenics*, **26**, 9 (1986).

(18) Gallinaro G.: *Il Nuovo Saggiatore*, **2**, 49 (1986).

(19) Mc Cammon D., Moseley, S. H., Mather J. C. and Mushotzky R. F.: *J. Appl. Phys.*, **56**, 1263 (1984).

(20) Douglas D. M., Falicov L. M.: *Prog. Low Temp. Phys.*, **4**, 97 (1964).

(21) Wood G. H. and White B. L.: *Can. J. Phys.*, **51**, 2032 (1973).

(22) Kurakado M. and Mazaki H.: *Phys. Rev. B.*, **22**, 168 (1980).

(23) Kurakado M. and Mazaki H.: *Nucl. Instrum. Methods*, **185**, 141 (1981).

(24) Kurakado M. and Mazaki H.: *Nucl. Instrum. Methods*, **185**, 149 (1981).

(25) Kurakado M.: *J. Appl. Phys.*, **55**, 3185 (1984).

(26) Barone A., Darbo G., De Stefano S., Gallinaro G., Siri A., Vaglio R. and Vitale S.: *Nucl. Instrum. Methods A*, **234**, 61 (1985).

DESIGN FOR 1-μm RESOLUTION X-Y SUPERCONDUCTING PARTICLE DETECTOR

O. LIENGME[1]

Argonne National Laboratory
Argonne, Illinois 60439
[1]Present address: 62A, Route de Vandoeuvres,
CH-1253 Vandoeuvres, Switzerland

1. INTRODUCTION

Position-sensitive particle counters[1,2] are currently used for X-ray diffraction experiments as well as high energy vertex detectors (e.g., wire chambers). In addition, such counters can be read electronically in real time. However they offer a poor spatial resolution, usually of the order of 100 μm, due to their fast signal speed. In some cases however, such as diffraction of synchrotron light or early event vertex detection for instance, a higher resolution of detection in conjunction with a high count rate is required, especially if the detector has to be placed close to the source of particles (for instance in the direct output beam of the synchrotron or close to the impact volume for electron-positron annihilation).

Metastable superconducting microstrips[3] were recently suggested for this purpose. Based on a hotspot propagation mechanism, it is shown that a thermal approximation of the complex nonequilibrium phenomena involved is relevant. In this paper, the feasibility of such detectors is investigated, where the detector geometry is scaled with the size of the interaction event for very low particle interaction energies. A practical design is suggested for high count rate and high resolution, in a way that solves elegantly the so-called N^2 problem. The cases of soft X-rays and high energy particles are emphasized, where a 1 μm resolution is predicted.

First we show which requirements should be fulfilled for achieving these detectors. Such devices consist of a particle-nucleated heat wave propagating slowly (10^3 m/s) along a metastable superconducting strip (propagating hotspot). In contrast, electric pulse propagation speeds are usually closer to that of light, resulting in a relatively poor spatial resolution even with subnanosecond time resolution. In our case of propagating heat pulses, the working conditions in terms of speed, time resolution and dead time suggest an affordable 1 μm linewidth technology. Particle induced hotspots are locally preamplified due to the finite drive current I_d. Moreover, if the width of each detector is kept equal to the reading resolution along its length, the total count rate (proportional to the inverse propagation time for a single detector line) is constant for a detector array made of independently measured strips. VLSI techniques are shown to be adequate also for this purpose. A practical proposal is given in the second part of the paper. Each element consists of a 1 μm wide superconducting wire driven at a current close to the critical current. For large currents (as limited by the critical current imposing low enough reduced temperatures), the zero-voltage state is metastable with respect to the out-of-thermal equilibrium dissipative hotspot state[4]. The sensitivity of this system to external perturbations is well known[5]. In the proposed device a hotspot triggered by a particle interaction event (as already observed using α particles[6]) is amplified[7] using a large enough bias current and propagates along that given wire as a heat wave[8].

2. VLSI arrays

In this section we establish the correlation between high-resolution, high-flux particle detectors and VLSI techniques. The position measurement for particle counters is defined by the time lag, Δt, corresponding to particle-generated signals propagating both ways and at a finite speed, v, in one transmission line or channel[3]. In such homogeneous systems, the precision of spatial detection, δx, along the channel is usually set by the accuracy, δt, of time measurement. This time resolution can be much smaller than a nanosecond if superconducting electronics is used. For simplicity we will assume hereafter a time resolution $\delta t = 1$ ns that is still compatible with the inductance of VLSI detector lines. An X-Y array of such individual channels (or lines) will be used for position detection in a plane (e.g. wire chamber for vertex detection). The distance from channel to channel is most efficiently set equal to the spatial resolution $\delta x = v\, \delta t$ of the device, that is for a square detector of N_L channels of length L :

$$N_L = L/\delta x = L/v\delta t \qquad (1)$$

Thus a signal propagation speed as low as $v = 10^3$ m/s is required for a position resolution of $\delta x = 1$ µm and our reading time constant δt = 1 ns; and equation 1 leads to the anticipated VLSI density of channels.

Even if a second set of pulses could physically coexist in a detector line for a large enough particle flux, it should be discarded because of the resulting confusion in the measurement of the position of the corresponding events. Hence, latching detectors that reject subsequent detection events are preferred. In any case the data acquisition rate is correspondingly limited by the dead time Δt, which is the time lag needed for the pulse to propagate through the entire wire of length L. If the particle rate is larger than this acquisition rate, many hotspots could propagate in the same detector line, resulting in an error over the N sites of particle interactions (as read as a single particle event closer to the center of the wire length). For a negligible reset time (that can be kept of the order of the propagation time) it is :

$$\Delta t = L/v \qquad (2)$$

However, the experimentally important parameter for a useful detector is the maximum count rate Γ. For independent channels this is the sum of the individual count rates $R = \Delta t^{-1}$. Although Δt^{-1} decreases proportionally to the signal propagation speed v according to equation 2, the density of channels may be increased as $\delta x^{-1} = (v\delta t)^{-1}$. In other words under this condition the total count rate depends only on δt^{-1}, and for a square detector satisfying equations 1 and 2 we obtain :

$$\Gamma = N_L R = \delta t^{-1} \approx 10^9 \text{ particles/sec} \qquad (3)$$

Let us illustrate this concept using some numbers corresponding to the above example. If we take an array of channels of length L = 10 cm, the speed v = 10^3 m/s gives us a dead time $\Delta t = 10^{-4}$ s and a maximum count rate per channel R = 10^4 s^{-1}. A square array of such individual channels would give $\Gamma/A = N_L R/A = 10^7 (cm^2 s)^{-1}$ where A is the area of the device. Only the maximum total count rate Γ is an intensive variable that does depend on the time resolution δt. This remarkable result means that such VLSI slow-signal detectors can be achieved without the expense of a lower count rate, provided each detector line is measured independently.

As for any detector, the counter saturates for particle rates larger than Δt^{-1}. In a X-Y array of undiscernable detectors however, if N particle events are to be read within the propagation time Δt, N^2 sites may be equivalently attributed to the N measured particles (so-called N^2 problem). N^2 is the number of possible configurations, that is, the sum of all the permutations over the N coordinates of the N particles interacting with that detector within the time Δt.

In our case of N_L independently probed detector lines, we have only M possible sites in each line for M particles impinging within the time lag Δt. Here, at most one particle should be read per channel over the time Δt. In addition, since each line saturates independently, at most one coordinate will be read per line, and that event will be correct even if subsequent events occur in the already latched areas. Accordingly, the use of independent reading channels solves that N^2 problem for such latching detectors.

3. HOTSPOT SUPERCONDUCTING DETECTORS

The principle of a hotspot superconducting detector is fairly simple, since the zero voltage state of a superconducting strip driven at a large current is in a metastable state. An initial hotspot, such as that created by an impinging particle, is amplified and allowed to grow at constant speed through a homogeneous superconducting microstrip until it reaches its ends where the temperature is detected, for instance by means of tunnel junctions (see Fig. 1). Its resistance, i.e the small voltage associated with the reading current I_d will drop when the temperature at the junction exceeds T_c. Then, the current bias I_b should be reset to allow the strip to recover the superconducting state and the next particle event to be observed. In this section we outline the working principles of such a device using results from the previous section. Practical examples are given in terms of dimensions, drive and choice of materials for the superconducting strip.

Figure 1 : *Metastable NbN strip with thermometer junctions at each end (shaded zones) to detect the propagating hotspot.*

Superconducting wires submitted to a large enough current density can be switched to the normal state under the influence of the excitations due to a particle interaction[6,7,8] because of the following mechanisms. A given amount of energy provided by the particle is released into the superconducting material breaking a corresponding amount of Cooper pairs. This energy spike should relax in a way similar to thermal excitations since the initial excitation energy is much larger than the superconducting gap energy[3]. After a time comparable with the thermal relaxation time (typically 0.1 ns), an initial hotspot of diameter D at the superconducting transition temperature T_c results. This thermal event, occurring in a very small sampled volume, may trigger the nucleation of a hotspot across the entire width w of the strip : if the critical current density J_c is reached in the remaining cross-section of the film, the whole film section will start dissipating. This behavior will occur for currents larger than the following current threshold[9] I_p :

$$I_p = I_c\,(1-D/w), \qquad (4)$$

where $I_c = J_c wd$ is the total critical current and d is the film thickness. In other words an amplification factor w/D of the hotspot size can be obtained by using a finite current drive I_p/I_c. Indeed, the time constant τ_0 of the superfluid electrons[10] is much faster than the thermal propagation time :

$$\tau_0 = m\sigma/n_s e^2 \approx 0.5\ \text{ps}/[Tc-T][K]. \tag{5}$$

Therefore the supercurrent redistributes itself quasi-instantaneously around the initial growing hotspot until the critical current density is reached, allowing hotspot amplification[3,9] (see Fig. 2).

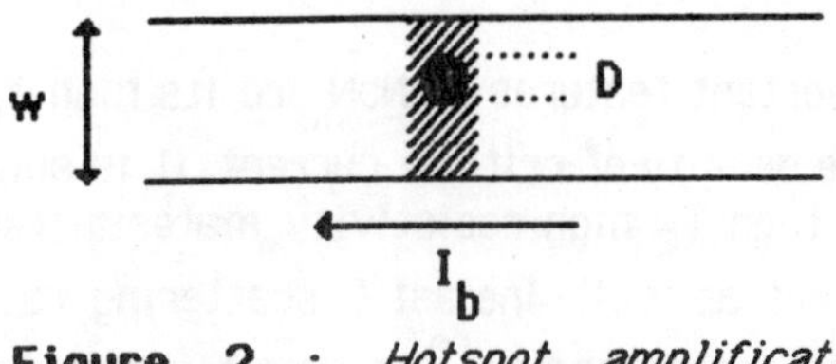

Figure 2 : *Hotspot amplification in a current driven superconducting wire of width* W. *Notice the smaller diameter* D *of the primary hotspot.*

It should be noticed that equation 4 applies strictly only to a cylindrical track of ionization such as that released by a high energy particle. In the case of X-rays however, the heated region presumably surrounds the atomic site emitting a core photoelectron. Nevertheless, we believe that in the interesting case of film thicknesses comparable with the diameter of the particle-induced hotspot, equation 4 should remain a good approximation. Furthermore, if we assume that, after the initial relaxation, all the particle energy is transferred into the electronic system, we can use as a first approximation the equilibrium thermodynamic relationship :

$$\partial U/\partial t = V^{-1}\ \partial E/\partial t = \gamma T. \tag{6}$$

This relationship should be valid for large enough integration time. Here, U is the internal energy of the system that (as incremented by the particle interaction event), V is the interaction volume and γ is the electronic specific heat.

For NbN operated at 4.2 K with $\Delta T = T_C - T \sim 10$ K, we find an initial hotspot diameter D = 0.25 µm for 6 keV X-rays, using[10] $\gamma = 1.66\ 10^3$ ergs/cm^{3}K^2. According to equation 4, the 1 µm wide microstrip would then have a threshold current $I_p = 0.75\ I_c$. Although substantially larger amplification factor could be obtained for a single strip driven at higher currents[4], that line would be correspondingly more sensitive to fluctuations, such as those induced by thermal switching of adjacent strips. In that case, the excessive thermal coupling from strip to strip would trigger the adjacent strip, allowing the hotspot to grow transversally as well. Therefore, thermal coupling to the substrate is important for proper operating conditions. We will see below that the thermal coupling to the substrate also affects the maximum speed of thermal growth along a given detector line. The use of low reduced temperatures where the critical current saturates might help with that respect too. Operating at 4.2 K would be adequate[5] for NbN from that point of view also.

Actually, the most important features of NbN are its high T_c, high resistivity and its reasonable density of critical current. It is suggested that working with with any high T_c high resistivity material (such as ceramics) should be convenient as well. Inelastic scattering rates are expected to be extremely fast, corresponding[12] to even higher calculated T_c^* and faster τ_0.

The second role of a large current drive is to make the zero voltage state metastable and to let an initially localized hotspot grow along the sample length, until it is detected at each end of the microstrip by means of the superconducting junctions (see Fig. 1). The one-dimensional wire can be considered as an open system fed by the energy of the current and triggered by the small particle induced hotspot. If the current is too small, the hotspot will collapse because not enough energy will be dissipated in the film as compared to that extracted by the thermal bath. For a low coupling with the substrate, the Joule generated heat is predominantly removed along the wire itself by metallic conduction. Under this condition, and for large enough current densities, that hotspot may grow with a speed comparable to that of sound[4]. A simple balance equation for the dissipated energy determines the growth speed of the hotspot, as given for large dissipation and negligible heat transfer to the substrate[8] :

$$v^2 = \rho_n J^2/(C_v^2(T_c-T_b)/K) \qquad 7)$$

where ρ_n is the normal state electrical resistivity of the film, C_v is the specific heat per unit volume and K is the thermal conductivity of the sample. Although the values of transport parameters at T_c describe well the time dependence of the normal-superconducting (NS) boundary, the nonequilibrium temperature inside of the hotspot can be quite high so that high melting-temperature superconductors like Nb-based materials or ceramics are indicated.

If the substrate is thermally insulating or decoupled enough from the film, then the Wiedemann-Franz law[6] (valid at low temperature) can be used in the equation 7. It leads to a velocity of the NS boundary that only depends on the critical current density and the specific heat, and not on the normal resistivity ρ_n of the sample :

$$v = [L/(1-T_b/T_c)]^{1/2} \, J / C_v \qquad 8)$$

Here L is the Lorentz number and we take $T = T_c$ at the moving NS interface.

Comparing equations 4) and 8) we notice that for submicron detectors the most restrictive condition is that of obtaining sufficient velocity according to 8), whereas for much wider detector lines the initial hotspot amplification determined by equation 4) sets the current drive. Hence a micron scale device is not only feasible using the now standard VLSI technology, but to some extent imposed by the dynamics of the detector itself.

NbN is known for its relatively high critical current density (typically over 10^6 A/cm^2) so that measured NS boundary speeds[4] exceed a few 10^3 m/s. However its large resistivity might perhaps lead to an excessive heat dissipation at the 4.2 K operating temperature, that could induce the transition of adjacent detector lines in case of excessive thermal coupling. This is of even more concern with ceramics, although one might consider working with higher reduced temperatures and somewhat lower net currents in this case. To that respect, NbN is a flexible material of adjustable ρ_n and J_c, a feature that might be found useful for first attempts.

Notice that the transport properties of concern in equations 4 and 7 are only reduced quantities, allowing the use of low thicknesses for a lower net dissipation. Although this will be at the expense of some efficiency of the counter in the case of X-ray detection, this feature may turn to be very useful for some very high intensity synchrotron applications.

So far only relative positioning of the particle event has been mentioned. Since the velocity of the pulse depends on the current density, absolute accuracy is restricted by cross section variations leading to systematic errors. For example, a 0.1 % uncertainty in cross section area would result as an uncertainty in the particle position of 0.1 % of the length of the detector strip. Even worse are errors that accumulate, such as systematic thinning or narrowing of the strip at one end for example. As is well known, thinning can be avoided by mounting the substrate onto an epicicloïd of revolution. However, for a regular distribution of small variations in thicknesses and in widths only small systematic errors would contribute, errors that should statistically compensate for each other.

4. DETECTOR LAYOUT

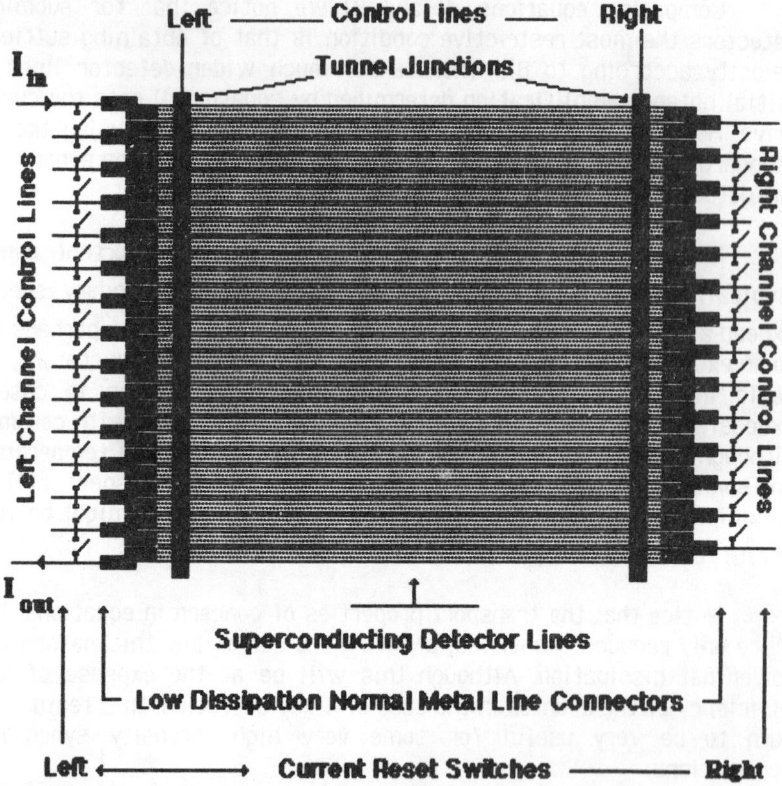

Figure 3 : *Layout example of a completed square detector consisting of 28 detector strips. See the text for details on the various parts of the detector.*

Figure 3 shows an example of a detector layout for a square detector consisting of 18 detector strips (lightly shaded horizontal lines). At each end of the strip, Giaever junctions detect the hotspot. These are interconnected by a single strip (control line) per sample side, that is the base line for those junctions. Individual detector strips are interconnected by low resistance low dissipation normal metal contacts, allowing the same current bias for all the strips but preventing the hot-spot of one given strip to propagate into the adjacent line. In this design, the resistance between the control line and a given (left or right) control line, as connected to the corresponding normal link, monitors the temperature of two adjacent detector strips. If both of those detector lines flip to the normal state, the resistance is dropped to the normal state value $R_N/2$; whereas when one junction is flipped only, a higher voltage value is read.

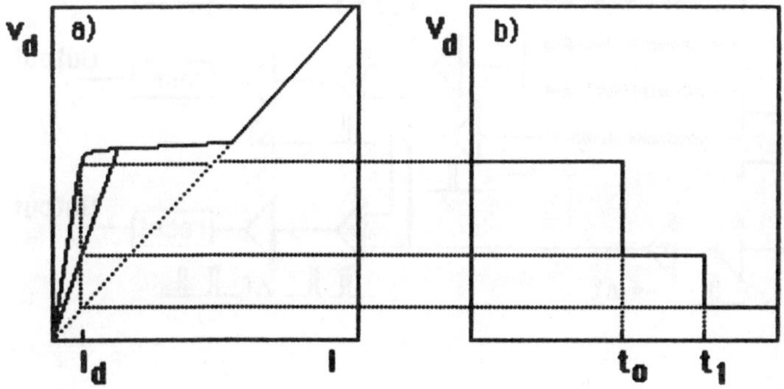

Figure 4 : *Three state level of the channel outputs, for current biased Giaever junctions at a current* I_d. *a) For maximum voltage difference, this current corresponds to a subgap voltage in the superconducting state. b) Voltage output changes at the first hotspot arrival* (t_0) *and at the second hotspot arrival* (t_1).

Clearly, this is a three-state system, as shown on Fig. 4. A pulse generator triggered at a high enough voltage would detect a hotspot at either of the two lines, leading to a 2N problem. Then, the pulses coming from one side (say, for instance, the left channel control lines) have to be delayed by a time lag Δt. The position measurement is that of the time lag between the right channel and left channel pulses, respectively, where each right channel pulse from the j right channel may correspond to a pulse from either the j-1 or the j left channel, the combination of the two informations giving the correct detector address of the corresponding particle event. Since the maximum time lag between two pulses from a single strip does not exceed Δt, as set by the propagation speed v, a simple demultiplexer would check for the presence of such two pulses within Δt.

At this point, the latching property of the detector can be used to reset the detector only once the particle record has been processed. The corresponding pair of detectors have to be shorted by either one of the adjacent switches, as shown on the associated electronic schematics (Fig. 5).

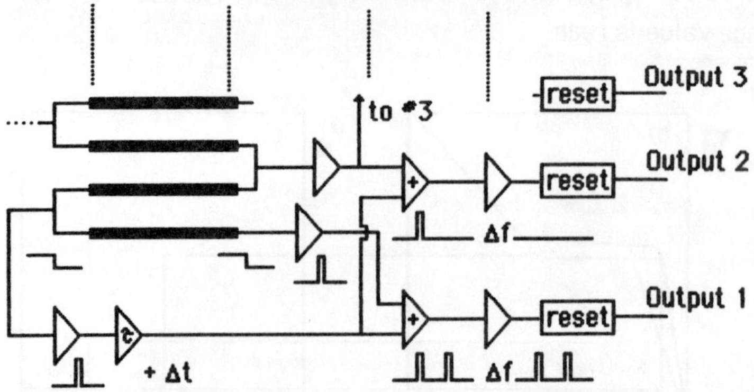

Figure 5 : *Electronics associated with the detector array shown on Fig. 3. The 4 bottom detector strips are schematized as bold lines, and pulse shaping is performed at each end of each strip. Delay lines are introduced on the left side channels, signals from both ends of each strip are summed (+) and demultiplexed (Δf) before the output. Each output front signal resets the corresponding line by means of the corresponding switch (not shown on this figure, see on Fig. 3).*

However, local encoding of the detected particle addresses would be appropriate in any application where the number of detector strips is very large. For instance, a local memory storing bits at detected particle positions would be very large, i.e. $10^8 = 10^4 \times 10^4$ bits cm^{-2}. Sending a fully multiplexed data stream with that number of bits at the reading rate of 10^8 data/s would impose a tremendously fast data transfer rate in a single line, namely 10^{16} bits cm^{-2} s^{-1} or about 10^{15} bauds.

CONCLUSION

High resolution high data acquisition rate superconducting X-Y position detectors are feasible using hotspots propagating in individually probed detector strips. This new class of detectors show advanced features, such as local signal preamplification, very high resolution and very high data acquisition rates.

Working conditions are explicitly formulated in expressions that agree with those preliminary data. Accordingly, the linewidth of each detector line should be matched to the size of the primary hotspot within the preamplification factor. A design is proposed that replaces the usual N^2 problem by an affordable 2N problem, at the expense of some local electronics.

REFERENCES

1) C. J. Borkowski and M. K. Kopp, Rev. Sci. Instr. **39**, 1515 (1968)
2) A. Gabriel, Rev. Sci. Instr. **48**, 1303 (1977)
3) O. Liengme, Proceeding of the 1987 Max-Plank-Institute Workshop on "Low Temperature Detectors for Neutrinos and Dark Matter", pp. 44-52, ed. K. Pretzl, N. Schmitz, and L. Stodolsky, Springer-Verlag.
4) W. J. Skocpol, M. R. Beasley and M. Tinkham, J. Appl. Phys. **45**, 4054 (1974)
5) K. E. Gray, R. T. Kampwirth, J. F. Zasadzinski and S. P. Ducharme, J. Phys. F : Met. Phys. **13**, 405 (1983)
6) D. E. Spiel, R. W. Boom and E. C. Crittenden, Appl. Phys. Lett. **7**, 292 (1965)
7) K. E. Gray and D. E. Fowler, J. Appl. Phys. **49**, 2546 (1978)
8) R. F. Broom and E. H. Rhoderick, Br. J. Appl. Phys. **11**, 292 (1960)
9) A. Barone, S. de Stefano and K. E. Gray, Nucl. Instr. and Meth. **A235**, 254 (1985)
10) M. Tinkham, Introduction to Superconductivity (McGraw Hill, New York, 1975)
11) C. Geibel, H. Reitschel, A. Junod, M. Pelizzone and J. Muller, J. Phys. F : Met. Phys. **15**, 405 (1985)
12) P. A. Lee, Phys. Rev. Letts. **58**, 2691 (1987)

SUPERCONDUCTING THIN FILM VERTEX DETECTOR FOR SSC APPLICATIONS

R. G. WAGNER, K.E. GRAY

Argonne National Laboratory,
Argonne Illinois 60439, USA

Most large solid angle detectors discussed at Superconducting Super Collider (SSC) workshops over the last few years have included a device for performing high precision tracking close to the bunch crossing point (within a few tens of centimeters). The main purpose of these microvertex detectors is to provide "close-in" track reconstruction with point-to-point spatial resolution of a few microns that allows precision location of the primary event vertex, of secondary vertices from decays of c and b quarks or τ leptons, and of vertices unrelated to the event of interest from other beam-beam interactions. In addition, the event vertex constraint and the track segments close to the interaction point improve the overall tracking momentum resolution provided by the large outer tracking chamber.

The available alternatives for a microtracker at the SSC are severely limited by the high charged particle multiplicity within events, by the requirement of fast detector response arising from the 16 ns bunch spacing, and by the high yearly radiation doses a detector would receive near the beam pipe. A silicon microstrip device has generally been the detector of choice since it represents proven technology and adequately meets the above requirements except for radiation hardness. With relation to their sensitivity to radiation damage, Si strip detectors are only marginally acceptable for long term operation at the SSC. A Si strip device has been made that functions after a 1 MRad dose, although with a greatly increased leakage current [1,2]. This is about the yearly dose expected at 10 cm radially away from the interaction point due to minimum ionizing particles from beam-beam interactions [3]. Other more speculative detectors such as scintillating fibers or semiconductor pixel devices either exhibit some change in their operating characteristics after MRad doses or have unknown radiation hardness. Obviously, one would prefer to have a detector that would show no measurable change in its operating characteristics even after several years operation at the SSC. We will describe a possibly very-radiation-hard alternative based on superconducting thin film strips. There exists materials whose superconducting characteristics have been shown to be orders of magnitude more radiation resistant than the properties determining the response of semiconductor devices. The possibility that strip detectors sensitive to minimum ionizing particles might be fabricated has led us to study their feasibility for use as a microvertex detector.

There are two key elements of the operation of the thin film detector: the interaction with charged particles and the required readout for high spatial resolution. This paper deals mainly with the former, and we consider the feasibility of extending previous work on α-particle detectors to the more stringent requirements for minimium ionizing particles. For the readout we briefly outline a novel approach to the problem [4] utilizing the slow phase velocity of electromagnetic waves possible in superconducting waveguides[5].

INTERACTION WITH CHARGED PARTICLES

Historically, the idea of using a superconducting film as a particle detector in this manner was first proposed 25 years ago by N.K. Sherman [6]. The device was envisioned as a detector of fission fragments or α–particles. Experimental observation of voltage pulses from superconducting-to-normal (S-N) transitions caused by alpha particles of about 5 MeV kinetic energy incident on thin indium, tin

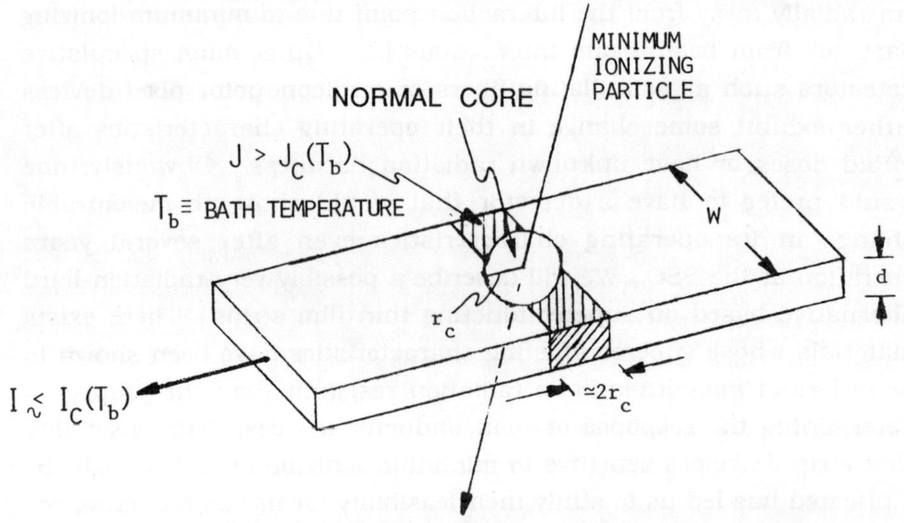

Fig. 1. Schematic diagram of the interaction of a minimum ionizing particle with a thin-film superconducting strip.

and aluminum films has been reported in Refs. [7-9]. Briefly, the device acts as a bolometer. Ionization from the passage of a charged particle through a strip, carrying a current bias that is nearly critical, serves to heat the region around the particle path above its critical temperature. The resulting S-N transition propagates across the width of the strip and the resistive region thus created produces a detectable voltage pulse.

<u>Detector operation.</u> The detailed operation of the detector is illustrated in Fig. 1. A thin film strip of width, w, and thickness, d, carries a current, I, close to but below the critical current, I_c (T_b), where T_b is the bath or substrate temperature at which the strip is maintained. Energy loss by the incident particle produces a local heating in the cylindrical region centered on the particle trajectory. Within a radius r_c, which is determined by the material's specific heat and the deposited energy from the particle, the temperature

rises above its critical value forcing the current into the areas between this region and the film edges. This reduces the effective width by $2r_c$, thereby decreasing the critical current proportionately:

$$I_c'(T_b) = I_c(T_b)\,[w-2r_c] \,/\, w. \qquad (1)$$

The current bias I is set to be greater that this reduced value, $I_c'(T_b)$, so that the whole cross-section of the film becomes normal. Clearly since the uniformities of w, d and $I_c(T_b)$ are limited by practical considerations, w cannot be too much larger than $2r_c$.

The normal radius, r_c, can be estimated from the molar enthalpy change, Δh, of the superconducting material in going from T_b to its critical temperature T_x. Assuming that the ionization heating is uniform and confined to the normal region, the result is

$$r_c = \sqrt{[(dE/dx) \,/\, \pi\,\Delta h]}, \qquad (2)$$

where dE/dx is the energy loss per unit distance for the particle. Thus a small value of the specific enthalpy is highly desirable. For example, the very large specific heat near 100 K of the high-T_c oxide superconductors precludes their use here since the normal region created by minimum ionizing particles would be negligibly small. The film widths used in the earlier work [7] ranged from 5 to 50 μm and the region heated above the critical temperature had a radius of 2.5 to 9 μm around the trajectory of the α−particles. Spiel, et al. [7] showed that the radius of the normal region agreed well with estimates given by a heat diffusion model. Since 5 MeV alpha particles deposit approximately 200 times the energy of minimum ionizing particles, these data and Eq. 2 imply that high energy particles would produce a normal region of a few tenths μm in radius.

The normal region will have a length along the strip of about $2r_c$ so that the instantaneous potential drop, assuming I is constant, is given by

$$V = 2\, r_c\, \rho\, J \leq 2\, r_c\, \rho\, J_c, \qquad (3)$$

where ρ is the normal state residual resistivity of the strip material and $J=I/dw$ is the current density. This voltage pulse can be observed across an external resistor, and is close to the unattainable limit given when $J=J_c$ for the reasons given after Eq. 1. The normal region thus created, can either collapse or propagate depending on whether the Joule heating in the region exceeds the cooling capacity of the substrate and/or bath. Reference [7] observed both modes in their work with α-particles. In the case where the normal region collapsed and the strip recovered its superconductivity, pulses of 10 ns risetime and 70 ns decay were observed. For use as a fast recovering detector for high multiplicity of events, operation in the collapsing mode would be preferable. For the present case of minimum ionizing particles with smaller energy loss, this should be the easier of the operating modes and one might also expect the signal decay time to be shorter. The reasoning is that α-particles deposit significantly more energy in the substrate than in the sample since dE/dx is steeply rising at this kinetic energy. The energy deposited in the substrate also contributes to heating of the strip. For minimum ionizing particles, dE/dx is essentially uniform throughout, and little additional heating of the strip can be expected.

Optimizing the material. The question of feasibility for minimum ionizing particles then rests on obtaining a value of V that is sufficiently above the noise level and a value of r_c that is large enough that film widths, which can be no wider than a few times r_c, can be made reliably and uniformly over the large detector areas envisioned. As can be seen from Eqs. 2 and 3, these depend on materials properties of the superconductor, i.e., ρ, J_c and Δh. An additional requirement of radiation hardness for a practical detector leads us to consider NbN as a suitable candidate for initial tests. A considerable literature exists for NbN, and in thin films J_c exceeds 10^6 A/cm^2 for zero field [10,11] and is [12] above 2×10^5 A/cm^2 for fields of 10 T. Superconductivity persists to fields above 20 T. High values of ρ (>200 $\mu\Omega$cm) are routinely found in sputtered films due to the granular microstructure and high concentration of defects [13]. Studies of NbN irradiated [14] with fast neutrons (> 1 MeV) have observed only a 1% change in T_c for doses up to 10^{19} n/cm^2. Recently the dependence of J_c on neutron irradiation has also been measured. For this test, samples of NbN were irradiated at room temperature by neutrons of energy >0.1 MeV at a test reactor at Oak Ridge National Laboratory. The J_c at 4.2 K in a 10 T field was measured before and after irradiation and no change was observed[15], within the measuring error, up to a fluence of 2.5 x 10^{17} n/cm^2. For twice this dose, which corresponds to at least 100 MRads, a decrease of about 10% was observed in most samples.

Clearly the largest values of ρ and J_c are highly desirable but also r_c, which comes from Eq. 2. For minimum ionizing particles, dE/dx varies roughly with the mass density so that differences between various superconducting materials will be small. However, Δh will depend on the specific heat and the critical temperature T_x under the operating conditions of the device. Large values of T_x imply large Δh and hence small values of r_c. An estimation of T_x comes from the measured dJ_c/dT of the material using

$$T_x = T_b + \varepsilon\,[J_c/(dJ_c/dT)]_{Tb}, \qquad (4)$$

where $\varepsilon = 1 - J/J_c$. For NbN, typical values [11] of $[J_c/(dJ_c/dT)]_{Tb}$ are 10 to 15 K for T_b=4.2 K and about 5 K for T_b=8 K. In the latter case J_c is 60% smaller than at 4.2 K. Thus choosing a superconductor with T_c (at zero current and field) much greater than T_b gives almost the maximum value of J_c for the material, which increases V through Eq. 3, but may result in a small value of dJ_c/dT leading to large T_x. Because of the strong dependence of r_c on T_x, through Δh, this is disadvantageous. Therefore T_c values closer to T_b=4.2 K than the 16 K of NbN may be desired. Experimental switching tests may be necessary to adequately determine the best T_c. If lower T_c is desired to reduce T_x, the compounds $Nb_{1-x}V_xN$ seem promising since these may maintain the high values of ρ and J_c found in NbN. For x=0, T_c is about 16 K, but it decreases [16] to about 3.5 K for x=0.4 and then rises to 9 K for x=1.

Conclusions. Using the measured [16] specific heat of NbN and anticipated values of T_x~6.2 K, based on ε=0.2 and $[J_c/(dJ_c/dT)]_{Tb}$ =10 K, one finds r_c~0.2 μm and that w must be of order 2 μm in order to achieve switching from minimum ionizing particles. Then if ρ=200 μΩcm and J_c=2 x 10^5 A/cm^2 one finds V=1.6 mV from Eq. 3, and this is comparable to values found in wire chambers.

READOUT FOR HIGH SPATIAL RESOLUTION

The energy loss for minimum ionizing particles requires strip widths of 1-2 μm in order that the initially created normal region is a sufficient fraction of the width that the current density is increased above the critical value in the remaining, superconducting portion of the strip. Such strips can be fabricated, but it is improbable that a detector composed of many parallel strips of this width could be made, or would even be desirable given the enormous number of channels it would contain. Resolution which is 5 times better than Si microstrip detectors can be achieved, for example, by fabricating 1-2 μm wide strips where each strip forms a zigzag pattern of width of 10 μm and several cm in length. The number of channels and the problem of connecting them to the readout electronics would be similar to that for a Si device. It should be noted that the device would be a replacement for the Si-based detecting medium only. Since it relies on the observation of voltage pulses produced by a S-N transition, its utility as an SSC detector depends ultimately on the availability of radiation hard readout electronics.

The arrival time of the voltage pulse from the S-N transition, contains information on the position of the particle interaction along the strip. However, if this travels with the speed of light in vacuum, the desired spatial resolution of 10 μm places impractical limits on the time resolution. However, the phase velocity of electromagnetic waves propagating in superconducting thin film waveguides has been shown [5] to be considerably less than the free space value of c=3 x 10^8 m/s. The explanation is based on the different penetration depths of electric and magnetic fields into superconductors. While the magnetic field can penetrate a depth of 50 to 500 nm, electric fields are strictly excluded from the superconductors and are confined to the insulator between them which can be only 1-2 nm. The phase velocity depends roughly on the square root of this ratio and can be as low as 0.05c in Josephson junctions[5] assuming a dielectric constant of 4. Thus, we propose a superconducting waveguide comprised of the thin film strip detector with a superconducting ground plane separated by a thin insulating layer. The strip-to-strip spatial resolution is 10 μm, but using the above estimates, one finds a spatial resolution along the strip length to be 2 mm for an electronic time resolution of 1 ns. Although this is inferior to conventional detectors, it is sufficient to unravel the N^2 problem if two detectors are placed at right angles with one above the other. Simply stated, with a multiplicity of N events per bunch, the 2 mm resolution along the strip is sufficient to identify which events in the two detectors correspond to the same particle track, and thus provide spatial resolution of 10 μm in each orthogonal direction. Without this information, it is impossible to identify which of the N events in each detector correspond to the same particle, unless stereo angles less than 90° are used in the detectors, resulting in a considerable loss of resolution and unambiguous identification.

ACKNOWLEDGEMENTS

One of the authors (KEG) thanks Ray Hagstrom for informative discussions. This work supported by the U.S. Department of Energy, Divisions of High Energy Physics and Basic Energy Sciences-Materials Sciences under contract #W-31-109-ENG-38.

REFERENCES

[1] T. Kondo, M. Nakamura, K. Niwa and Y. Tomita, Proc. of the 1984 Summer Study on the Design and Utilization of the SSC, edited by R. Donaldson and J. Morfin, p. 612.

[2] T. Kondo, KEK preprint 86-100, to be published in the Proc. of the 1986 Summer Study on the Physics of the SSC.

[3] M.G.D. Gilchriese, Proc. of the 1984 Summer Study on the Design and Utilization of the SSC, edited by R. Donaldson and J. Morfin, p. 607.

[4] K.E. Gray, patent disclosure.

[5] M. Tinkham, Introduction to Superconductivity, (McGraw-Hill, New York, 1975), p. 207.

[6] N.K. Sherman, Phys. Rev. Lett. **8**, 438 (1962).

[7] D.E. Spiel, R.W. Boom and E.C. Crittendon, Jr., Appl. Phys. Lett. **7**, 292 (1965); E.C. Crittendon, Jr. and D.E. Spiel, J. Appl. Phys. **42**, 3182 (1971).

[8] N. Ishihara, R. Arai, T. Kohriki and N. Ujiie, Japan. J. Appl. Phys. **23**, 735 (1984).

[9] K.W. Shepard, W.Y. Lai and J.E. Mercereau, J. Appl. Phys. **46**, 4664 (1975).

[10] J.R. Gavaler, M.A. Janocko, A. Patterson and K.C. Jones, J. Appl. Phys. **42**, 54 (1971).

[11] R.T. Kampwirth and K.E. Gray, IEEE Trans. Magn. **MAG-17**, 565 (1981).

[12] J.W. Ekin, J.R. Gavaler and J. Gregg, Appl. Phys. Lett. **41**, 996 (1982); R.T. Kampwirth, K.E. Gray, D.W. Capone,II, J.M. Murduck and J.F. Lingevitch, Proc. 1987 ICMC.

[13] H.L. Ho, R.T. Kampwirth, K.E. Gray, D.W. Capone,II and L.S. Chumbley, Ultramicroscopy **22**, 297 (1987).

[14] D. Dew-Hughes and R. Jones, Appl. Phys. Lett. **36**, 856 (1980).

[15] H.W. Weber, private communication.

[16] C. Geibel, H. Rietschel, A. Junod, M. Pelizzone and J. Muller, J. Phys. F:Met. Phys. **15**, 405 (1985).

A REVIEW OF SUPERHEATED SUPERCONDUCTING GRANULES AS A DETECTOR FOR DARK MATTER, SOLAR NEUTRINOS, MONOPOLES AND DOUBLE BETA DECAY

KLAUS P. PRETZL

Max-Planck-Institut fur Physik und Astrophysik
Munchen, Germany

1. Introduction

The interest in superconducting devices for particle detection is based on the very small quantum energies involved (it takes about 1meV to break up a Cooper pair) as compared to conventional ionization (~ 20eV) and semiconductor detectors (~ 1eV). Consequently, one hopes to reach very much lower energy thresholds and better energy resolutions with such devices.

In the following, the discussion is focused on superheated superconducting granules (SSG) as a possible particle detector. The idea to use SSG as a particle detector goes back to 1967[1]. Since then, progress in this field has been very slow. This is probably due to the enormous problems one had to face like the production of small size granules (several μm in diameter), the development of sensitive read out electronics, and the necessary extravagant cryogenics. Nevertheless, in recent years this technique has regained considerable interest for the detection of dark matter, solar neutrinos, monopoles, and double beta decay. However, the proposed applications are extremely ambitious, and there is still a long way to go before any of these ideas can be implemented.

This paper is organized as follows: The SSG detector principle is described in chapter 2. Possible applications are discussed in chapter 3. Chapter 4 gives an account of the present state of SSG detector development. Conclusions and a future outlook are presented in chapter 5.

2. Principle of Detection

A SSG detector typically consists of tiny spherical granules (Fig. 1) which are cooled down below their critical temperature T_c and are mounted in an applied magnetic field. The granules are made from type I superconductor material, i.e. they have a superheating (H_{SH}) and supercooling (H_{SC}) phase transition as shown in Fig. 2. The choice of the superconducting material depends on the physical application. In table 1 some of the well known type I superconductors are listed.

The grains serving as target and detector material at the same time are usually embedded in some dielectric material (paraffin) with a typical filling factor of 30% in volume. An interaction of an incoming particle within a grain leads to a temper-

TABLE 1. Some well known type I superconductors with their critical temperature T_c and critical field H_c (at $T = 0$) are listed.

Metal	T_c (K)	H_c (Gauß)
Pb	7.2	803
Sn	3.7	305
In	3.4	293
Re	1.7	200
Al	1.2	99
Ga	1.1	51
Mo	0.92	98
Zn	0.85	54
Zr	0.61	47
Cd	0.56	30
Ru	0.49	69
W	0.015	1

ature change ΔT which can suffice to flip a grain from the superconducting to the normal state. In a uniform heating model the temperature change experienced by the granule is

$$\Delta T \sim \frac{3\,\Delta E}{4\pi \cdot c \cdot \rho \cdot R^3} \tag{1}$$

with ΔE, the energy release of the particle in the grain, c, the specific heat, ρ, the density and R, the radius of the grain. The flip of the granule can be detected by a pick–up loop which measures the flux change due to the disappearance of the Meissner effect (Fig. 3)

$$\Delta\phi \sim \frac{H \cdot R^3}{D} \tag{2}$$

with D, the diameter of the loop, and H, the applied magnetic field. Thus, the average signal voltage is

$$\Delta V \sim \frac{\Delta\phi}{\tau} \tag{3}$$

with τ, the time the external magnetic flux needs to penetrate the grain (the flipping time of the granule). This flipping time is related to the decay time of the eddy–currents in the diamagnetic granule and should follow[2]

$$\tau \sim \frac{R^2}{\rho'} \tag{4}$$

with ρ', the normal state resistivity of the grain material. To our knowledge the flipping time has not yet been experimentally measured. For a better understanding of the flux penetration dynamics this measurement is of great interest.

As an example, Fig. 4 shows a possible pick–up loop arrangement for a SSG detector[3]. The current loops can be made on printed circuit boards with wire distances of the order of 50 μm or less. A layer of granules diluted in paraffin is surrounded by a X,Y–coordinate system of current loops. An event can be localized in space by a coincidence of two loops perpendicular to each other. As explained below, this spatial information will be used in monopole and double beta decay detectors. For a full–size detector, many of these layers can be combined and sandwiched together. The number of read out channels for such a device can be quite large depending on the required sensitivity and spatial resolution of the detector. Considerable progress in the design of electronic read out circuitry has recently been made[4]. With a newly developed wide band amplifier, single flips of granules with diameters of $\leq 10\mu$m were measured.

A SQUID (superconducting quantum interference device) system has recently been employed to measure phase transition signals from 5μm grains[5]. With commercially available dc–SQUID's, read out loops of serveral cm in diameter can be envisaged in the future. It would allow to measure single grain flips within a detector volume of several cm^3. This would simplify the read out problem considerably, at least for detectors where spatial resolution is not of importance (dark matter, reactor neutrino and neutral current detectors).

Minimum ionizing particles can be recognized by causing many granules to flip. Their energy loss in the detector is proportional to the number of flipped grains.

SSG detectors with spatial and time resolution can be used as real time tracking devices (monopole experiments).

Another principle using SSG has been proposed by A. Drukier and L. Stodolsky[6] for the detection of neutrinos. In this case the neutrino is measured by its neutral current interaction with a nucleus inside a granule (coherent neutrino nucleus scattering). The nuclear recoil energy transformed into heat can cause a phase transition of the granule. In this application the neutrino, as distinct from a minimum ionizing particle, would only flip one single granule. This signature can be used for background rejection of cosmic rays and natural radioactivity in the detector material.

The cross section for coherent neutrino nucleus scattering can be calculated from the standard electroweak model:

$$\frac{d\sigma}{d(cos\theta)} = \frac{G_F^2}{8\pi}\left[Z\left(4\,sin^2\theta_w - 1\right) + N\right]^2 E^2\left(1 + cos\,\theta\right) \quad (5)$$

with Z, the number of protons, and N, the number of neutrons in the nucleus. Small incoherent contributions due to the axial vector current interactions for nuclei with spin are neglected in (5). For $sin^2\theta_w \simeq 0.22$ (with θ_w, the weak interaction Weinberg angle) and $G_F = 1.166 \cdot 10^{-5}\ \mathrm{GeV}^{-2}$ (the weak interaction Fermi coupling constant) the cross section

$$\sigma \simeq 2.5 \cdot 10^{-18}\,\frac{N^2E^2}{A}\left[\frac{cm^2}{kg}\right] \quad (6)$$

is essentially proportional to N^2 and E^2, the energy of the incident neutrino measured in MeV. The validity of (5) and (6) applies to the coherence condition $R_A \cdot \Delta \leq 1$ where R_A is the radius of the nucleus and Δ the momentum transfer of the neutrino to the nucleus. For Sn granules it is fulfilled for neutrino energies up to $E \simeq 40$ MeV.

The advantage of this method is twofold:

1. The cross section for neutrino nucleus scattering is about a factor 1000 larger as compared to inverse beta decay reactions[7] (see also chapter 3.2.). Thus, a SSG detector with an effective mass of about 100 kg will measure the same solar neutrino event rate as a multiton detector based upon inverse beta decay reactions.

2. The SSG detector is sensitive to all neutrino flavors (neutral current reactions). However, the principal difficulty with this method is the detection of the small nuclear recoil energy. This will be discussed further in chapter 3.

The transformation mechanism of energy into heat inside a grain is experimentally unexplored and will be one of the main subjects to be investigated. A whole catalogue of questions needs to be studied in the future: Are thermal (global heating) or nonthermal effects (local effects on the surface of the grain) or both responsible for the phase transition of a grain? How much energy will be used up by permanent lattice dislocations in the grain material? What happens with lattice defects at very low temperatures? Do they release enough energy to flip a grain? Do granules with identical dimensions flip at the same energy threshold? Do neighboring granules influence each other diamagnetically? Some of these effects have already started to be investigated and experimental results are discussed in chapter 4.

An SSG detector measures in principle only an energy threshold. However, it has been speculated whether it is also possible to extract an energy information from a phase transition of a granule by using the so-called "flip-flop"[6] effect. This effect works as follows: The grain sits in the phase diagram at a temperature $T(< T_c)$ and an applied field $H(< H_{SC})$ (Fig. 2). Due to the energy transfer from an interacting particle the grain heats up, becomes metastable and eventually flips into the normal state. After this, it loses its heat slowly to the surroundings (due to the Kapitza resistance effect). By choosing the appropriate surrounding material, the cooling time of the grain can be adjusted to be longer than its flipping time. After cooling down, it flops back into the superconducting state. Thus, the measurement of the flip-flop time is a measurement of the energy deposited inside the grain. This flip-flop effect may be realizable for grains with small difference between superheating (H_{SH}) and supercooling (H_{SC}) fields. For example, by adding impurities to the superconducting material, the field penetration depth can be enlarged and, consequently, the metastability of the grain reduced (see Ref. 8 and chapter 4).

3. Proposed Applications

3.1. Dark matter detection

The first hint for the existence of invisible matter came from measurements performed by F. Zwicky[9] in 1933. He determined the kinetic energy of galaxies in large galactic clusters using red shift analysis. His measurements showed that the kinetic energy of the galaxies is much larger than expected from the virial theorem ($E_{kin} = \frac{1}{2} E_{pot}$ for a mechanically stationary system), indicating the existence of large amounts of invisible matter. More recently, measurements of the rotation velocity of visible matter in spiral galaxies have shown that the velocity stays constant at large distances from the galactic center in a region where the visible matter density is decreasing[10](Fig. 5). These measurements (more than 60 galaxies have been investigated), originally performed with visible light, have also been confirmed with 21 cm radiowaves[11]. The latter technique allowed to extend the measurements to distances more than twice the visible radius of the galaxies. These results are in apparent conflict with Kepler's law and call for large amounts of invisible matter in galaxies. Approximately a factor 10 more matter than visible is needed to explain the rotation velocity results.

What is dark matter? Is it of baryonic or non–baryonic nature? The strong constraints put on the baryon density in the Universe by the observed helium and deuterium abundance and the standard model of nucleo–synthesis disfavor a large amout of invisible baryonic matter. The baryon density in the Universe cannot account for more than 4% of the density that would be required for closing the Universe. Thus, the search for dark matter concentrates more on non–baryonic particle candidates such as massive neutrinos, axions and particles like photinos and sneutrinos postulated by supersymmetric theories. For more details on the subject of dark matter the reader is referred to some excellent review articles[12].

Weakly interacting dark matter particles can be detected with SSG using the coherent scattering method. Goodman and Witten[13] have calculated the event rates per kg per day with various detector materials and various mass values m_x of the unknown dark matter particle. Their results, listed in table 2, show that an SSG detector with an effective mass of 1 kg is quite sensitive to particles within a large mass range.

TABLE 2. Event rates in $\mathrm{kg}^{-1}\,\mathrm{day}^{-1}$ *for coherent weak interactions with various SSG detector materials and various values of the mass* m_x *of the unknown particles. A correction to account for the loss of coherence at high momentum has been included (taken from Ref.* 13*).*

m_x(GeV)	Al	Ge	Sn	Pb
1	27	90	170	370
10	150	700	1400	3.4×10^3
10^2	120	1400	4700	6.0×10^3
10^3	18	340	1700	6.0×10^2
10^4	2	40	200	58
10^5		4	21	5.8
10^6			2	0.6

The maximum nuclear recoil energy needed to flip a grain is

$$E_{max}^{recoil} = \frac{2\,m_x^2\,M}{(m_x + M)^2} \cdot v^2 \geq E_{thresh} \tag{7}$$

with M, the mass of the nucleus in the grain, and v, the velocity of the dark matter particle. Assuming Sn grains and $v/c \sim 10^{-3}$, for a typical velocity of objects in our galaxy the recoil energies range from 16 eV to 50 keV for particles with 1 GeV to 100 GeV mass, respectively. As can be seen, the low mass cut–off depends strongly on the minimum threshold energy E_{thresh} of the SSG detector.

The other prominent dark matter candidate, the photino, interacts with nuclei via spin dependent forces. Thus, grain material with a large nuclear spin (Al or Ga) should be chosen. A list of suitable materials with their quality factors is given in Ref. 14. By using SSG detectors with different materials, one can distinguish between spin–dependent (photinos) and spin–independent (massive neutrino, sneutrino) interactions of dark matter particles. The event rate due to photino interactions in a detector using Ga grains is 100 times larger than in one using Zn grains, for example. The estimated event rate for photino masses of $\sim$ 2 GeV is 0.1 – 1 events/kg day.

The mass of the dark matter particle can be estimated from (7) by changing the threshold energy of the detector and assuming a fixed value of v (the rotation velocity of objects in our galaxy, for example). Compared to the large expected event rates from the dark matter particles, the background from solar neutrinos and neutrinos from nearby nuclear power stations is small[15]. An additional significant signature of dark matter detection is an expected signal modulation due to the motion of the earth around the sun[14] (Fig. 6).

Although many fundamental problems still have to be solved, a 1 kg SSG detector with SQUID read out can perhaps be aimed for in the not too far future. A working device of this kind would be able to set interesting limits on dark matter particle fluxes.

3.2. Solar neutrino detection

Davis and his collaborators[16] found that the solar neutrino flux is about a factor three smaller than estimated from the standard solar model[17]. This discrepancy between theory and experiment is known as the so–called "solar neutrino puzzle." It caused a variety of speculations about an explanation.

A mechanism has been proposed by Mikheyev and Smirnow[18] which adiabatically converts electron neutrinos ν_e, created in the interior of the sun, into muon neutrinos ν_μ through charged current interactions with electrons of high density. Since the Davis experiment was only sensitive to electron neutrinos which were measured via the inverse beta decay reaction

$$\nu_e + Cl^{37} \rightarrow Ar^{37} + e^- \tag{8}$$

electron neutrinos converted into muon neutrinos would remain undetected in this experiment. Thus, a neutral current SSG detector being sensitive to all neutrino flavors would be of great importance for the investigation of the above mechanism.

Other possible explanations for the solar neutrino puzzle include the decay of massive neutrinos[19], the effect of a neutrino magnetic moment[20], and a possible reduction of the interior solar temperature due to dark matter particle interactions in the sun[21].

A new solar neutrino experiment (GALLEX)[22] is planned for the Gran Sasso

Laboratory (Italy) employing about 60 tons of Gallium in the reaction

$$\nu_e + Ga^{71} \rightarrow Ge^{71} + e^- . \tag{9}$$

Like the Davis experiment, it is a pure neutrino counting experiment based upon radiochemical techniques. Since the GALLEX experiment has a lower neutrino energy threshold than the Davis experiment, it has the advantage of being sensitive to the most abundant neutrinos from the $p+p \rightarrow D + e^+ + \nu_e$ processes in the sun (Fig. 7). However, in both experiments the neutrino energy cannot be measured and, therefore, no detailed information on the solar neutrino energy spectrum can be obtained.

A rather ambitious SSG experiment

$$\nu_e + In^{115} \rightarrow Sn^{115} + e^- + \gamma + \gamma \tag{10}$$

using Indium granules has recently been proposed[4]. This reaction was originally advertised by R.S. Raghavan[23] because it not only has a low neutrino threshold energy of $E_\nu = 128$ keV but also offers a possibility to measure the solar neutrino energy spectrum. The neutrino energy can be extracted from the measurement of the recoil e^- spectrum. One of the problems is the rejection of radioactive background from other Indium isotopes. Monte Carlo studies have shown that an In SSG detector needs an energy resolution of 50 keV around 100 keV and 100 keV around 500 keV and a spatial resolution of $\Delta V = 1\text{cm}^3$ to obtain a signal–to–background ratio of 10. Four tons of In grains are necessary to yield one solar neutrino event per day. To obtain the required energy resolution, grains with a size of about 2μm in diameter have to be used. The 4 ton detector has to be cooled down to $T = 200$mK. Considering the technological effort to fabricate 4 tons of granules with identical dimensions, to cool them to 200 mK, and to electronically read them out, puts the hopes for an implementation of such a project into the very far future.

Because of the larger cross section, a SSG detector based upon coherent neutrino scattering can be of much smaller weight. The expected solar neutrino event rates for a neutral current SSG detector are given in table 3.

Using SQUID read out, the number of read out channels can also be kept to a manageable quantitiy. The principal difficulty of this method is the detection of the

TABLE 3. Expected solar neutrino event rates for a neutral current detector for various SSG materials assuming a neutrino flux of $2 \times 10^{10}/\mathrm{cm}^2$ *sec for* $E_\nu > 300$ *keV*

	Al	Cd	Sn	Pb
events/kg year	1.6	9	10	19

very small nuclear recoil energy E_A. The average recoil energy

$$\overline{E}_A = \frac{2}{3}\,\frac{E_\nu^2}{A}[\mathrm{keV}] \tag{11}$$

with E_ν, measured in MeV, and A, the atomic weight of the grain material, is calculated for various neutrino energies and grain materials and listed in table 4.

TABLE 4. The average nuclear recoil energy $\overline{E}_A$ *in eV is listed for various grain materials and different neutrino energies* E_ν.

E_ν (MeV)	Al	Cd	Sn	Pb
0.3	2.25 eV	0.6 eV	0.5 eV	0.3 eV
0.4	4 eV	1 eV	0.9 eV	0.5 eV
1.0	25 eV	6 eV	5.6 eV	3.2 eV
3.0	222 eV	53 eV	50 eV	29 eV
8.0	1578 eV	377 eV	355 eV	206 eV

When choosing the detector material, one has to compromise between a cross section increasing with N^2 and a recoil energy decreasing with A. Assuming global heating, the sensitivity in detecting very small recoil energies depends on the specific heat and the size of the grains. For example, Sn grains of 2μm diameter have to be cooled down to $T = 50$mK to be sensitive to solar neutrinos with $E_\nu = 400$ keV. The fundamental problems to be solved before the realization of a 100 kg detector are still enormous. However, a neutral current detector being sensitive to all neutrino flavors would be an interesting complementary device to the GALLEX detector which will soon be set up in the Gran Sasso Laboratory.

3.3. Monopole Detection

In 1931 Dirac proposed that monopoles with a magnetic charge of

$$g = \frac{\hbar c}{2e} = \frac{137e}{2} \tag{12}$$

might exist. Recently, the interest in monopoles was revived since their existence is a natural consequence of the basic ideas of Grand Unified Theories (GUT). They are predicted to have a mass $M \sim 10^{16}$ GeV/c^2 near the grand unification mass and velocities of $10^{-3}c$. They have been searched for in a number of experiments[24] which either looked for a flux change $\Delta\phi = 4\pi g$ in a superconducting coil or tried to detect the ionization of atoms caused by through–passing monopoles. Assuming the above monopole mass and velocity and a uniform monopole density equal to the critical density to close the Universe one can derive an upper limit on the monopole flux of $\sim 6 \times 10^{-15}$ cm^{-2} sr^{-1} s^{-1}. Another limit, the so–called "Parker limit", with $\sim 10^{-15}$ monopoles cm^{-2} sr^{-1} s^{-1}, gives the maximum tolerable monopole flux for not destroying the galactic magnetic field of 3 – 5 μ Gauß. The lowest experimental upper limit of the order of the Parker limit has been obtained by the Baksan experiment[25].

A SSG monopole detector has been proposed by Gonzales–Mestres and Perret–Gallix[3]. The flipping mechanism caused by a monopole interaction with the granule is not due to a thermal effect in this case but rather due to a surface effect. The monopole when passing through a granule leaves behind a flux tube $\phi = 2\phi_0$ (with $\phi_0 = \frac{hc}{2e} = 2.067 \cdot 10^{-7}$ Gauß cm^2, the flux quantum) about the size of a Cooper pair coherence length. The flux tube creates a nucleation center at the surface of the grain which breaks the superconductivity and causes the grain to flip. Since this flipping mechanism is a surface effect, it is independent of the grain size. Large grains can be used in such a detector, which have the advantage of delivering large signals. The use of large grains also provides an effective background rejection. The energy needed for a minimum ionizing particle to flip a grain is very much larger than the energy needed for a monopole to flip a grain.

For a given time resolution of the detector the speed of a slow moving monopole could in principle be determined from the measurement of the monopole crossing at several well spaced detector planes. In this case the SSG detector could be used as a real time track detector. However, when using large grains their flipping time (4) and consequently the time resolution of the detector is slow. Perhaps the flipping time of large size grains can be shortened by adding small amounts of impurities to the grain material.

To obtain 1 monopole event per year at the Parker limit, 3000 m^2 sr of SSG detector planes are necessary. Considering the gigantic technical and cryogenic problems still to be solved, an SSG monopole detector has a difficult time competing with the MACRO Detector[26] (with a sensitivity of 4 events per year at the Parker limit and $\sim$ 12000 m^2 sr detector planes) shortly to be in operation in the Gran Sasso Laboratory.

3.4. Double beta decay detection

Up to now we do not know whether the neutrino and the antineutrino are different (so–called Dirac neutrinos) or identical particles with different helicity states (so–called Majorana neutrinos) (helicity is the projection of the spin on the direction of the particle motion). The experimental investigation of this problem has been concentrated in the last years on the search for neutrinoless double beta decay reactions ($\beta\beta\, \emptyset\, \nu$) of the type

$$(A, Z) \rightarrow (A, Z + 2) + 2e^- \tag{13}$$

As shown in Fig. 8a this lepton number violating process can only occur if the neutrino is a Majorana particle. Lifetime limits of larger than $5.3 \cdot 10^{23}$ years have recently been established[27] for the neutrinoless double beta transition $Ge^{76} \rightarrow Se^{76}$.

Pacheco and his coworkers[28] have investigated the possibility of using SSG for the detection of double beta decay reactions. They have found a number of interesting candidate materials for the grains with superconducting properties: Zr^{96}, Mo^{100}, Cd^{106}, Cd^{116} Sn^{124}, Ru^{96}. In order to discriminate $\beta\beta\, \emptyset\, \nu$ from $\beta\beta\, 2\nu$ (Fig. 8) reactions, one needs to measure the electron energies with high accuracy. They have performed a Monte Carlo simulation of the transport of electrons, positrons and bremsstrahlungs photons generated either in the $\beta\beta\, \emptyset\, \nu$ or $\beta\beta\, 2\nu$ case on the example of ^{124}Sn. The results show (Fig. 9) that, in addition to a very small grain size ($\sim 1\mu$m), a very good spatial resolution of the detector is needed to distinguish genuine ($\beta\beta\, \emptyset\, \nu$) from the ($\beta\beta\, 2\nu$) background. However, the fundamental problem with double beta decay experiments is the suppression of background radioactivity in the material. It remains to be seen whether the required energy resolution can be obtained with SSG to be competitive with other techniques like Ge–detectors for example.

4. Status of Detector Development

In this chapter a short account of the latest developments in detector studies is given. For more details the reader is referred to Ref. 29 or to the contributions of this conference.

4.1. Production of the Granules

Today it is possible to industrially fabricate large quantities of granules[30]. The production technique is similar to the one used in commercial humidifiers. A thin film of liquified grain material is brought onto the surface of an ultrasonic transducer head. The granules evaporate from the head with an average diameter of

$$\overline{d} \sim \left(\frac{\sigma}{\rho}\right)^{1/3} f^{-2/3} \qquad (14)$$

with f, the operation frequency of the ultrasonic transducer, σ, the viscosity, and ρ, the density of the grain material. They obtain their spherical shape from the surface tension which they experience during their flight and cool down time in an inert gas before being collected in a funnel. The sphericity and surface quality of these grains can be seen in Fig. 1. However, the grain diameter distribution is very large ($\frac{\Delta d}{d} \sim 50\%$ full width) and requires further grain size selection after the fabrication of the grains. Several possibilities are conceivable like molecular sieves, sedimentation or mass spectroscopic selection. However, no technically sound solution has been found so far which would select small size grains in large quantities with a diameter variation of a few percent. Up to now small grains with $\overline{d} \sim 25\mu m$ have been industrially fabricated and there is hope that grains with $\overline{d} \sim 5\mu m$ can be produced in the near future.

4.2. Superconducting properties of granules

Early measurements with Sn and Cd granules[31] in a varying external magnetic field have shown a washed out phase transition $\frac{\Delta H_{SH}}{H_{SH}} \sim \frac{\Delta H_{SC}}{H_{SC}} \sim 20-30\%$ [Fig. 10]. The effects leading to such a behavior were recently studied in experiments with individual granules[32]. The experiment was performed with Sn granules of various diameters $20\mu\text{m} \leq d \leq 112\mu\text{m}$ in a temperature range $1.4\ \text{K} \leq T \leq 3.26$ K. The granules were rotatable around an axis perpendicular to the applied magnetic field. By cycling the field at fixed temperature, H_{SH} and H_{SC} have been measured as a function of the rotation angle (Fig. 11). It was found that all measured grains exhibited variations $\frac{\Delta H_{SH}}{H_{SH}}$ and $\frac{\Delta H_{SC}}{H_{SC}}$ which were approximately of the same magnitude as the washed out phase transitions found in the early measurements with groups of grains.

Another effect contributing to the smearing of the phase transition has been discovered by the MPI Munich group[32] on measurements with single grains. For some grains, flip signals were observed which were very much smaller than expected for a phase transition of an entire grain. It was also found that the superheating phase transition occurred successively with increasing external field leading to several H_{SH} field values for one grain. These observations are consistent with what one would expect if the granules were subdivided into superconducting and normal zones (formation of an intermediate state). In these cases, the superconductivity of the grain is only partially broken, up to the point where $H_{\text{eff}} < H_c$ (with H_{eff} being the effective field at the equator of the grain where the field lines are compressed, Fig. 3). Since the probability of the intermediate state formation is decreasing with decreasing grain size, this effect may no longer be of importance for very small grains.

Nucleation centers due to surface defects or internal structures (crystalline structure of granules) can be made responsible for the observed H_{SH} and H_{SC} variations as a function of the rotation angle. An analysis of the crystalline structure of the grains showed several monocrystalline regions with random orientations[33]. Experiments with single superconducting Sn crystals have shown that the field penetration depth exhibits a considerable anisotropy with respect to the crystallographic orientations[34]. Furthermore, since the supercooling phase transition (being insensitive to surface defects) shows a similar variation as the superheating phase transition, the crystalline structure is probably the dominant effect leading to the above behavior. It is conceivable that by adding some impurities the crystal formation in the grain can be influenced such that the variation of the phase transition diminishes (many nucleation centers in one grain smoothen the phase transition variations[8]).

The understanding of these problems and their solution are of utmost importance for the construction of a SSG detector with uniform response and good energy resolution.

While studying the behavior of SSG at lower temperatures (T<350 mK), the TU–Garching group[35] discovered an instability in the phase transition for Cd grains. By lowering the temperature, they found that a larger and larger fraction of the grain sample undergoes a phase transition at the same external magnetic field (Fig. 12). Diamagnetic influence of neighboring granules cannot account for this effect since the used SSG sample had a very low filling factor (~ 3.4 %). This so-called "avalanche effect" may be due to the release of latent heat during the phase transition of a grain. If this heat is large enough to induce flips of neighboring grains, a chain reaction starts leading eventually to a complete phase transition of the entire grain sample. Attempts to reproduce this effect with Sn granules have failed so far[36]. In any case, this behavior needs further study since it may constrain seriously the use of SSG at very low temperatures.

Calculations[35] based on a uniform heating model for various grain materials have shown that highest sensitivities can be obtained with low T_c materials (Cd or W) and operation temperatures near T_c. Under these conditions, however, the external magnetic field is very low and consequently the phase transition signal small. This underlines again the necessity to develop very sensitive read-out electronics.

4.3. Irradiation experiments

Considerable effort has been made to study the behavior of SSG when irradiated by external sources. These studies range from electron beam experiments[37] to α, β, γ ray detection[5,32,35,36] using radioactive sources. The aim of these experiments was to test the sensitivity of SSG and to investigate the relevant heating processes (global heating versus local effects) in the grain. One of the difficulties with these measurements is that the amount of energy lost by a particle in a grain is unknown. Even with a monoenergetic α–source (the absorption length of a 5.5 MeV α–particle in Sn is $\lambda_a \sim 14\mu$m) the particles lose some energy on their way to the grain traversing liquid helium ($\lambda_a \sim 300\mu$m) and the support material of the grain.

From the many irradiation experiments only two examples are discussed here:

1. Sn granules with diameters 10μm $< d < 25\mu$m were irradiated with a Fe^{55} (E_γ = 6 keV) source[36]. Near the phase boundary a certain energy deposition ΔE in the granules leads to a temperature rise ΔT which corresponds to a magnetic field jump ΔH (Fig. 2). Thus, the sensitivity of SSG can be expressed as $\Delta H/H$. The dependence of the number of grain flips on $\Delta H/H$ after 1 hour of γ irradiation is shown in Fig. 13. In a global heating calculation the value $\Delta H/H \simeq 0.05$ corresponds to $\Delta E = 15$ keV for 10μm grains and to $\Delta E = 250$keV for 25μm grains. Since the observed sensitivity of the grains is larger than expected for a global thermalization of 6 keV, it is conceivable that a mechanism other than global heating also takes place. A similar behavior has been observed when irradiating the same sample with other sources like Tc^{99} ($E_\gamma = 140$ keV), Cl^{36} ($E_\beta < 714$ keV), and Am^{241} ($E_\alpha = 5.5$ MeV). It is very difficult in these experiments to obtain information on the flip efficiency of the granules.

2. Individual Sn grains with 56μm, 100μm, and 30μm diameter have been irradiated with an Am^{241}–source[32]. As schematically shown in Fig. 14, the measurement was performed at various irradiation angles. For mono–energetic α's, a sharp peak at $H_{SH} - \Delta H$ would be expected. However, because of the finite energy loss of α's in the helium bath and non–central collisions with the grain (only for "head on" collisions the α is totally absorbed in the grain; if the α hits the grain more peripherally, for example, only part of the α energy is absorbed in the grain), this distribution is broadened. The observed change of the center of gravity of the distributions at 80° to lower values of H (Fig. 14) indicates that the grain is more sensitive if an α–particle hits near the equator (80°), where H_{eff} is largest, than at the pole (0°). It was also found in this experiment that ΔH did not vary with the grain size like $\Delta H \sim 1/R^3$ as expected from global heating.

These observations suggest that a local process takes place leading to a fast break up of Cooper pairs, which apparently occurs before the grain is globally heated by phonons. Since the energies involved are well above the ionisation energy of the grain material, lots of electrons will be released which break up Copper pairs and destroy locally the superconductivity of the grain. This effect makes SSG more sensitive for particle detection than expected from pure thermal considerations. However, in coherent neutrino scattering the recoil energies can be well below the ionization energy in which case the global heating model may apply. Future SSG tests with reactor neutrinos will hopefully clarify these points.

5. Conclusions

Low temperature detectors will be an important new tool in future astro–particle physics research. The described SSG is just one of many other possible low temperature devices presently under study[29] (like bolometers, ballistic phonons in connection with tunnel junctions and trapping, transition edge thermometers and rotons in superfluid helium). The future of SSG depends strongly on the progress towards solving the related problems of homogeneity of response and energy resolution problems. New ways have to be found to produce grains with either very many or no nucleation centers at all. Perhaps the adding of small amounts of impurities to the grain material could play an important role. This clearly needs further investigation. Furthermore, the behavior of very small size ($< 5\mu$m) granules is still unexplored. For this more sensitive electronics have to be developed. However, SQUID systems being sensitive to single flux quanta are very expensive, but useful devices for this exploration. A future technological challenge will be the fabrication of small and equal size granules. If all these problems can be solved in the near future, the realization of a prototype SSG detector (about 1 kg of grain material) is in sight. This prototype could, for example, be tested at a reactor station with a high intensity neutrino flux. It also could be used as a dark matter detector in an underground laboratory.

REFERENCES

1. H. Bernas et al. Phys. Lett. A, 24 (1967) 721 and C. Valette: Thesis Orsay (1971)
2. T.E. Faber: Proc. R. Soc. London Ser. A, 219 (1953) 75 and D. Huebner: Thesis Orsay(1980)
3. L. Gonzales–Mestres and D. Perret–Gallix: Il Nuov. Cim. Vol.9 (1986) 573
4. A. de Bellefon et al., Report GPS–ENS 87 007 (1987) Faisabilite de la detection de neutrinos solaires aveç l'indium
5. A. Kotlicki et al: Proceedings of the workshop on Low Temperature Detectors for Neutrinos and Dark Matter, at Ringberg Castle 1987, edited by K. Pretzl, N. Schmitz, L. Stodolsky (published in Springer Verlag Heidelberg), page 37
6. A. Drukier and L. Stodolsky: Phys. Rev. D30 (1984) 2295
7. J. N. Bahcall: Rev. Mod. Phys. Vol.50 (1978) 881
8. see P. Freund these proceedings
9. F. Zwicky: Helv. Phys. Act. 6 (1933) 110
10. V.C. Rubin et al., Astrophys. J. Lett. 225 (1978) L107, and V.C. Rubin Proceedings of the 117th Symposium of the international astronomical union held in Princeton, New Jersey (1985), page 51
    V.C. Rubin: Scientific American (June 1983), page 88
11. R. Sancisi and T.S. van Albada: Proceedings of the 117th Symposium of the international astronomical Union held in Princeton, New Jersey (1985), page 67

12. J.R. Primack, Proceedings of the second ESO–CERN Symposium (1986), page 193
    L.M. Krauss, Scientific American (Dec. 1986) page 50
    J. Silk, Proceedings of the 23 International Conference on High Energy Physics in Berkeley Vol. 1 (1986) 465

13. M.W. Goodman and E. Witten: Phys. Rev.D. Vol 31. No. 12 (1985) 3059

14. A. Drukier et al. Phys. Rev.D. Vol. 33, No. 12 (1986) 3495

15. P.O. Lagage Nature Vol. 316 (1985) 420

16. R. Davis et al. Phys. Rev. Lett. 20 (1968) 1205
    J.N. Bahcall and R. Davis in Essays in Nuclear Astrophysics edited by Ch.A. Barnes et al. (Cambridge Union Press, Cambridge, England 1982) page 243

17. J.N. Bahcall Rev. Mod. Phys. Vol. 50 (1978) 881
    J.N. Bahcall et al. Rev. Mod. Phys. Vol.54 (1982) 767

18. S.P. Mikheyev and A.Y. Smirnov, Nuovo Cim. C9 (1986) 17
    H.A. Bethe, Phys. Rev. Lett. 56 (1986) 1305

19. J.N. Bahcall, N. Cabibbo and Y. Yahil, Phys. Rev. Lett. 28(1972) 316

20. V.B. Voloshin, M.I. Vysotskii and L.B. Okun, Sov. Phys. JEPT 64(3) (1986) 446

21. J. Falkner, Moriond Proceedings on massive neutrinos, edited by J. Tran Than Van (1987)

22. T. Kirsten et al. Moriond Proceedings on massive neutrinos, edited by J. Tran Than Van (1986), page 119

23. R.S. Raghavan, Phys. Rev. Lett. 37 (1976) 259

24. G.E. Groom, Physics Reports 140 (1986) 323
    G. Giacomelli, La Riv. del. Nuovo Cim. Vol.7 (1984) 1

25. E.N. Alexeyev et al.: Proceedings of the 19th International Cosmic Ray Conf., La Jolla (1985) page 250

26. MACRO: Il. Nuov. Cim. 9c, No.2 (1986) 281

27. D.O. Caldwell, Moriond Proceedings on New and Exotic Phenomena, edited by J. Tran Than Van (1987)

28. A.F. Pacheco et al.: Proceedings of the workshop on Low Temperature Detectors for Neutrinos and Dark Matter, at Ringberg Castle 1987, edited by K. Pretzl, N. Schmitz, L. Stodolsky (published in Springer Verlag Heidelberg), page 65

29. Proceedings of the workshop on Low Temperature Detectors for Neutrinos and Dark Matter, at Ringberg Castle 1987, edited by K. Pretzl, N. Schmitz, L. Stodolsky (published in Springer Verlag Heidelberg)

30. EXTRAMET, Zone Industrielle de Crauves–Sales, F–74380 Bonne–sur–Menoge (Haute–Savoie) France

31. F.v. Feilitzsch et al.: Il. Nuov. Cimento, Vol. 9C, N2 (1986) 598

32. K. Pretzl et al.: Proceedings of the workshop on Low Temperature Detectors for Neutrinos and Dark Matter, at Ringberg Castle 1987, edited by K. Pretzl, N. Schmitz, L. Stodolsky (published in Springer Verlag Heidelberg), page 30
see also J. Feder and D.S.Mc Lachlan, Phys.Rev. 177, (1969) 763

33. J. Paul and S.Mc Kay: Internal Report, Max–Planck–Institut für Metallforschung, Stuttgart (1986)

34. P.C.L. Tai et al.: Phys. Rev. B Vol. 11, N1 (1975) 411

35. W. Seidel et al. Rev. Sci. Instr. 58(8) (1987) 1471

36. L. Gonzales–Mestres and D. Perret–Gallix: Proceedings of the workshop on Low Temperature Detectors for Neutrinos and Dark Matter, at Ringberg Castle 1987, edited by K. Pretzl, N. Schmitz, L. Stodolsky (published in Springer Verlag Heidelberg), page 9

37. A. de Bellefon et al.: Proceedings of the workshop on Low Temperature Detectors for Neutrinos and Dark Matter, at Ringberg Castle 1987, edited by K. Pretzl, N. Schmitz, L. Stodolsky (published in Springer Verlag Heidelberg), page 59

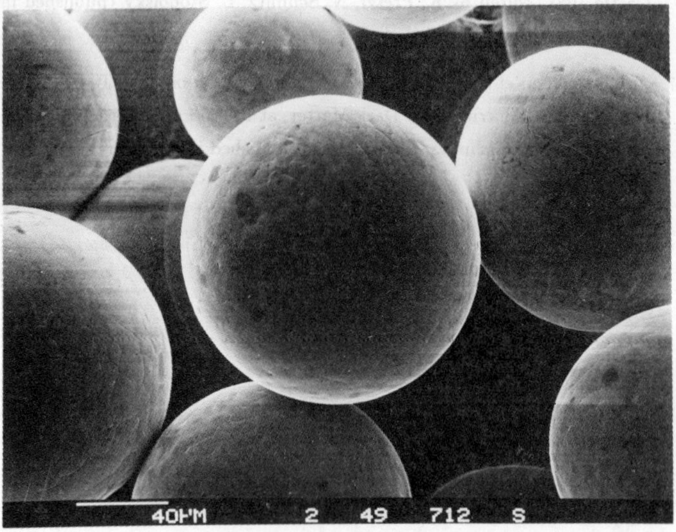

Fig. 1 : Tin granules (~ 120μm) as shown by an electron microscope.

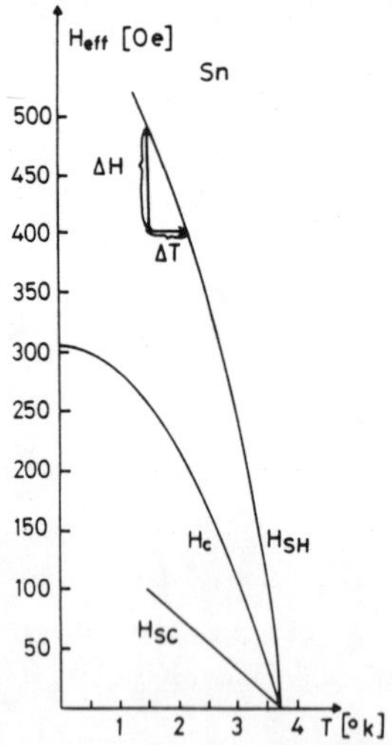

Fig. 2 : Phase diagram of tin granules with H_{SH} = superheating field, H_{SC} = supercooling field, H_c = critical thermodynamic field.

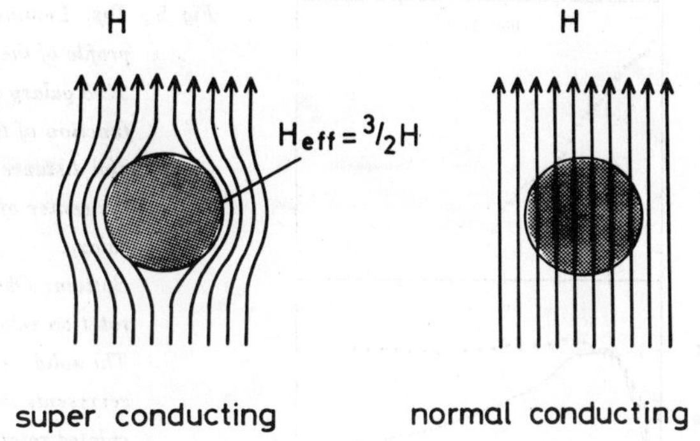

Fig. 3 : Meissner effect of a superconductor.

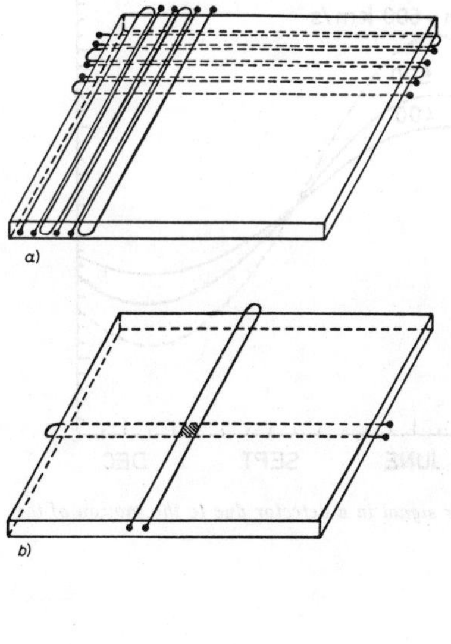

Fig. 4 : a.) A foil of grains surrounded by a (X, Y)-coordinate system of pick up loops.
b.) A localized event detected by a coincidence of two loops perpendicular to each other.

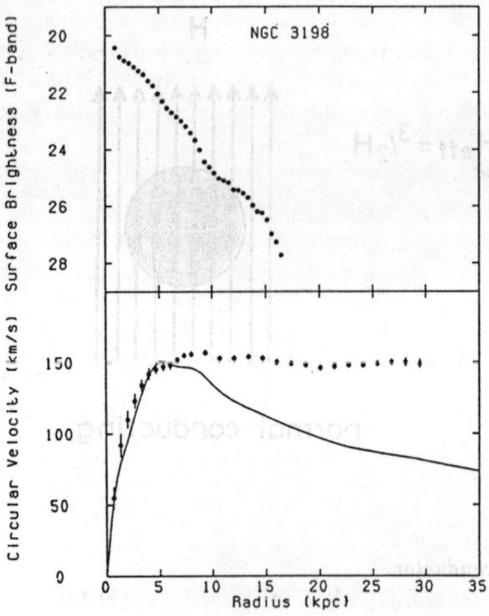

Fig. 5 : Top: Luminosity profile of the NGC 3198 galaxy as a function of the radial distance from the center of the galaxy. Bottom: Observed rotation velocities. The solid curve represents the calculated rotation velocity from the luminous matter profile.

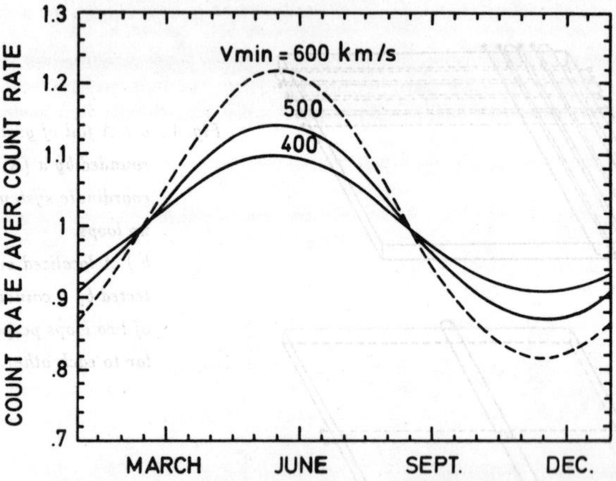

Fig. 6 : Modulation of the dark matter signal in a detector due to the motion of the earth around the sun.

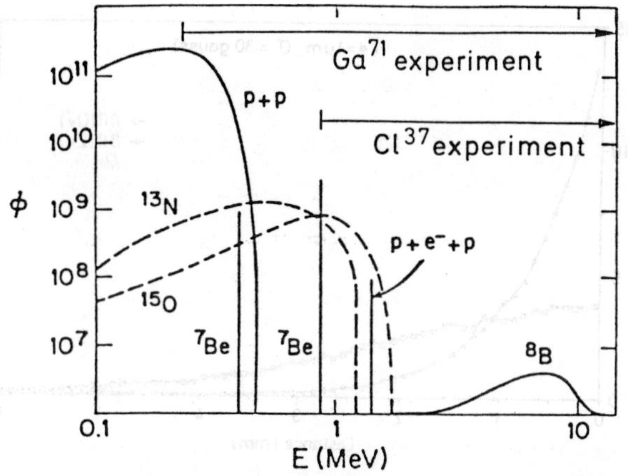

Fig. 7 : Solar neutrino spectrum. The energy range covered by the GALLEX (Ga^{71}) and the Davis (Cl^{37}) experiments are also shown.

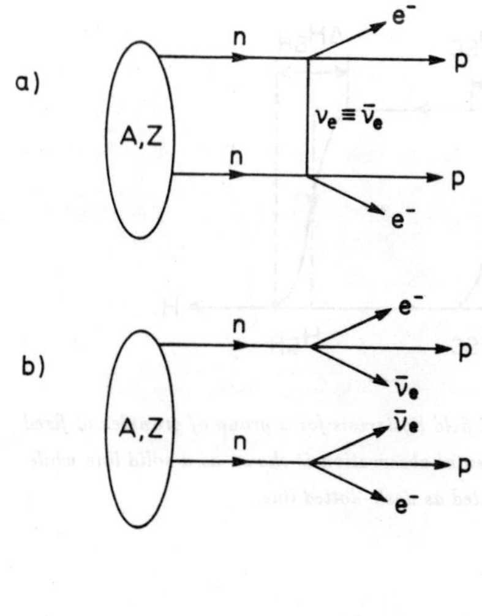

Fig. 8 : a.) Neutrinoless double beta decay. b.) Two neutrino double beta decay.

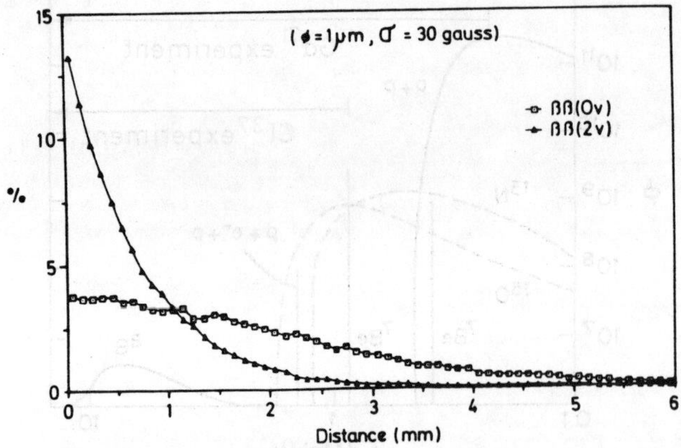

Fig. 9 : The spatial distribution of flipped granules for the $\beta\beta$ $\not{0}$ ν and $\beta\beta$ 2 ν decay modes are shown.

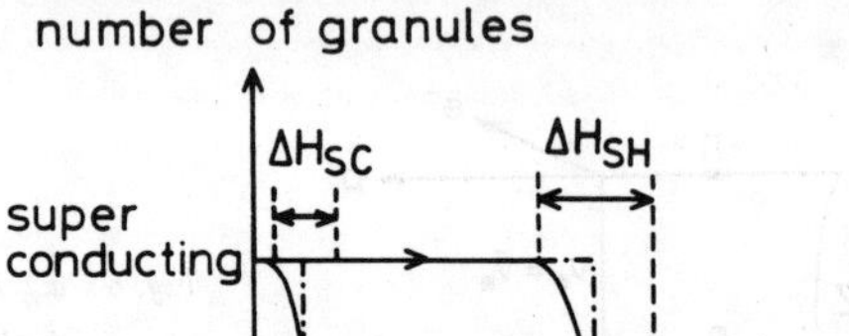

Fig. 10 : Schematic view of external field hysteresis for a group of granules at fixed temperature. The experimental observation is shown as a solid line while the ideal behavior is indicated as dash–dotted line.

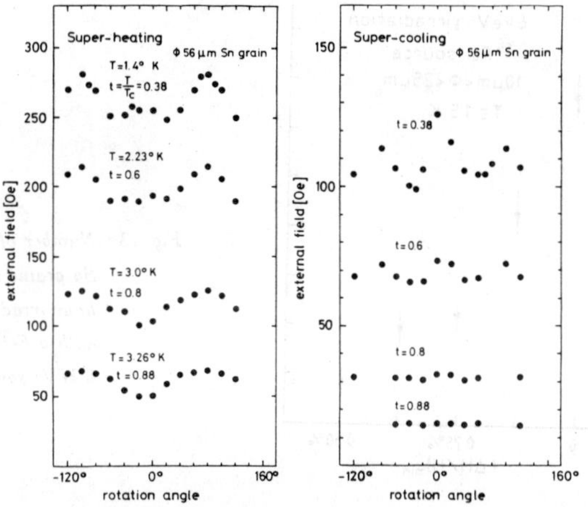

Fig. 11 : Superheating and supercooling field for a Sn grain with 56 μm diameter at various temperatures.

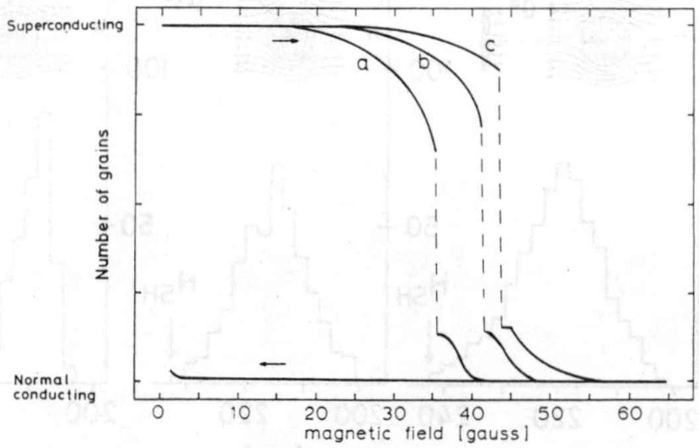

Fig. 12 : Hysteresis curves for Cd grains showing an avalanche effect at different temperatures; a.) T = 300 mK b.) T = 200 mK and c.) T = 135 mK.

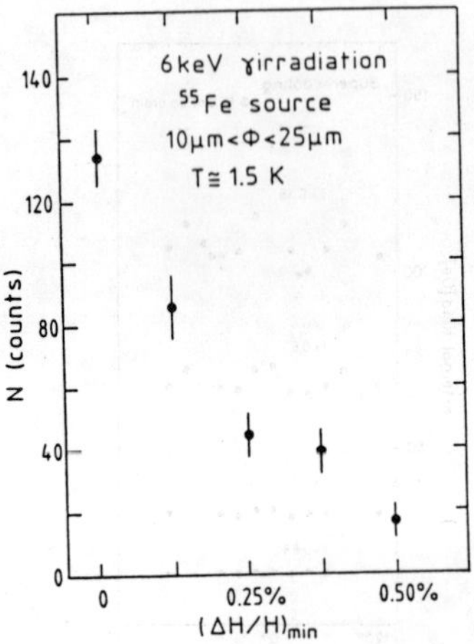

Fig. 13 : Number of flipped Sn grains after 1 hour irradiation with a Fe^{55} (E_γ = 6 keV) source.

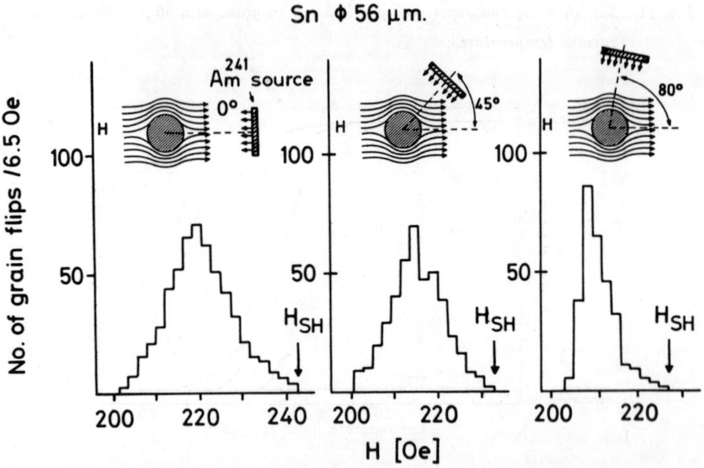

Fig. 14 : The phase transition field H of a single Sn grain irradiated by α–particles at different angles is shown. The temperature was T = 1.4 K.

SUPERHEATED SUPERCONDUCTING GRANULES

A. DE BELLEFON°, D. BROSKIEWICZ°, R. BRUERE-DAWSON°,
P. ESPIGAT°, B. METTOUT*, N. PERRIN*, D. LIMAGNE*,
L.C.L. YUAN[1], G. WAYSAND*

*Groupe de Physique des Solides de l'Ecole Normale Supérieure,
Université Paris 7, Tour 23,
75251 Paris Cedex 05, France
° Lab. de Physique Corpusculaire, College de France,
Place M. Berthelot
75005 Paris, FRANCE
[1]Brookhaven National Laboratory,
Upton Long Island (N.Y.)

1- INTRODUCTION

In this workshop other papers deal with superheated superconducting granules as detectors. Unfortunately, the speaker for this contribution had not been able to attend their presentation. It is hoped that the overlapping with other contributions is not too large.

The purpose of this paper is to underline new approaches that we are developing in Paris, rather than repeating the basic presentation of metastable superconductivity or reporting results of other participants to this workshop.

Ideally, superheated superconducting granules are a collection of perfect spheres dispersed in a dielectric medium. Here lies already a problem. Not only are the granules not ideal, but they are evidently a macroscopic disordered material. Very often, we are asked why are not we using a quasi-crystalline repartition of the granules? The answer is that it is worthless to try to make a crystal structure if the grains are not perfect and identical. In addition to this, if one imagines experiments involving large quantities of granules, it is surely better to study actual dispersions. Last, but not least, for those of us which are solid state physicists, it will be foolish not to take advantage of a disordered medium in which each element can be distinguished from the others thanks to the signal provided by a phase transition.

The former considerations explain in part the organization of this talk. We will start from a discussion of the phase diagram of type I superconductors. Then we will look into questions related to macroscopic disorder and magnetic nucleation. For most of the applications thermal nucleation is involved. In a section devoted to thermal nucleation, we present a description of the creation of quasiparticles by an incident radiation. In the final section we discuss some applications based on thermal nucleation. That is to say: X-ray imaging, transition radiation detection, and solar neutrino detection with indium granules.

2-PHASE DIAGRAM OF TYPE I SUPERCONDUCTORS

First order phase transitions are characterized by the possibility of occurence of superheated and supercooled metastable states. Type I superconductors such as Sn, In, Hg, Ga, belong to this category. However, it is only when they are divided in microspheres that the theoretical limits for superheating can be easily reached. The phase diagram in the H,T plane is represented below.

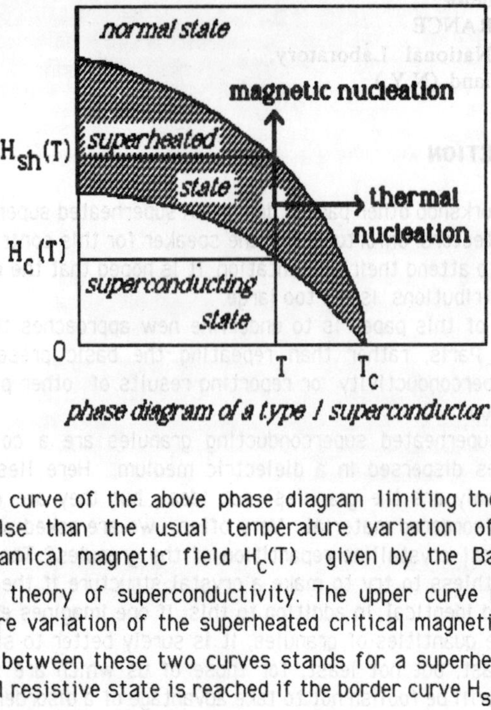

phase diagram of a type I superconductor

The lower curve of the above phase diagram limiting the grey area is nothing else than the usual temperature variation of the critical thermodynamical magnetic field $H_c(T)$ given by the Bardeen-Cooper-Schrieffer theory of superconductivity. The upper curve represents the temperature variation of the superheated critical magnetic field $H_{sh}(T)$. Any point between these two curves stands for a superheated situation. The normal resistive state is reached if the border curve $H_{sh}(T)$ is crossed from a starting point into the superheated zone.

Experimentally there are two possibilities for such a process:

-Magnetic nucleation: the temperature T is kept constant and the magnetic field is increased up to the value where at least one point of the surface of the granule senses a local field equal to H_{sh} (T).

-Thermal nucleation: at constant applied field H,the temperature is raised by ΔT, such that $H_{sh}(T+\Delta T)$ becomes smaller than H, the applied magnetic field.

Implicitely, the above diagram is valid only for a metal occupying half of the space and with a flat interface, parallel to H, separating the metal from the vacuum. This is true in practical cases, since the lengths characteristic of the superconducting state are small compared to the radius of a granule. The London penetration depth - the distance over which the magnetic field is screened- is a few hundred Å , except the immediate vicinity of T_c. The coherence length- the spatial extension of the two correlated electrons forming a Cooper pair - is a few thousand Å in the same temperature range .

However, if the granule is a sphere, rigourously speaking, it cannot be represented by a single point in the phase diagram. The local value of H is different from point to point at the surface of the granule. It is 3/2 H at the equator and zero at a pole. The whole granule , even if isolated, is represented by a vertical segment because the temperature is the same everywhere in the granule. The upper extremity of the segment represents the situation at the equator. The lower end must be on the axis H=0 because there is no field at the pole.

If we consider now a collection of granules, the diamagnetic interactions between them are responsible for the fact that the sample is represented by a vertical linear segment. Another cause of complications is: the granules are never perfectly spherical. Their surfaces are not defect-free; this modifies the length of the representative segment in the phase diagram.

The thermal nucleation is naturally controled by the horizontal distance of the upper part of the segment from the $H_{sh}(T)$ curve, because the magnetic field is weaker at the poles.

3- PRODUCTION OF THE GRANULES

The production of granules is always based on a single principle which is called the Faraday effect in the anglo-saxon litterature, capillary waves in Landau-Lifshitz, and has been recently renamed crispation by physicist in search of linguistic innovation. A thin layer of liquid metal is deposited on top of an ultrasonic transducer. The vibration creates wrinkles on top of the liquid metal. If enough ultrasonic power is transmitted to the liquid film, nebulisation occurs. If, σ is the viscosity of the liquid metal, ρ its density, f the ultrasonic frequency, ϕ the mean diameter of the granules is given by the formula: $\phi=(\sigma/\rho)^{1/3}\, f^{-2/3}$. This gives granules around 30µm for a 20kHz ultrasonic vibration.

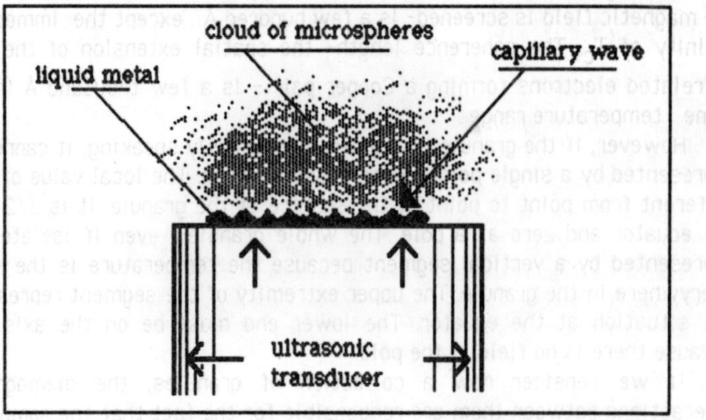

production of microspheres by generation of capillary waves on top of a thin liquid metal layer

The size distribution of the granules is far from being monodisperse, which is not surprizing for such a highly non-linear mode of excitation. The $f^{2/3}$ dependence makes it difficult, for the time being, to obtain smaller sizes in large quantities. However, in the old technique used in our lab in which molten metal is ultrasonorized in tensioactive oil, we have been able to produce 2µm granules. For such a purpose, the 20kHz ultrasonic quarter wavelength resonator sonorized molten tin into a trucated conical bucket. This amplifies high harmonics of the ulrasonic wave and gives smaller granules. Their shape is however far from regular.

So-called atomization techniques provide powders with very irregularly shaped granules and of smaller sizes. Since a granule of a few thousand Å cannot be placed in a superheated state, the production of small granules remains a problem. For granules with Ø>50μm, the production of droplets by a shower of liquid metal gives good spherical shapes (producer: Extramet 74-Annemasse -France).

Except for the "atomized powder", *even when the shapes seem perfectly spherical*, surfaces of granules exhibit many defects: surface dips, surface fractures are the most frequent surface defects. This is due to the cooling process: the external enveloppe of the granule solidifies first, while the interior remains liquid. The dilatation for the liquid phase is larger than for the solid phase.The solid external crust must relax surface tensions. This is done by the creation of one or a few dips and of many surface fracture lines. These lines are very similar to those created at the surface of a drying clayey soil. For the granules, the result of these geometrical alterations is a local enhancement of the magnetic field, making the magnetic nucleation more easy. Though it is possible to measure in-situ, by a self-consistant method, the surface density of the effective defects[1], we are presently not able to correlate experimentally these defects to well-defined geometrical imperfections.

In experiments where a single particle flips only one granule, this question of effective defects is of the utmost importance. It is secondary if the energy resolution relies on the change of state of many granules under a single impact. In any case we have to be sure that we deal with a true random dispersion if we want to obtain reproducible results.

4- GRANULES DISPERSIONS AS A MACROSCOPIC DISORDERED MATERIAL

4-1 Baker's theorem:

When the granules are in close contact, they cannot exhibit a superheated behaviour. The point of contact is a nucleation center for the normal state, for both granules, as this will be discussed later in this paper. Therefore clusters of granules, as a rule, must be avoided. This is why we use paraffin wax at 42C to disperse the granules. Furthermore, in spite of the fact that we do not have a crystalline structure, we want to have reproducible samples. We want their behaviour in magnetic and thermal nucleation to be the same when the filling factor in granules is the same. If no precautions are taken for the mixing of the powder of granules within wax, the samples are not equivalent, even if they have the same filling factor. The recipe is to make use at this point of a topological theorem. This theorem is : Two points in a continuously deformable volume can be separated by a given distance if the volume is submitted N times to the following cycle:

step 1: flatten the volume to get a pancake

step 2: fold this pancake around its diameter parallel to the Y axis.

step 3: refold the result of step 2 around a perpendicular direction passing by the center

step 4: do step 1.

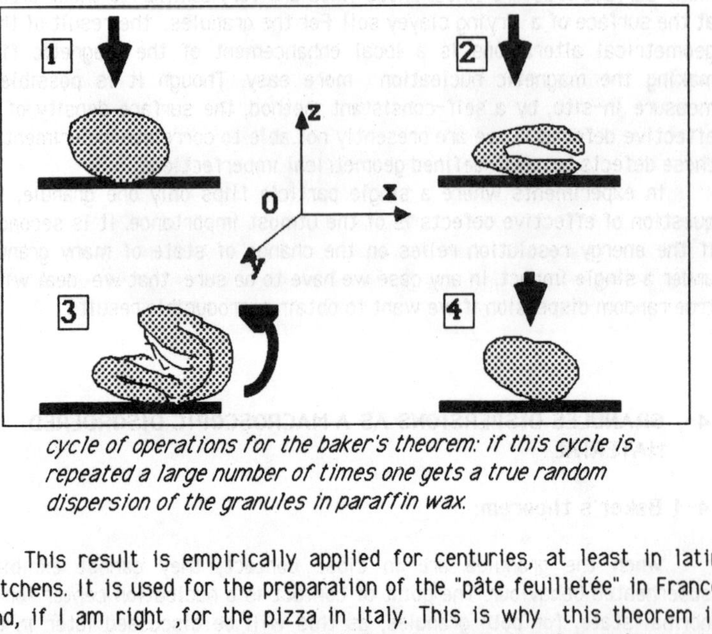

cycle of operations for the baker's theorem: if this cycle is repeated a large number of times one gets a true random dispersion of the granules in paraffin wax.

This result is empirically applied for centuries, at least in latin kitchens. It is used for the preparation of the "pâte feuilletée" in France and, if I am right ,for the pizza in Italy. This is why this theorem is called the baker's theorem. The result for our composite of granules can be seen in the next view. Tin granules have been dispersed in paraffin wax kept at 42 C following the baker's recipe. The granules are effectively randomly dispersed, though a layered structure is apparent in the wax .

For a given filling factor and the same number of cycles, 40 for $1 cm^3$, we get identical superheated curve (see below the definition of this curve which is characteristic of a composite of metastable granules).

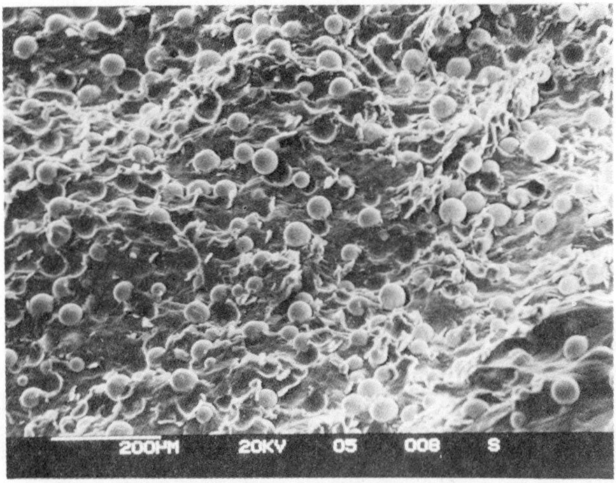

picture taken with an electron microscope, of tin Extramet microspheres randomly dispersed in wax by the baker's method. $33<\varnothing<40$ µm, filling factor 28,5%

4-2 Dielectric and diamagnetic N-body interactions:

A suspension of randomly dispersed spheres is not something new in physics. More than a century ago Clausius and Mossotti calculated the dielectric constant of such a suspension of spheres. If we replace the electric field by the magnetic field, and the metal by a superconductor which is perfectly diamagnetic , their problem is our problem. As in their case it is not sufficient to associate each sphere with a diamagnetic dipole to get a valid description of their interactions [2][3].

Just to help the reader realize how much this is important, we would like to emphasize that, because of these interactions, even if the volume fraction in granules is of the order of a few percent,it is impossible to find two granules flipping at the same applied field. By scanning up the magnetic field and counting the number of granules N(H) which have already flipped up to H, one gets a sigmoïdally shaped curve called the superheated curve. The detailed shape of this curve is characteristic of the disorder of the collection of granules and of their surface defects. Its general aspect is drawn below.

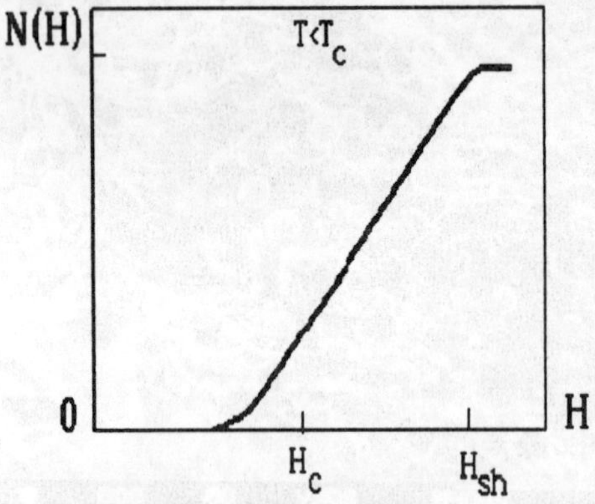

typical aspect of a superheated curve

The results of references [2][3] are very far from this type of variation.The discrepancy cannot be rejected on the surface defects, because using the method defined in [1] we are able to obtain a "defect-free" superheated curve for comparison with theory.

The detailed description of this macroscopically disordered situation has been drastically improved , using a multipolar expansion, by P.Mazur and U.Geigenmüller [4][5][6]. Their method gives an "exact" expression of the local field at the surface of a granule interacting with others. The simplest way to appreciate this approach is to apply it to the computation of the dielectric constant of the dispersion of granules at various filling factors. The result is compared to the classical results of Clausius-Mossotti and Bruggeman in the following plot.

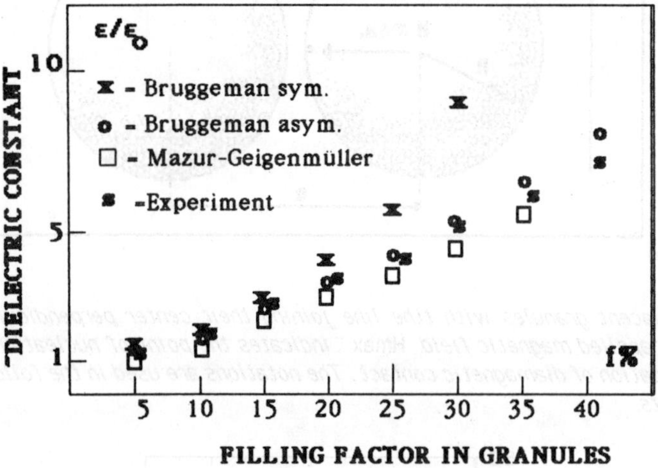

Dielectric constant as a function of the filling factor in granules for various theoretical treatments. The experimental points are in agreement with Geigenmüller approach. We thank A.Broniatowski and Ph. Dumas for the communicaton of these results before publication.

It is possible to use the Mazur-Geigenmüller method for the analysis of a simple situation: the description of the diamagnetic interaction between a pair of adjacent granules. When the line joining their centers is perpendicular to H, the local magnetic field between these granules is strongly enhanced. To compute this field enhancement one has to go to a high order in the multipolar expansion as it is shown in the next graphs.

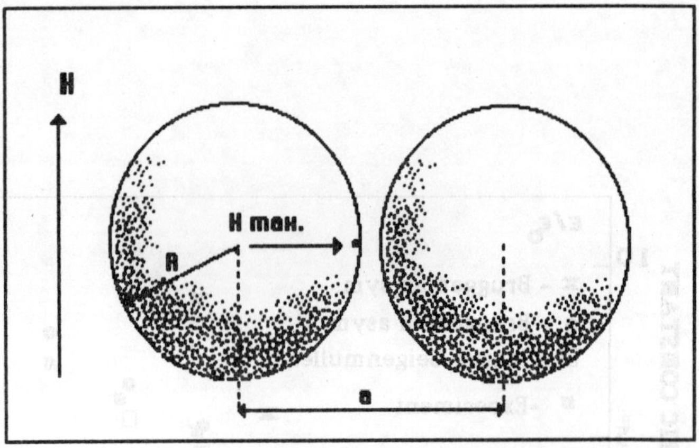

Adjacent granules with tthe line joining their center perpendicular to the applied magnetic field. Hmax *indicates the point of nucleation in a situation of diamagnetic contact. The notations are used in the following plots.*

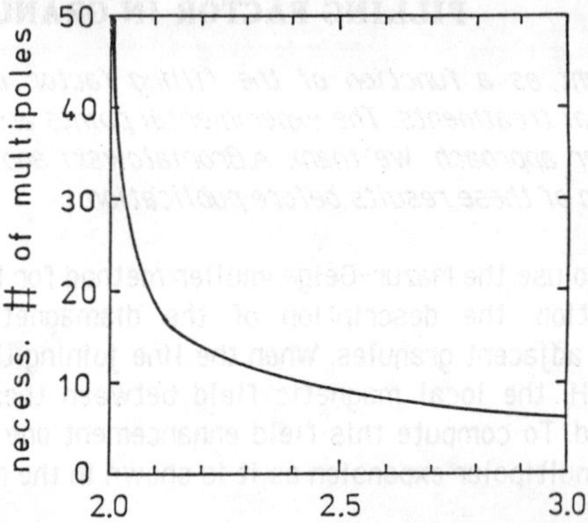

Number of terms to take into account for the computation of field enhancement as a function of a/R . R *is the radius of the granules,* a *the distance of their centers. (from reference 6)*

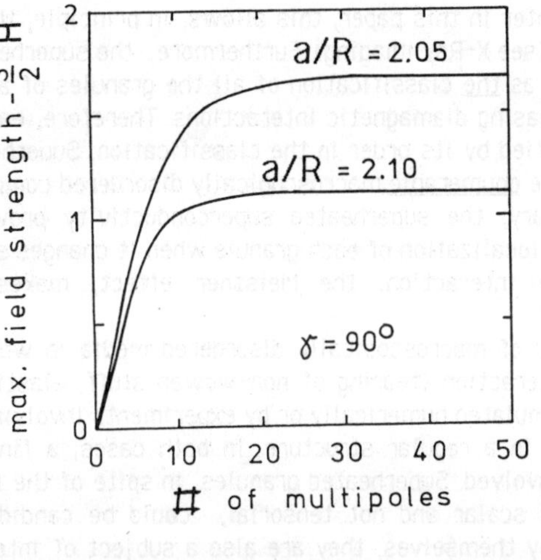

Computed magnetic field enhancement . This function is plotted for different spacing between two identical spheres with their line or centers perpendicular to the applied magnetic field. R *is the radius or the spheres,* a *is the distance between their centers. (from reference 6)*

If this applied field is smaller than $H_c(T)$, this creates a situation of diamagnetic contact. Even if the two granules are only in quasi-contact, their strong proximity allows a magnetic nucleation, in one of them, at $H<H_c(T)$. The propagation of the normal state stops when the field becomes equal to H_c at the interface between the two phases. Therefore, in the superheated curve, the points of $H<H_c$ correspond to signals smaller than the signal for the volume of a granule [7]. If one wants to make a detector with a reasonable energy resolution, this situation must be avoided.
The same diamagnetic situation exists also in the new granular high T_c oxide superconductors [8].

4-3 Simulation of others macroscopically disordered material ?

The singularity of a dispersion of superheated granules comes from the fact that each elementary entity within can provide an individual signal at its transition from the normal to the superconducting case. As we will see later in this paper, this allows, in principle, the localization of each event (see X-Ray imaging). Furthermore, the superheated curve can be considered as the classification of all the granules of a dispersion by order of decreasing diamagnetic interactions. Therefore, each granule can also be identified by its order in the classification. Superheated granules dispersions are enumerable macroscopically disordered composites.

In summary, the superheated superconductivity provides, first, a signal for the localization of each granule when it changes state. Secondly a long range interaction, the Meissner effect, makes the system enumerable.

The study of macroscopically disordered media in which there is a long range interaction (tearing of non-wowen stuff, elastic percolation) are usually simulated numerically or by experiments involving elementary cells, usually in a regular structure. In both cases, a finite number of elements is involved. Superheated granules, in spite of the fact that their interaction is scalar and not tensorial, could be candidates for such simulations. By themselves, they are also a subject of interest from the point of view of disorder.

5- THERMAL NUCLEATION

5-1 Creation of quasiparticles:

When a nuclear radiation, photon or electron, pass through a superconductor, metastable or not, the main process of energy loss is quasiparticle production by electromagnetic interaction. However, in the process of losing energy, quasiparticles are entering in a range where they can deliver a fraction of their energy to the lattice. The energy gap of the Cooper pairs Δ, is smaller than ω_D the Debye frequency which is the highest possible energy for phonons. This allows Cooper pair-breaking by phonons. To the authors' knowledge, the most detailed description of this cascade of events has been provided by Kurakado and Mazaki [9] A simplified, and less accurate summary of their classification is provided in the table on the next page.

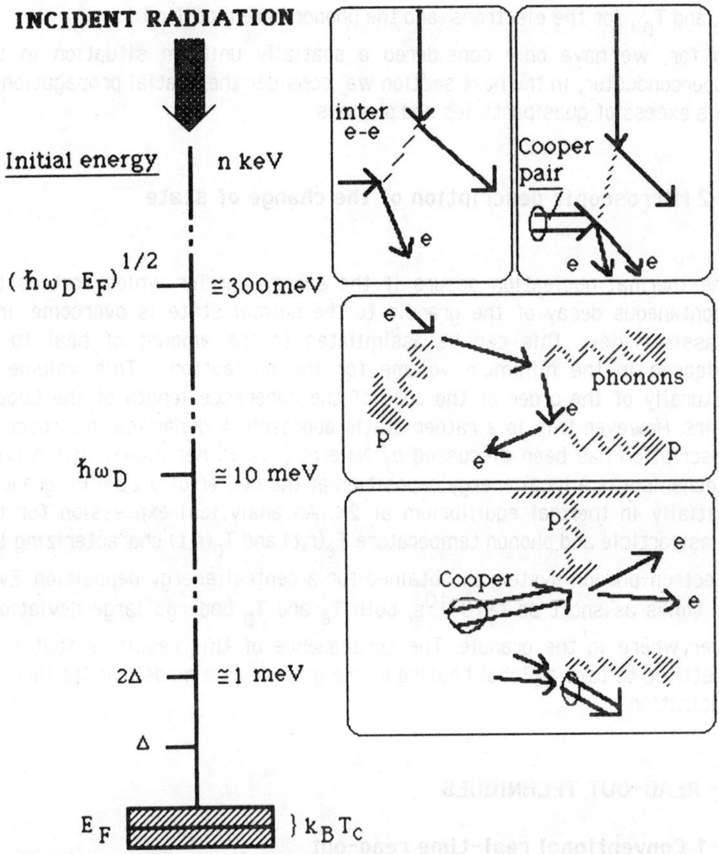

Simplified representation of the cascade of events following the arrival of an incident radiation in a superconductor. Electrons are represented by arrows, Cooper pairs by double arrows with an ellipse around the tails, phonons by wavy lines.

Above an energy $(\hbar\omega_D E_F)^{1/2}$, electrons-electrons interactions are dominant [10]. Under this energy and the Debye limit, energy decay by creation of phonons is possible. Under the Debye energy and above $2\Delta = 3{,}5\ k_B T_C$, phonon pair-breaking takes place.

Finally sub-gap phonons are ineffective for pair-breaking. The overall result is that, in a very short time, we have plenty of quasiparticles, just above the gap. These quasiparticles are candidates for pair-breaking or, eventually, recombination by phonon absorption. The average temperature T_e and T_p, for the electrons and the phonons are modified.

So far, we have only considered a spatially uniform situation in the superconductor, in the next section we consider the spatial propagation of this excess of quasiparticles and phonons.

5-2 Microscopic description of the change of state

The thermal nucleation occurs if the energy barrier which forbids the spontaneous decay of the granule to the normal state is overcome. In a classical view, this can be assimilated to the amount of heat to be released in the minimum volume for the nucleation . This volume is naturally of the order of the cube of the coherence length of the Cooper pairs. However this is a rather static approach. A dynamical microscopic description has been discussed by N.Perrin[11]. In her model, the driving mechanism is a local energy deposition at the center of a 2µm In granule, initially in thermal equilibrium at 2K. An analytical expression for the quasiparticle and phonon temperature $T_e(r,t)$ and $T_p(r,t)$ characterizing the electron-phonon system is obtained for a central energy deposition. Even at times as short as $t=10^{-10}$s, both T_e and T_p undergo large deviations everywhere in the granule. The consequence of this result is that it is justified to take a global heating of the granule as a model for its thermal nucleation.

6- READ-OUT TECHNIQUES

6-1 Conventional real-time read-out

It is the breakdown of the Meissner effect which allows the detection in real-time of the change of state. The conventional technique, now used by all the teams involved in the study of superconducting granules is the one which we have worked out over the years. It is a charge amplification set-up equivalent to a ballistic fluxmeter operating in the microsecond range [12] [13]. The flux change is detected by a pick-up coil surrounding a collection of granules. This coil is connected to the amplifier input by two pulses transformers in series. The first one in the helium bath, the second at room temperature, just in front of the input FET of the amplifier. Both transformers are step-up in voltage. The connecting line from the low-temperature transformer to the high-temperature transformer is not matched in impedance. The best performance obtained with an hybrid amplifier is the read-out of a 8μm granule in a 500μm wide, U shaped, 81mm long pick-up coil. In this type of set-up, the collected voltage pulse at the output of the charge amplifier is proportionnal to the change of flux and therefore to the volume ($4\pi R^3/3$) of the granule in transition to the normal state.

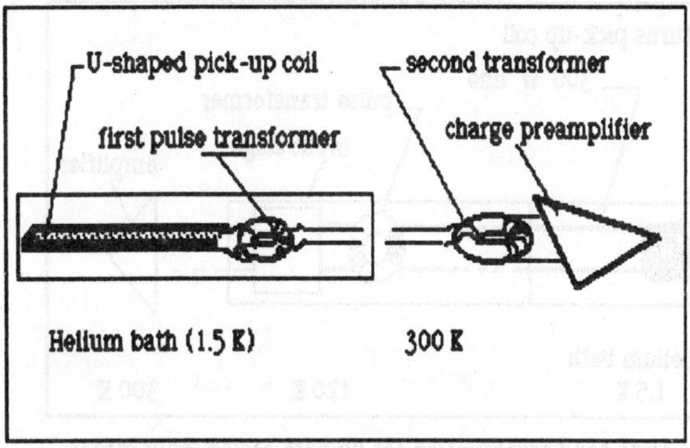

Charge amplification set-up

With this set-up magnetic nucleation as well as thermal nucleations can be studied.

6-2 Wideband amplification with cooled input stage

The source of the signal when the nucleation occurs is nothing else than the decay of the eddy currents opposing the penetration of the magnetic flux.
At the very beginning, when the energy barrier has been overcome, these

eddy currents are inexistent, therefore the velocity of flux penetration is very large, and so is $\partial \emptyset / \partial t$. If an amplifier has a wideband large enough we can take advantage of this effect. One essential condition is a very careful impedance matching from the pick-up coil to the amplifier. A second feature is that by cooling the input stage of the amplifier an additional benefit in terms of signal/noise ratio is obtained. The actual realization is a 10 turn solenoidal pick-up coil directly connected to a 300Ω transmission line. This feeds a single pulse transformer. The secondary of this transformer is itself connected to the input stage of the preamplifier. This input stage and the transformer are cooled at 120K.

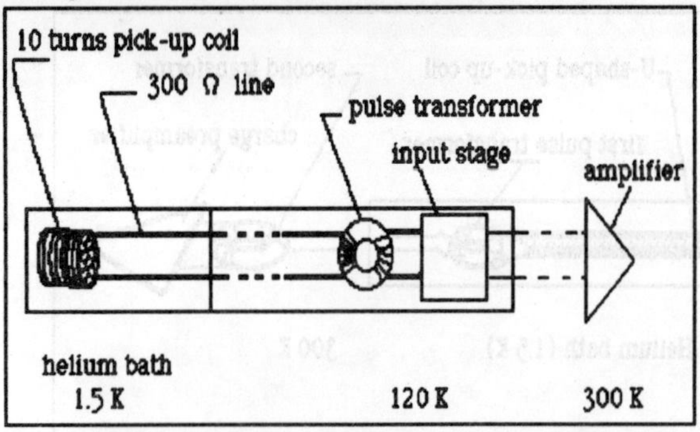

wide band amplification set-up with cooled input stage

With a 2 Mhz wideband amplifier a granule smaller than 8 μm in diameter has been read. Furthermore, with this wideband solution, the amplitude of the signal varies like R^2. The benefit is that if we are using 2μm granules we will be able to read 16 granules, provided they are flipped at the same time and that we do not loose the R^2 dependence at higher frequencies. With the classical charge amplifier, for the same performance we will have to read $8^3/2^3 = 64$ granules.

7-1 X-ray imaging detector

In this application, the suspension of granules in paraffin wax is sandwiched between two families, X and Y, of U-shaped pick-up coils, each orthogonal to the other. The incident X-rays are converted into photo-electrons within the suspension. The energy losses of these electrons, when they are large enough, are responsible for the transition of a granule to the resistive state. This transition is detected as a coincidence between two coils X_i and Y_j, thus providing the localisation of the event.

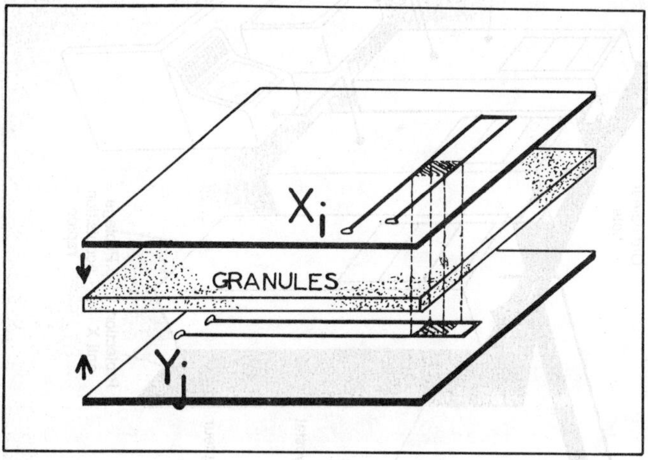

To get a picture it suffices to pile up the events at a given address. A practical realisation is shown in the following figures. The detecting plate itself is 350 µm thick, its area is $81*81 mm^2$, covered as explained above by 128 X wires and 128 Y wires.The filling factor in 10-15 µm tin granules is 20%. Thus the pitch, which is the intrinsic spatial resolution of the system is 600µm.

The overall view of the device is shown below. The X-ray source is a conventional industrial tube operating at 150kV and 4mA. The complete monitoring of the experiment is realised by CAMAC modules controled by a CAB500 microcomputer in the CAMAC crate itself.

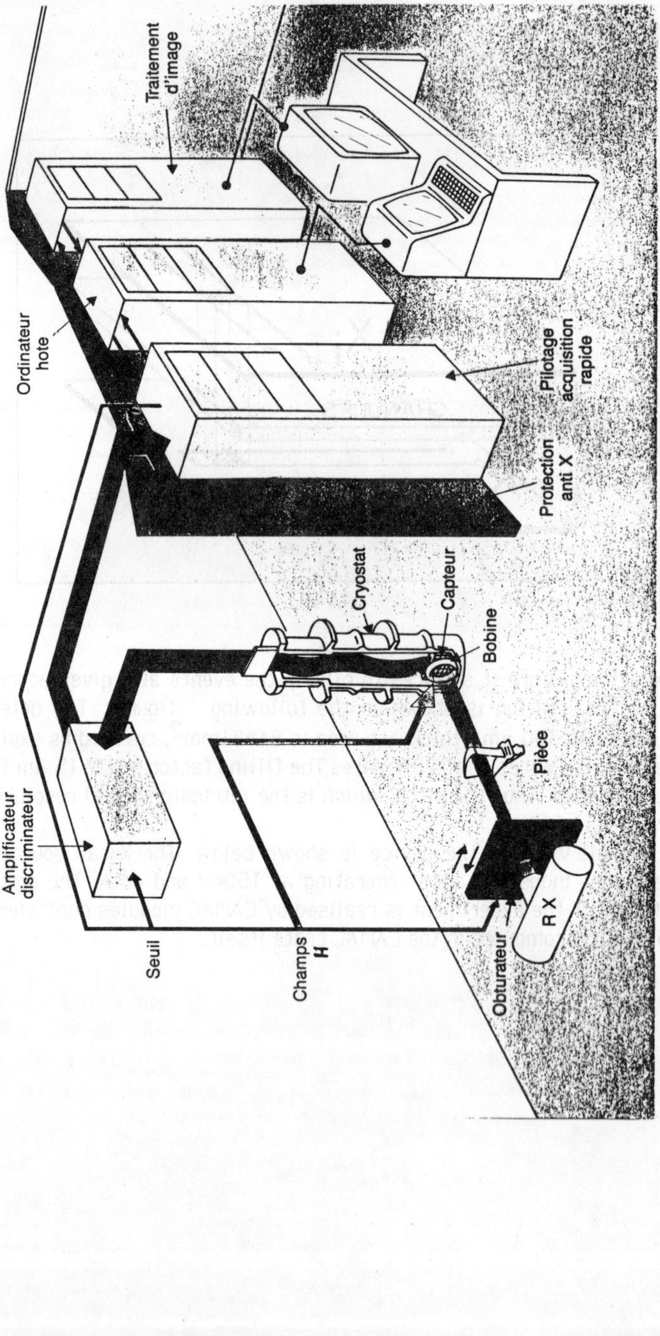
Amplificateur
discriminateur
Seuil
Champs
H
Obturateur
R X
Pièce
Bobine
Capteur
Cryostat
Protection
anti X
Pilotage
acquisition
rapide
Ordinateur
hote
Traitement
d'image

A practical result of what can be obtained is represented in the next graph. It is one square centimeter of the detector (16*16 pixels) partially covered from the X-ray irradiation by 3mm of lead.The number of counts at each pixel is represented by the height of the box which has this pixel as its base. At addresses where the detector is behind the lead foil ,only a few diffused photons are detected, whereas much more are counted in the zone under direct irradiation. About 1200 photons are counted in this "picture". No corrections and no filtering were done, thus the flucuations from pixel to pixel represents the inhomogeneity of the beam and the detector.

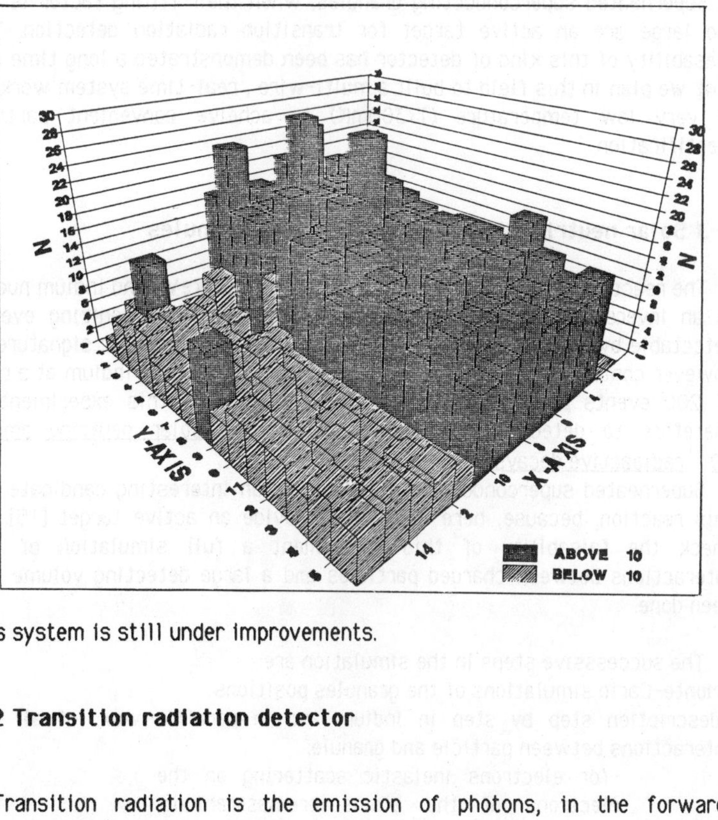

This system is still under improvements.

7-2 Transition radiation detector

Transition radiation is the emission of photons, in the forward direction, at the interface between a dielectric and a normal metal when this interface is crossed by a charged relativistic particle. A crude explanation is that, when crossing the interface, the group velocity of the particle adjusts itself to the new conditions of propagation it is experiencing .The probability of emission is 1/137 therefore many interfaces are required for a photon to be emitted. In spite of this , the importance of transition radiation lies in the fact that the energy of the emitted photon varies linearly with the ratio E/mc^2 of the particle and not its $ß=v/c$. At high energies, all the ß's of the various particles are near one and high magnetic fields are required for particle identification; on the other hand, the ratios E/mc^2 remains very different. This is a strong argument for transition radiation detectors in future beams.

Superheated superconducting granules, when their filling factor is not too large are an active target for transition radiation detection. The feasability of this kind of detector has been demonstrated a long time ago [14]. We plan in this field to built a multi-wire , real-time system working at very low temperature (T<300mK) to acheive convenient particle identification.

7-3 Solar neutrino detection with indium granules

The reaction between solar neutrinos, with E>128keV, and indium nuclei is an inverse double beta decay which produces three ionizing events detectable by a triple delayed coïncidence. This very specific signature is however conterbalanced by the spontaneous beta decays of indium at a rate of 200 events per kg per second. The challenge of this experiment is therefore to detect in 4 tons of indium one solar neutrino among 10^{11} radioactive decays.

Superheated superconducting granules are an interesting candidate for this reaction, because, here also, they provide an active target [15]. To check the faisability of this experiment a full simulation of the interactions between charged particles and a large detecting volume has been done.

The successsive steps in the simulation are:

-Monte-Carlo simulations of the granules positions

-description step by step in indium and in paraffin wax, of all the interactions between particle and granule:

-for electrons inelastic scattering on the electrons of the atoms bremsstrahlung, multiple scattering on the nuclei.

-for photons Rayleigh scattering,Compton effect, photoelectric effect.

The results are represented in the next table for a filling factor in granules of 20%.

	T=1K		T=0.2K	
	E=100 keV	E=500 keV	E=100 keV	E=500 keV
Ø=5µm ΔE=→	N=6.5 k=0.12 100 keV	N=68.5 k=0.06 200 keV	N=6.5 k=0.3 50 keV	N=68.5 k=0.2 110 keV
Ø=2µm ΔE=→	N=14.9 k=0.45 25 keV	N=168 k=0.27 70 keV	N=14.9 k=0.63 15 keV	N=168 k=0.46 50 keV

Ø=granule diameter
N=number of granules crossed by an electron of energy E
k=efficiency of flipping
ΔE= energy resolution

It is clear that the indium experiment must use small granules Ø=2µm at low temperature (16). This last requirement is related to the T^3 dependence of the phonon term in the specific heat of the metal.

The authors express their thanks to the numerous collegues for many discussions and suggestions, specially to C.Caroli, Pr. P. Mazur, Pr. M. Froissart, J. Rich, J.P. Maneval, A.Broniatowski, Ph. Dumas, N.Booth , K.E. Gray and all the people involved in the study of superheated granules who have ,for themselves ,verify that, according to Cesare Pavese, whose life has been so influenced by the city of Torino, " Lavorare stanca". Finally, the speaker would like to thank Pr. R.Bland of San-Francisco State University for his hospitality during the preparation of this talk.

REFERENCES

1- D. Hueber, C. Valette, G. Waysand, J.Physique Lettres **41** (1980) L-611
2- C. Valette, G. Waysand, D. Stauffer Solid State Com. **41**(1982) 305
3- Hillel, D. Stauffer Solid State Com. **43** (1982) 305.
4- U. Geigenmüller, P. Mazur Physica **136A** (1986) 316.
5- U. Geigenmüller, P. Mazur Physica **138A** (1986) 269.
6- U. Geigenmüller Thesis, April 1987 (Leyden)
7- B. Mettout and al. Solid State Com **64** (1987) 1069
8- G. Waysand "Diamagnetic interactions in granular high T_c superconductors" (to be published in Europhysics Letters).
9- M. Kurakado and H. Mazaki, N.I.M. **185** (1981) 141.
10- S.B. Kaplan, C.C.Chi, D.N. Langenberg, J.J. Chang, S. Jafarey and D.J. Scalapino, Phys. Rev. **B14** (1976) 4854.
11- N. Perrin, J.Physique **47** (1986) 1939
12- R.L. Chase, Ch. Gruhn, A. Hrisoho,C.Valette, G.Waysand.Proceedings of the second Ispra nuclear symposium. Stresa 29 ,(1975).
13- A. Hrisoho,G.Waysand NIM, **214**, (1983) 415,
14- A.K. Drukier, C. Valette, G. Waysand, L.C.L.Yuan, E. Peters Lett. al Nuovo Cimento **14**, (1975) 300.
15- G.Waysand in "Astrophysique et interactions fondamentales" edited by J.L.Basdevant et J.Audouze CNRS ,16(1983). see also " Moriond Workshop on Massives Neutrinos in Astrophysics and Particle Physics" edited by J.Tran Thanh Van, éditions Frontières, Gif (1984),319.
16- A. de Bellefon, P.Espigat,L.Gonzalez-Mestres,R.Kuenzler, B.Mettout D.Perret-Gallix,N.Perrin,G.Waysand, Report to the direction of the CNRS on the faisability of the detection of solar neutrinos with indium. GPS-ENS 87-007, (1987)

RECENT RESULTS AND PROSPECTS ON SUPERHEATED SUPERCONDUCTING GRANULES DETECTORS

L. GONZALES-MESTRES and D. PERRET-GALLIX

L. A. P. P.
B.P. 909, Chemin de Bellevue,
Annecy-le-Vieux,
74019 Cedex France

The basic properties of superheated superconducting granules (SSG) detectors are reviewed. Recent results are presented and discussed.

Granule samples of different sizes ($10\ \mu m < \phi < 400\ \mu m$) and impurity content ($6\times10^{-8}\ \Omega.cm < \rho$, residual $< 10^{-6}\ \Omega.cm$) exhibit encouraging sensitivity to low energy β and γ sources (down to ^{55}Fe 6 keV γ's). All tests were performed with real-time read-out electronics, detecting single granule flips under the action of individual particles. Sensitivity is shown to depend on normal state resistivity. Irradiation of very large tin granules ($45\ \mu m \leq \phi \leq 400\ \mu m$) with α particles (^{241}Am, $E \simeq 5.5$ MeV) shows evidence for local heating, where the observed energy threshold is far below the one predicted by equilibrium thermodynamical calculations.

Tests made at lower temperatures ($T \geq 450$ mK) show the absence of avalanche effect (seen by other authors in different conditions) for several samples of tin granules. A theoretical discussion of the avalanche effect is presented. The understanding of the role of heat exchanges in the composite medium leads to the concept of "localized micro-avalanche" and opens the way to drastic improvements of SSG performance for particle detection. Such a phenomenon should be obtained by a better thermal matching between dielectric and granules, working at temperatures where the released latent heat is slightly positive. Estimates of the behavior of the detector at very low T are also given, where a thin layer of normal electrons near the surface is shown to contribute to the heat capacity of a superheated granule.

We discuss the main points to be studied in the near future, and give a brief evaluation of the present status of feasibility investigation for several proposed experiments (solar neutrinos, monopoles, dark matter, double β,...). An updated working program for SSG development is proposed.

1. INTRODUCTION

In a superconductive material, the interface energy between superconducting and normal zones at H_o (applied magnetic field) = H_c (thermodynamical critical field) is positive for κ (Ginzburg – Landau parameter) < $1/\sqrt{2}$ [1]. Materials satisfying this condition are called type I superconductors, and exhibit metastable phases. Such states (superheating and supercooling) are known experimentally since more than thirty years [2]. Supercooling occurs when a superconductive sample remains normal for H_{sc} (supercooled critical field) < H_o < H_c , whereas superheating appears when the superconducting state is preserved for H_c < H (external magnetic field on the surface of the specimen) < H_{sh} (superheated critical field). The corresponding phase diagram is shown in Fig.1.

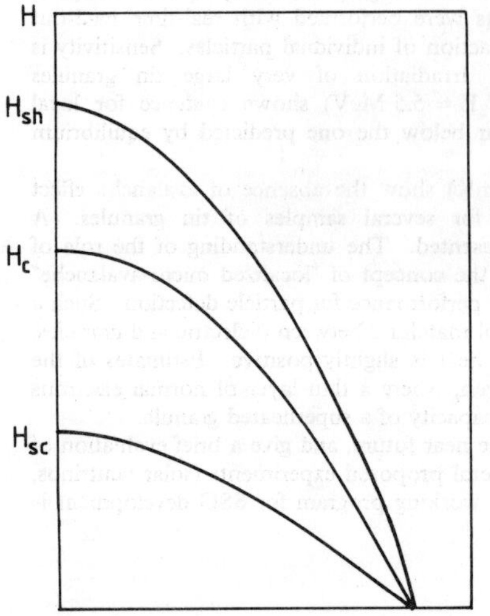

Fig.1 : Temperature dependence of H_{sh}, H_c, and H_{sc} for a type I superconductor.

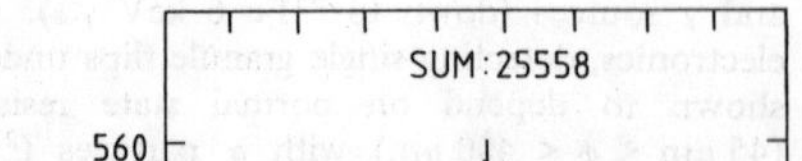

Fig.2 : Superheating curve for a collection of tin granules produced by EXTRAMET, and sieved at 10 μm < ϕ < 25 μm. dN/dH_o indicates the number of counts per unit interval in applied magnetic field when H_o is raised from 0 to 500 G. ΔH_{min} is the sweep in magnetic field usually made in irradiations at $H_o = H_{test}$.

Defects and inhomogeneities act as nucleation centers of the thermodynamically stable phase. Twenty years ago, solid state physicists developed a considerable amount of work to obtain metastable states with small spherical samples [3]. Suspensions of microspheres into some dielectric material were thus used to determine κ for several pure metals, through the formulae:

$$H_{sh} \simeq 2^{-1/4} \kappa^{-1/2} H_c \qquad \{1a\}$$

$$H_{sc} \simeq \sqrt{2}\, \kappa\, H_c \qquad \{1b\}$$

obtained [4, 5] within the framework of the Ginzburg-Landau (GL) approximation [6] for $\kappa \ll 1$. Other approximations, such as the extreme anomalous limit (EAL) [7] have also been used. Samples other than spheres (e.g. wires [8, 9] or films [10]) can also reach the superheated state, but will not be discussed here.

For a spherical sample in the presence of an applied field $\vec{H}_o$, the magnetic field on the surface is given by:

$$H \simeq 3/2\ H_o\ |\sin\theta| \qquad \{2\}$$

where θ is the angle between $\vec{H}_o$ and the position vector $\vec{r}$ taken from the center of the sphere. In the case of a perfect sphere, the superconducting to normal or intermediate state phase transition would happen at $H_o = 2/3\ H_{sh}$. However, because of defects, inhomogeneities or anisotropies, and diamagnetic interactions, each granule changes state at a different value of H_o (H_{sh}^{eff}, the effective superheated critical field). For a given isolated granule, H_{sh}^{eff} has been shown [11, 12] to smoothly depend on its orientation with respect to $\vec{H}_o$. Real collections of granules give a superheating curve (Fig.2), where dN/dH_o is the number of flips observed per unit interval in H_o.

The idea of using superheated superconducting granules (SSG) [13] for particle detection [14] is due to the Orsay group [15]. In a very elegant experiment, performed twenty years ago, it was shown that 2 μm diameter granules doped with radioactive gold were sensitive to the emitted β rays. Evidence for irreversible superconducting to normal phase transitions under elementary particle interactions was thus reported for the first time. A series of tests with electron beams was made in Rennes [16] a few years later. X-ray imaging [17] or transition radiation detection [18] led to the first attempts to build prototypes of photon detectors. More recently, the growing interest in high sensitivity devices for Astrophysics and Particle Physics has triggered new developments on SSG and other cryogenic detectors.

Solar neutrinos can possibly be detected with SSG devices, either through the nucleus recoil energy [19] or, using indium granules [20], through Raghavan's reaction [21]. Magnetic monopoles are expected [22, 23] to destroy the superheated state independently of the size of the grains. Thus, a large area detector may be feasible if large granules of good quality are available (good background rejection, better signal over noise ratio). Cold dark matter [24, 25, 26] and solar axions [27] may be detectable with suitable materials such as Al or Ga, if sensitivity to energies below 1 keV is demonstrated. Finally, double β experiments [28] could be performed with cadmium or molybdenum. Molybdenum granules (if available, and if they can reach the superheated state) would be a most interesting material because of the high value of Q (3.03 MeV for ^{100}Mo).

SSG devices (SSGD) provide a superconducting active target with a large mean density. Suitable materials are most type I superconductors (pure metals, such as: Sn, In, Hg, Cd, Ga, Mo,...), usually embedded into paraffin or epoxy. Physical applications are closely related to the choice of the superconductive material. The energy threshold depends on the granules size (in a global heating situation), the operating temperature and the gap in magnetic field. By choosing the appropriate material, and tuning these parameters, one expects to find solutions for specific detection purposes, when conventional devices fail. SSG detectors can be viewed as a "metallic, preformed bubble chamber", where a metastable phase is disrupted by particles which, instead of forming gas bubbles, destroy the superconductivity of some metal grains. Except for magnetic monopoles [22,23], the phase transition is due to a thermal deposit of energy that heats one or several granules. Since superheating happens in the presence of an applied magnetic field, the read-out technique is based on the disappearance of the Meissner effect, where current loops placed around the granules detect a local variation in magnetic flux.

SSG detectors are still at the level of a feasibility study. Present experimental tests are at the moment limited by the difficulty to produce very small granules (ϕ < 5 μm) with the required size homogeneity. But there are also obstacles related to the basic properties of the detector. One major drawback is the spreading of the superheating curve, i.e., the dispersion in the effective superheated critical field H_{sh}^{eff}. This implies a loss in efficiency for a required sensitivity. Another difficulty may be the existence of an avalanche effect [29, 30] that, in some cases, seems to prevent from operating the detector at temperatures far below T_c (the critical temperature), where the fast decrease of superconducting specific heats would provide higher sensitivity. Finally, the use of very small granules may lead to signals beyond the reach of present day electronics, especially if good time resolution is simultaneously required.

Chapter 2 proposes a review of the basic properties of SSG.

Experimental results on irradiation with α, β and γ particles are presented in chapter 3. Results obtained with large granules (up to 400 μm diameter) irradiated with α particles, bring for the first time quantitative information on the mechanism of phase transition by a local deposit of energy. Irradiation of 200 μm < ϕ < 400 μm with 6 keV photons provides a dramatic confirmation of local heating mechanism. Irradiation of smaller granules (10 μm < ϕ < 25 μm) with γ and β^- sources (including ^{55}Fe 6 keV γ's) brings encouraging results as far as sensitivity is concerned. In all cases, flips have been observed, and for granules of equal sizes sensitivity appears to depend on normal state resistivity.

Chapter 4 presents results and estimates on the behavior of the detector at very low temperature, in particular: a) The observed absence of avalanche effect for several samples of Sn granules at T $\geq$ 450 mK; b) A detailed description of the avalanche effect explaining existing data; c) The heat capacity of small superconducting spheres in the presence of an applied magnetic field; d) Numerical estimates concerning some of the proposed experiments.

Chapter 5 develops the new concept of localized thermal micro-avalanche, where latent heat is used to increase the amount of flipping grains for a given deposit of energy. Electronic signal and energy resolution should thus be considerably improved. Using well suited dielectric materials, heat exchanges in the composite medium seem to be fast enough to allow for the use of micro-avalanche phenomena in experiments with $\approx 10^{-7}$ sec time resolution.

Present status of the feasibility study for several proposed experiments is reviewed in chapter 6, whereas chapter 7 presents a working program for the near future.

The concepts and notations used throughout this contribution are the similar to those of references [14, 31], where other details can be found.

2. REVIEW OF BASIC PROPERTIES.

2.1 GL description of superheating.

Superheating is a surface phenomenon, where an energy barrier protects the superconducting state on the surface of the specimen. For a sphere of radius R $\gg \xi$ (coherence length), with $\kappa \ll 1$ and H large enough, the GL approximation gives [4,32]:

$$F \simeq - H^2\lambda/8\pi f(0) + 2\sqrt{2}\, H_c{}^2\lambda/8\pi\kappa\,[2/3 - f(0) + 1/3\, f(0)^3] \qquad \{3\}$$

F is the difference in free energy per unit surface between normal and superconducting state, $\lambda = \kappa\xi$ the London penetration length, f(0) the value of the real GL order parameter [6] on the surface of the specimen and H the value of the external field on each point of the surface. At fixed θ, one has: $f = f(z)$ where z is the depth from the surface towards the center of the sphere: $0 \leq f(0) \leq f(z) \leq 1$, and $f(z) \rightarrow 1$ far inside (up to finite size corrections that we neglect in what follows). F is given by the integral of the free energy density over the z coordinate. As a function of f(0), expression {3} has:

- An absolute minimum at $f(0) = 0$ ($F \rightarrow -\infty$) corresponding to the normal state.
- A local maximum at $f(0) = f_{max} = 1/\sqrt{2}\,(1-x)^{1/2}$
- A local minimum (the superheated state) at $f(0) = f_{min} = 1/\sqrt{2}\,(1+x)^{1/2}$

where:
$$x = (1 - H^2/H_{sh}{}^2)^{1/2} \qquad \{4a\}$$

$$H_{sh} \simeq 2^{-1/4}\,\kappa^{-1/2}\,H_c \qquad \{4b\}$$

Then, the superheated state is protected by an energy barrier per unit surface [23,32]:

$$\Delta F = F(f_{max}) - F(f_{min}) \simeq 4/3\,\sqrt{2}\,(1-(1-x^2)^{1/2})\,[(1+x)^{1/2}-(1-x)^{1/2}]\,H_{sh}{}^2\lambda/8\pi \qquad \{5\}$$

At finite κ, numerical calculations [33] give values of H_{sh} somewhat larger than those predicted by the low κ expression. In particular, $H_{sh}/H_c > 1$ at $\kappa = 1/\sqrt{2}$ in agreement with experimental results [34]. Solving numerically the GL equations at $\kappa = 0.16$ [23, 32], corrects the approximation {5} by only 20%, if the actual value of H_{sh} at finite κ is used.

2.2 EAL estimates.

The question of the stability of theoretical estimates for different approximations has to be addressed. Strictly speaking, the GL approximation holds only for λ, $\xi \gg \xi_o$ (the Pippard coherence length). For a type I superconductor, the condition $\lambda \gg \xi_o$ is quite far from reality. A better approximation is the so called "extreme anomalous" limit [7], that holds for $\lambda \ll \xi_o \ll \xi$. For $\kappa \ll 1$, one gets:

$$H^2/H_c{}^2 \simeq \varepsilon\,\lambda/\delta\kappa\, f(0)\,(1 - f(0)^2) \qquad \{6\}$$

where δ is the effective penetration length.
In the most naive description, [6,7]: $\varepsilon \simeq 3\sqrt{2}$ and $\delta \simeq \lambda_{eff}{}^{WF}\, f(0)^{-2/3}$, where:

$$\lambda_{eff}^{WF} \simeq (\sqrt{3}\, \xi_o\, \lambda_L^2 / 2.66\pi)^{1/3} \qquad \{7\}$$

WF stands for "weak field" and λ_L is the value of λ for pure material. An approximate model for the dependence of λ, δ and κ on T and ℓ (the mean free path of conduction electrons, that accounts for impurities and defects) is [6,7]:

$$\lambda_L \simeq \lambda_o\, (1 - t_r^4)^{-1/2} \qquad \{8a\}$$

$$\lambda \simeq \lambda_L\, (1 + \xi_o/1.33\ell)^{1/2} \qquad \{8b\}$$

$$\phi_o \simeq 2\pi\, \sqrt{2}\, H_c\, \lambda\, \xi \qquad \{8c\}$$

$$H_c \simeq H_c(0)\, (1 - t_r^2) \qquad \{8d\}$$

$t_r = T/T_c$ is the reduced temperature and $\phi_o \simeq 2.06 \times 10^{-7}$ Gauss.cm^2. From {6} and {8}, we see that the value of H_{sh} is obtained for $f(0)^2 = 5/11$, and:

$$H_{sh} \simeq H_c\, \kappa_o^{-1/3}\, (1 + \xi_o/1.33\ell\,)^{-1/4}\, (1 + t_r^2)^{5/12}\, (1 - t_r^2)^{-1/12} \qquad \{9\}$$

The energy barrier ΔF can be computed [32] in a similar way to the GL approximation. For H close to H_{sh} ($x \ll 1$), the result is:

$$\Delta F \simeq 0.66\, x^3\, H_{sh}^2\, \lambda_{eff}^{WF} / 8\pi \qquad \{10a\}$$

to be compared with the expression:

$$\Delta F \simeq 0.95\, x^3\, H_{sh}^2\, \lambda / 8\pi \qquad \{10b\}$$

obtained from {5}.

2.3 Role of impurities.

According to the Gorkov-Goodman relation [35], the dependence of κ on impurities and defects can be expressed as:

$$\kappa \simeq \kappa(\rho = 0)\, (1 + \xi_o/1.33\ell\,) \simeq \kappa(\rho = 0) + 7.5 \times 10^3\, \gamma^{-1/2}\, \rho \qquad \{11\}$$

where ρ is the normal state resistivity in Ω.cm, and γ the linear electronic specific heat coefficient in erg.cm^{-3}.K^{-2}. Then, from {11}, expressions {4b} and {9} can be easily cast in terms of ρ. An immediate consequence of the above formulae is that superheating can exist for a rather high impurity level. For $\ell \simeq 2\xi_o$, the ratio H_{sh}/H_c decreases by 14% according to {4b} and by 8% according to {9}. In both cases, large superheating is still present if the metal used is a good type I superconductor. Such a slow dependence of H_{sh}/H_c on the normal state resistivity has been checked experimentally by

H. Parr [34] for In-Bi alloys, and previously by C.Valette [15]. As we shall see later, normal state resistivity may be an important parameter for a realistic detector.

2.4 Electronic signal.

The read out system is made of current loops coupled to low noise pre-amplifiers. In the presence of $\vec{H}_o$ one has:

$$\vec{H} = \vec{H}_o + \vec{D} \qquad \{12a\}$$

$$\vec{D}(\vec{r}) = 1/2\, R^3/r^3\, (\vec{H}_o - 3(\vec{H}_o \cdot \vec{r}/r^2)\, \vec{r}) \qquad \{12b\}$$

$\vec{r}$ is the position vector with respect to the center of the sphere. The signal in magnetic field produced by the superconducting to normal phase transition ($H_o > H_c$) is:

$$\Delta\phi = -\int_{\text{loop surface}} \vec{D}(\vec{r}) \cdot d\vec{S} \qquad \{13a\}$$

For $2/3\, H_c < H_o < H_c$, the phase transition occurs from superconducting to intermediate state and the signal in magnetic flux is given by:

$$\Delta\phi_{int} = (3h - 2)\, \Delta\phi \qquad \{13b\}$$

where $h = H_o/H_c$. The signal in voltage $\varepsilon \simeq d\phi/dt$ is expected at small t (time) to follow the law [36,37]:

$$d\phi/dt \sim \rho\, t^{-1/2} \qquad \{14\}$$

and the total flipping time (extrapolated from cylinders [36] to spheres) is of the order:

$$\tau \sim \rho^{-1}\, \pi\, R^2\, 3/2\, H_o\, (3/2\, H_o - H_c)^{-1} \qquad \{15\}$$

Therefore, increasing ρ lowers the flipping time and improves the signal in voltage [23,32].

Various read out systems can be implemented depending on the nature of the experiment: from long hair-pin like loops forming a X-Y matrix and giving position information by time coincidence, to circular pad-like coils. Different types of pre-amplifiers are also being developed ranging from room temperature low noise silicon pre-amplifiers to low temperature Ga-As devices, able to work even at detector temperature. SQUIDS [38] are also being used [39] with rather encouraging results.

2.5 Working temperature

Some materials like Sn (T_c = 3.72 K) or In (T_c = 3.4 K) can be cooled with He^4 but Ga (T_c = 1.09 K), Cd (T_c = 0.56 K) or Zn (T_c = 0.875 K) require lower temperatures. In all cases, working at very low temperature ($T \ll T_c$) minimizes the heat capacity and improves the sensitivity of the detector. The specific heat of a superconductor is given by [40]:

$$c_s = c^{ph} + c_s^{el} \qquad \{16a\}$$

where the phonon term c^{ph} varies as:

$$c^{ph} \simeq a\,T^3 \qquad \{16b\}$$

and the superconducting electron term c_s^{el} is often parameterized [42, 43, 44] by:

$$c_s^{el} \simeq b\,\gamma\,T_c \exp(-f\,t_r^{-1}) \qquad \{16c\}$$

a, b and f are known experimentally for all relevant materials.

The above mentioned dispersion in H_{sh}^{eff} presented by existing collections of granules sets an intrinsic limit to SSGD performance [14, 41]. The thermodynamical superheated critical field is expected to follow, at low t_r, the law [45]:

$$H_{sh} \simeq H_{sh}(0)\,(1 - \alpha\,t_r^2) \qquad \{17\}$$

with $0.5 < \alpha < 1$, according to the approximations used. From both {4b} and {9}, one gets: $\alpha \simeq 0.5$, but the condition $\xi \gg \xi_o$ is not realistic at $T \ll T_c$. The crucial point, however, is the vanishing of dH_{sh}/dt_r at $t_r = 0$. In what follows, we assume that, for a given material, the T dependence of H_{sh}^{eff} is the same for all granules and is similar to {17} up to a T-independent demagnetizing coefficient.

Let us consider a granule flipping at some $H_{sh}^{eff}(T)$. If the applied field is smaller, $H_o = H_{sh}^{eff}(T) - \Delta H$, the granule would flip only if it is heated to a temperature $T + \Delta T$, where:

$$H_{sh}^{eff}(T + \Delta T) = H_{sh}^{eff}(T) - \Delta H$$

Then, relation {17} gives:

$$(t_r + \Delta t_r)^2 - t_r^2 = \alpha^{-1}\,\Delta H / H_{sh}^{eff}(0) \qquad \{18a\}$$

which, for $t_r \ll 1$, simplifies to:

$$t_r' \equiv t_r + \Delta t_r \simeq \Delta t_r \simeq [\alpha^{-1}\,\Delta H / H_{sh}^{eff}(0)]^{1/2} \qquad \{18b\}$$

In a global heating model, where we assume that the whole granule is thermalized before the phase transition starts, the amount of heat needed to flip the grain will be:

$$\Delta Q = \int_T^{T+\Delta T} C(T')\,dT' \qquad \{19a\}$$

where C(T) is the heat capacity of the granule. Writing $C \simeq c_s\,V$, ΔQ becomes:

$$\Delta Q \simeq V\,T_c \int_{t_r}^{t_r'} c_s(t_r''\,T_c)\,dt_r'' \qquad \{19b\}$$

From this analysis, and with the above hypothesis, we see the relevance of the parameter ΔH to estimate the sensitivity of each individual granule at a given value of H_o. If $t_r' \gg t_r$, the value of ΔQ

will be dominated by the contribution from the region $t_r'' \simeq t_r'$ to the integral in {19b}. In other words, the physically meaningful quantity is $c_s(t_r')$ rather than $c_s(t_r)$. Since t_r' is bounded below by {18b} at fixed ΔH, the possibility of taking advantage of very low T specific heats is limited by the realistic values of $\Delta H/H_{sh}^{eff}(0)$. In a practical detector, ΔH cannot be arbitrarily small for all granules at the same time due to the width of the differential superheating curve.

Performing a gaussian fit to dN/dH_o, we can choose two values of H_o, H_{max} and H_{min}, symmetric with respect to the mean value of the distribution, and such that $H_{min} = H_{mean} - \sigma = H_{max} - 2\sigma$, where H_{mean} is the mean value of H_o in the superheating curve and σ the standard deviation. We then require all granules flipping at $H_{min} < H_o < H_{max}$ to be sensitive to the deposit of energy under consideration. Sitting at $H_o = H_{min}$, the highest threshold in magnetic field for such granules will be $\Delta H_D = H_{max} - H_{min}$, which corresponds to grains with $H_{sh}^{eff} = H_{max}$. Then, ΔQ is obtained from {19b} replacing $\Delta H/H_{sh}^{eff}(0)$ by $\Delta H_D/H_{max}$. The value of $\Delta H_D/H_{max}$ tends to a constant for $t_r \to 0$, and provides an intrinsic parameter that characterizes the quality of a given collection of granules. The finite width ($\Delta H_D \simeq 2\sigma \simeq$ 15 to 20%) observed in all existing collections of granules is by now the strongest limitation to the performance of SSG.

2.6 Heat propagation and exchanges in SSG devices.

Three phenomena are involved:

a) To a first approximation, heat propagation inside the granules is governed by the heat diffusion coefficient of the superconducting material. However, since some Cooper pairs are broken at the surface of the granule due to the presence of an applied magnetic field, heat propagation by normal electrons through the surface layer may become important at very low temperatures, especially at the equatorial region of the sphere ($\theta \simeq \pi/2$), where pair breaking is maximum.

b) Heat exchanges between the granules and the dielectric are governed by the Kapitza resistance [46]. In a simple description we can write:

$$dQ/dt \simeq A\,\Delta T/\rho_K \qquad \{20\}$$

where dQ/dt is the amount of heat exchanged per unit time, ΔT the difference in temperature between the two media, ρ_K the Kapitza resistivity and A the surface concerned by heat exchange. Actually, expression {20} is only valid for $\Delta T \ll T$. Otherwise, the difference $T_1^3 - T_2^3$ appears on the right hand side of the equation, T_1 and T_2 being the temperatures of the two media.

c) Heat propagation in the dielectric material is altered by the presence of granules, which provide a random distribution of discontinuities. At the early stage of the diffusion process, when heat propagation is fast as compared to heat exchanges, the granules can be viewed as obstacles to heat propagation. When heat propagation slowers down, the situation may be different and a precise numerical simulation is required for each particular case. For a porous dielectric plunged in liquid helium, superfluid helium penetration can play an important role.

A detailed numerical study of heat propagation and exchanges in some precise scenarios is obviously required. Although paraffin or epoxy are the commonly used dielectric in SSG prototypes, other materials can also be considered. We have recently prepared good quality suspensions of granules in varnish GE 7031, known [47] to have a remarkably good thermal conductivity at very low temperatures. As will be seen later, work along this direction may open the way to fundamental improvements.

2.7 Industrial grain production.

Laboratory techniques to produce granules exist since the early days [3], but a particle physics experiment would require: perfectly reproducible parameters, small sizes, good size homogeneity, good sphericity and surface quality, and the possibility of producing a reasonably large amount of grains (from a few Kg to several tons). It is obvious that such a development can only be made at the industrial level.

Fig.3a shows a sample of tin spheres obtained by modern fabrication techniques (EXTRAMET [48], June 1987). Dispersion in size is apparent in the photograph. Surface defects, easily noticed on the largest granules, form a hexagonal honeycomb structure that may be related to the crystallographic properties. A granulometric study for the best present EXTRAMET production is shown in Fig.3b. 200 to 400 μm diameter BILLITON [49] granules used in our irradiation tests are shown in Fig.4a. The surface structure of these grains appears in Fig.4b, where a regular lattice of defects is again seen.

Progress made by granule producers [48,49] allows now to obtain collections of metastable spheres from 10 μm to 400 μm in diameter. Producing a large batch of small granules (ϕ < 10 μm) is

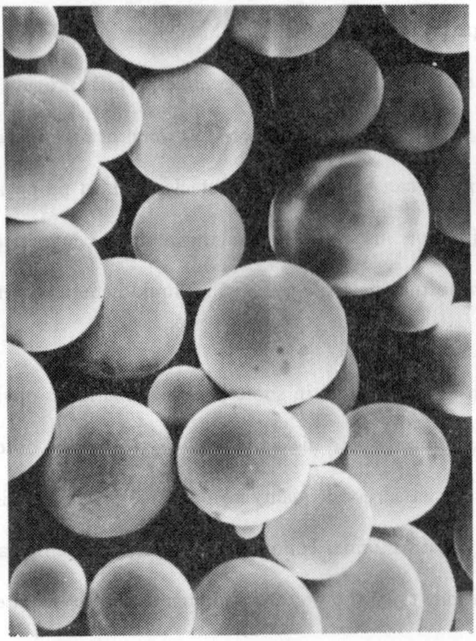

Fig.3a: Tin granules produced by EXTRAMET in July 1987. The sample is ≃ 125 μm × 168 μm.

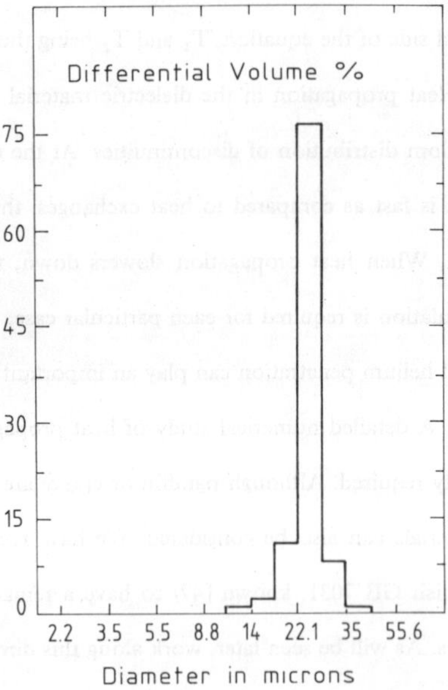

Fig.3b: Granulometric stuy of $\phi \leq 40$ μm tin granules produced by EXTRAMET in January 1987.

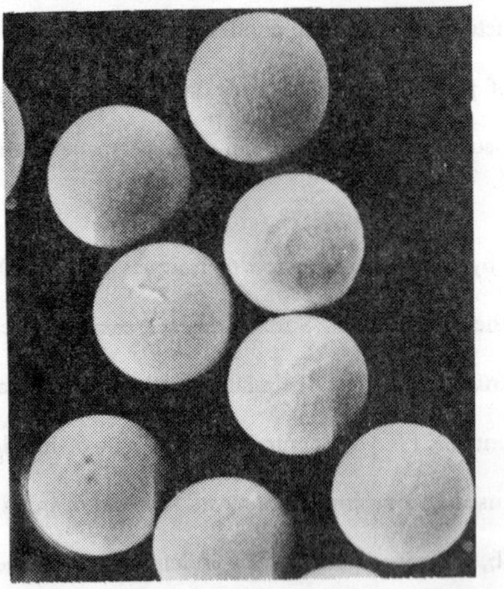

Fig.4a: Tin granules produced by BILLITON in November 1985 with an $Sn_{99}Sb_1$ alloy. The sample is ≃ 0.88 mm × 1.05 mm.

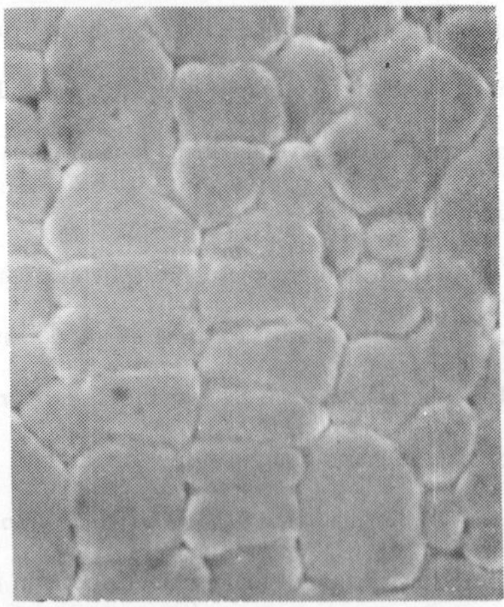

Fig.4b: Detail of the surface of a BILLITON granule. The sample is ≃ 56 μm × 68 μm.

less obvious, especially if good size homogeneity is required. This problem is being studied by EXTRAMET and other producers.

3. IRRADIATION RESULTS WITH REAL TIME READ-OUT SYSTEM.

The electronic read-out used in [15, 10] was not in real time and granules could not be seen to flip individually. A similar situation occurred in the Garching test [29, 30], where Cd grains of diameter 5 μm < ϕ < 25 μm were shown to be sensitive to low energy γ-rays. Results on γ irradiations using 10 μm Sn granules read by RF SQUID have recently been reported [39]. We present here results [50, 14, 31, 51] obtained with α, γ and β^- sources, where Sn granules are read individually by conventional fast electronics. The conclusion of our α irradiation test with large granules have recently been confirmed by the München group [11] with a set up involving single grain irradiation. Results obtained in Paris with an electron beam are reported elsewhere [52].

Our irradiation tests were performed as follows. The applied field H_o was first increased to some value H_{test}. Then, a sweep ΔH_{min} was performed creating a threshold $\Delta H \geq H_{min}$ for all granules that remained superconducting. As explained above, this procedure sets a minimal threshold ΔT in temperature and, in a naive model, can be directly related to an energy threshold ΔQ. By changing the value of ΔH_{min}, one can get information on the sensitivity of the irradiated granules. The threshold in magnetic field for each individual granule inside the collection is then:

$$\Delta H = H_{sh}{}^{eff} - H_{test}$$

The experimental results will be described in terms of the minimum of $\Delta H/H \equiv \Delta H/(H_{test} + \Delta H)$ i.e.:

$$(\Delta H/H)_{min} = \Delta H_{min}/(H_{test} + \Delta H_{min}).$$

3.1 α irradiation

We performed this test from February to July 1986 with the following collections of tin granules:

(a_1) EXTRAMET granules made in March 1985 with Billiton tin, ρ (measured at T = 4.2 K) $\simeq 6\times 10^{-8}$ Ωcm. After sieving, a collection of grains with diameters 45 μm < ϕ < 63 μm was obtained and used for our α irradiations.

(a_2) From the same set of granules, a collection with $125\ \mu m < \phi < 200\ \mu m$ was obtained by sieving and irradiated with α particles.

(b) EXTRAMET granules, made in January 1986 with Prolabo tin in inert atmosphere (helium). The measured value of ρ at 4.2 K was $\rho \simeq 1.5 \times 10^{-7}\ \Omega cm$. By sieving, a collection with $250\ \mu m < \phi < 300\ \mu m$ was extracted and used for α irradiations.

(c) BILLITON granules, $200\ \mu m < \phi < 400\ \mu m$, made in November 1985 with an alloy $Sn_{99}Sb_1$, ρ (estimated) $\approx 10^{-6}\ \Omega cm$.

Small prototypes were prepared with filling factors of 5 to 20% in volume. The multi-turn reading coils were in the range of 3 mm to 1 cm diameter. Because of the high values of ρ of the tin used, the flipping times were rather fast. Rise times in the range of 300ns to 1 μsec were seen at the output of a 142B ORTEC pre-amplifier. The prototypes were put in contact with an open ^{241}Am source (5.5 MeV α's, 4π activity $\simeq$ 5000 Bq) implanted on platinum. Some ^{241}Am atoms drifted into the detectors, that remained irradiated after the source was removed.

Individual granule flip counts were accumulated during finite time intervals. A typical irradiation curve is shown in Fig.5, where N(t) is the integrated number of counts after a time t. The nearly exponential decrease of dN/dt with time reflects the decrease of the number of remaining sensitive granules. We obtained such curves in all reported irradiation tests. A striking characteristics of α irradiations were the large values of $(\Delta H/H)_{min}$ for which flips were still seen. We found in all cases $(\Delta H/H)_{min} \geq 0.2$. The finite width of the superheating curve prevented us from neatly exploring larger values of ΔH_{min}. Fig.6 shows as a function of $(\Delta H/H)_{min}$ the number of flips after 10 minutes observed for collection (a_2) at $T \simeq 3.28$ K. The main result of our α irradiations is that large granules turned out to be sensitive to deposits of energy much smaller than those predicted by the expression {19a} taking for C(T') the heat capacity of the whole granule (global heating model).

3.2 140 keV γ irradiation

Our γ irradiation tests started in October 1986, but the results reported here were obtained with a collection:

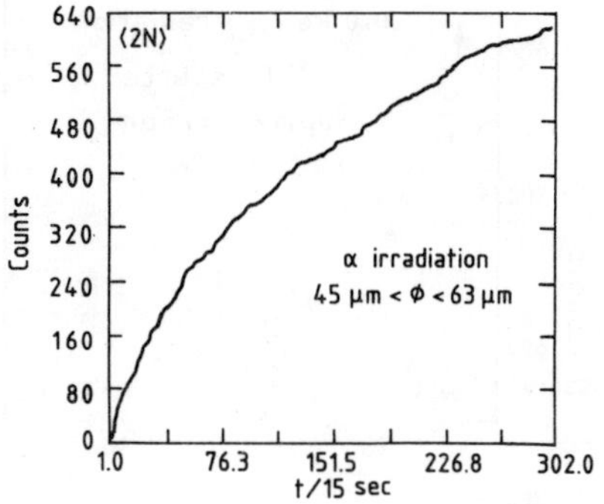

Fig.5 : Integral irradiation curve for sample a_1 at $T \simeq 1.5$ K (^{241}Am α's). The integrated number of counts, N is shown as a function of the irradiation time t. The results of two identical tests have been added in order to attenuate statistical fluctuations

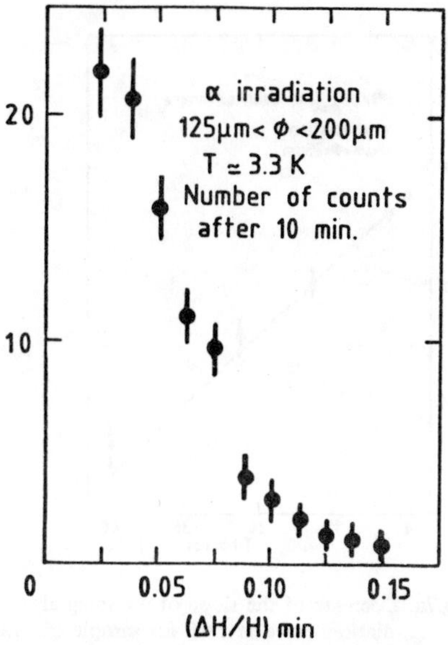

Fig.6 : Number of counts after 10 min for sample a_2 irradiated with ^{241}Am α's at $T \simeq 3.3$K, as a function of the relative gap in magnetic field $(\Delta H/H)_{min}$.

(d) EXTRAMET granules, produced in December 1986 and January 1987 with Metallum tin, ρ (estimated) $\approx 10^{-7}$ Ωcm. The new granules had smaller sizes than the previous collections (ϕ < 80 μm and ϕ < 40 μm). After sieving, collections of granules with 10 μm < ϕ < 25 μm were obtained and used for our γ irradiation test.

A medical source of ^{99}Tc (140 keV γ's, activity ~ 200 μC) was deposited on an absorbing texture, isolated and put inside the cryostat very close to a 2.5 mm diameter detector with a granule filling factor of about 15%. After 20 min of irradiation at $T \simeq 1.5$ K, the integrated number of counts amounts to about 6% of the total number of granules read in the detector. Asymptotically ($t \to \infty$), one can estimate the fraction of sensitive granules to about 8%. Flips were observed for $(\Delta H/H)_{min}$ larger than 2%. This is a rather good sensitivity as compared to other results on γ irradiation. The 6 hours life-time of the source was used to check that flips were due to individual particles. Shown in Fig.7a as a function of the source activity is dN/dt at the beginning of different irradiations at $(\Delta H)_{min} = 0$. Fig.7b shows the number of counts after 20 min. as a function of $(\Delta H/H)_{min}$.

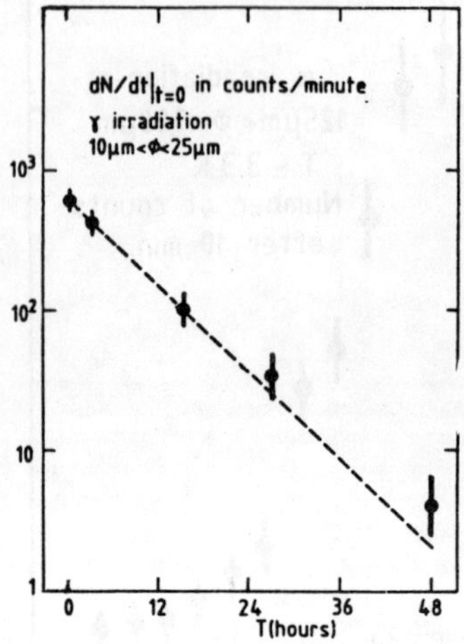

Fig.7a: Decrease of the slope of the integral irradiation curve at $t=0$ for sample (d) irradiated with ^{99}Tc 140 Kev γ's, as a function of the time T since the source started being used. Data are taken at $T \simeq 1.5$ K and $(\Delta H/H)_{min} = 0$. The dashed line reproduces the time dependence of the source activity.

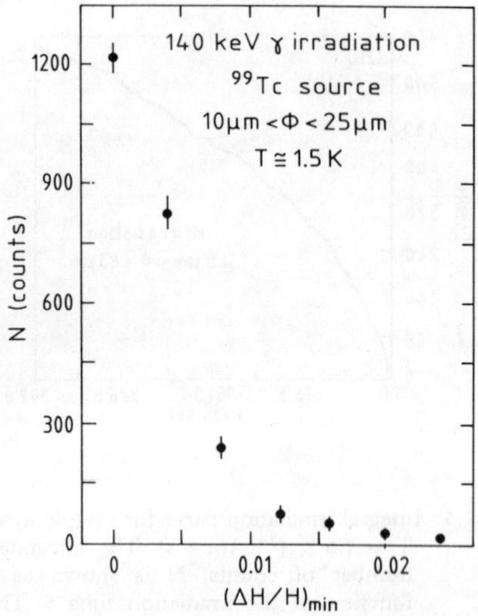

Fig.7b: Integrated number of flips observed after 20 min for sample (d) irradiated with ^{99}Tc γ's at $T \simeq 1.5$ K, as a function od $(\Delta H/H)_{min}$

3.3 β^- irradiation.

β^- irradiation tests were performed with granules of collections (d) and:

(e) EXTRAMET granules, produced in June 1987 with the standard BILLITON $Sn_{99}Sb_1$ alloy. From a production of $\phi < 40\mu m$, collections with $10\mu m < \phi < 25\ \mu m$ were extracted by sieving. Several filling factors were implemented, from 5 to 30%.

Two β^- sources were used:

- ^{36}Cl emitting β^- rays of energy $0 < E < 714$ keV (maximum probability at $E = 252$ keV).
- ^{14}C emitting β^- rays of $0 < E < 158$ keV (maximum probability at $E = 49$ keV).

The sources (very low activity, in the 100 Bq range) were mixed directly with the detector. For ^{36}Cl, a detailed study was performed leading to new interesting results. With sample (d) flips were observed [31] at $(\Delta H/H)_{min} \leq 1\%$. Sample (e) appears to exhibit a much better sensitivity, with flips at $(\Delta H/H)_{min} \leq 4\%$ and about 15% sensitive granules. Since samples (d) and (e) contain granules of the same sizes, produced in similar conditions, a tempting interpretation of such a result would be that sensitivity depends strongly on the purity of the material used. Such a strong dependence of sensitivity on normal state resistivity, if confirmed by other tests, would provide a direct evidence for local heating.

3.4 Tests with very low energy sources

Several low energy sources where used: ^{55}Fe (6 keV γ's); ^{3}H (β^-, $0 < E < 18$ keV, peaked at $E = 5$ keV); ^{65}Zn (γ, $E = 8{,}9$ keV). Granules of samples (d) and (e) were irradiated at several filling factors (5 to 30%). Again sample (e) seems to exhibit better sensitivity than sample (d).

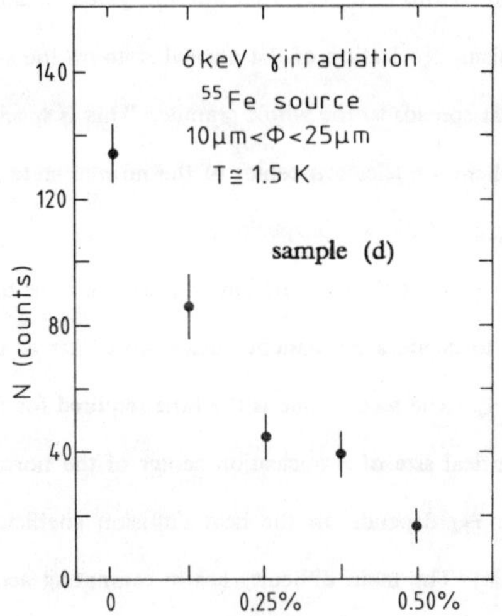

Fig.8a: Integrated number of counts observed after 1 hour for sample (d) irradiated with ^{55}Fe 6 keV γ's at $T \simeq 1.5$ K, in terms of $(\Delta H/H)_{min}$.

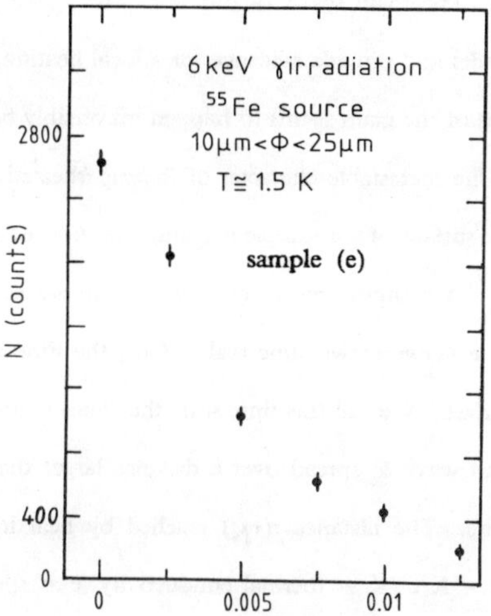

Fig.8b: Integrated number of counts observed after 20 min. for sample (e) irradiated with ^{55}Fe 6 keV γ's at $T \simeq 1.5$ K, in terms of $(\Delta H/H)_{min}$.

Fig.8 shows ^{55}Fe (about 1 μC) irradiation results for both samples in terms of $(\Delta H/H)_{min}$. About 5% of granules in sample (e) turn out to be sensitive to 6 keV γ's. For sample (e), in a global heating model, we should have $\Delta Q \simeq 25$ keV for 10 μm grains and $\Delta Q \approx 400$ keV for $\phi = 25$ μm. Again, sensitivity seems to depend on normal state resistivity and local heating may be at work.

3.5 Local heating mechanism and normal state resistivity

In a global heating model, the particle must deposit enough energy to raise the temperature of the whole sphere in order to induce a flipping process. Assuming uniform distribution of heat inside the granule, grains of sample a_1) were expected to be sensitive to individual 5.5 Mev α particles for $\Delta H/H < 0.016$ at $T \simeq 1.5$ K. For larger values of $\Delta H/H$, a sharp cutoff would be the normal behavior if global heating were the relevant mechanism. At $T \simeq 3.4$ K, the amount of heat ΔQ required for a 50 μm grain flip with $\Delta H/H = 20\%$ would be an order of magnitude larger than the energy of the α particles. At $T \simeq 1.5$ K, $\Delta Q \simeq 130$ MeV.

The main result of our tests with α particles is therefore that they rule out the global heating model and provide evidence for a local heating mechanism. Nucleation of the normal state on the surface of the grain seems to happen irreversibly before heat spreads to the whole granule. This is specific of the metastable character of the superheated state, where a nucleation center of the normal state on the surface of the sample is enough to induce the transition of the whole specimen.

A complete model of local heating has never been elaborated, but it mainly implies a precise balance between two time scales. One, the time required to create a permanent nucleation center at the surface. We call this time scale the "nucleation time" τ_N. The second one is the time required for the heat wave to spread over a distance larger than the critical size of a nucleation center of the normal state. The distance $r(\tau_N)$ reached by heat in a time τ_N depends on the heat diffusion coefficient $D = K/c$ (K = thermal conductivity, c = specific heat). The main difficulty lies in estimating accurately τ_N, in terms of H, ρ and the amount of energy ΔE deposited near the surface of the grain.

The penetration of the magnetic field into superconducting samples at $H_o > H_c$ has been carefully studied for cylinders [36, 37], and some of these results have been extrapolated to spheres [17] without further justification. We do not believe such an extrapolation to be entirely correct, since superheated spheres can go to the intermediate state, which is not the case for cylinders when the applied field H_o is parallel to the axis. But, locally, this description [36] and [37] is likely to give the correct order of magnitude for a small penetration around a nucleation center on the surface of the sphere. In the simplest model [36], the time evolution equation for penetration of a parallel magnetic field into a superconducting cylinder of radius R at $H_o > H_c$ is:

$$t(d_N) \simeq \rho^{-1} \pi [R^2 - (R-d_N)^2(1 + 2 \ln R/(R-d_N))] H_o/(H_o - H_c)$$

where d_N is the distance to which the magnetic field has penetrated. Then, an ansatz for τ_N, to get the right order of magnitude, could be to take $d_N \sim \xi$ and write:

$$\tau_N \sim 2\rho^{-1}\pi\xi^2 H/(H - H_c)$$

where H is the value of the field near the point of the surface where the deposit of energy took place. For tin ($\xi \simeq 2300$ Å) and $H \simeq 2 H_c$, one gets: $\tau_N \sim 6 \times 10^{-9}$ sec if $\rho \sim 10^{-9}$ Ω.cm and $\tau \sim 6 \times 10^{-12}$ sec if $\rho \sim 10^{-6}$ Ω.cm. In the last case, we reach the characteristic time scale of Cooper pairs.

Changing ρ also modifies the value of D, mainly through the electron contribution to the thermal conductivity K. For tin alloys, data at superconducting temperatures [53] show a linear T dependence consistent with the Wiedemann-Franz law, whereas the ratio K(superconducting)/K(normal) remains reasonably close to 1 in the whole range $0.4 < T/T_c < 1$. In what follows, we neglect microscopic phenomena and use the heat diffusion equation :

$$\partial \Delta T/\partial t = D \Delta' (\Delta T)$$

where, in the second term, Δ' means "laplacian". A rough estimate of the possibility of local heating phenomena can be given considering the case of an α particle releasing its energy on the surface of the grain, and approximating the distribution in r of ΔT by an isothermal half sphere of radius r(t). For r(t) we take the maximum of the distribution $\exp(-r^2/4Dt)$ that appears when considering the three dimensional isotropic diffusion equation. One then gets: $r(t) \simeq 2\sqrt{tD}$. The temperature at the center of the sphere varies as $t^{-3/2}$, as usually in three dimensional diffusion phenomena. To further simplify the calculation, we ignore effects due to the granule-dielectric interface and assume that the above formulae apply to a half sphere of radius $r(\tau_N)$. From data on thermal conductivity [53] and specific heat [42, 43], we can make a rough estimate of the energy required to heat a half sphere of radius $r(\tau_N)$ centered on a point of the surface of the grain and produce nucleation, for $\Delta H/H > 0.2$, at $T \simeq 1.5$ K. With the value $\rho \simeq 6 \times 10^{-8}$ Ωcm, one gets:

$$r(\tau_N) \sim 11\ \mu m \text{ and } \Delta Q \sim 3\ MeV$$

which is compatible with our experimental results.

We have performed an additional irradiation test to check the local heating mechanism, with sample (c). Due to the high residual normal state resistivity of such grains, local heating at the $\approx 1\mu m$ scale may be expected. The granules were irradiated with ^{55}Fe γ's ($E \simeq 6$ keV). Remarkably enough, flips were observed for $(\Delta H/H)_{min} \leq 0.01$. In a global heating model, for $\phi = 200\ \mu m$, this would correspond to an energy threshold of about 120 MeV, i.e. four orders of magnitude larger than the energy of ^{55}Fe photons. Such data lead to $r(\tau_N) < 5\ \mu m$. Fig. 9 shows the variation of the number of flips seen after 1 hour, as a function of $(\Delta H/H)_{min}$. In reasonable agreement with this result, data from sample (e) irradiated with 6 keV γ's give $r(\tau_N) < 3\ \mu m$. The dynamics of local heating is certainly a crucial item and deserves further study.

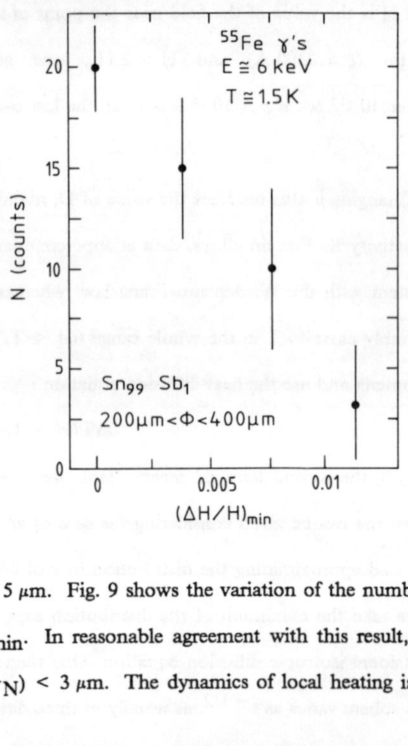

Fig.9: Counts after 1 hour for sample c) irradiated with 6 keV γ's versus $(\Delta H/H)_{min}$.

4. VERY LOW TEMPERATURE STUDIES

4.1 The avalanche effect

Equilibrium superconducting to normal state phase transition is endothermic, but in the case of superheating a positive latent heat is released at low t_T. The released latent heat per unit volume q^ℓ is given by [50,9] :

$$q^\ell = H_c^2 (8\pi)^{-1} (3/2\, h^2 - 1 - 4t_T^2/(1-t_T^2)) \qquad \{21a\}$$

for $H_o > H_c$ (transition from superconducting to normal state), and :

$$q^\ell = H_c^2 (8\pi)^{-1} [2 + 9/2\, h^2 - 6h - (12h-8)t_T^2/(1-t_T^2)] \qquad \{21b\}$$

for $2/3\, H_c < H_o < H_c$ (transition from superconducting to intermediate state). At $H_o = H_c$, one gets:

$$q^\ell > 0 \text{ for } t_T < 1/3$$

For Cd granules (where superheating is rather high), the Garching group [29] observed an "avalanche effect" at $T \le 350$ mK. The heat released by the flip of a single granule spreads to the surrounding granules and produces new flips, which in turn release more heat so that the whole detector collapses. We performed tests at $T \ge 450$ mK with several samples of Sn granules with about 10% dilution coefficient in volume, and using a He^3 cryostat where the sample was directly in contact with helium. No avalanche effect was seen and the detector could be normally operated in the whole range of T accessible to our cryogenics (between 450 mK and 1 K).

In order to understand the difference between both experiments, we study here more closely the condition for an avalanche to occur. In what follows we assume uniform heat propagation inside the detector, where only the granules and the dielectric material are present (detector in vacuum). The situation may be more complex in the presence of superfluid helium, which is a good thermalizer.

Assume that, at some fixed T and H_o, an avalanche propagates carrying a uniform increase in temperature ΔT and flipping granules at $\Delta H \le \Delta H(\Delta T)$. We then write a consistency condition requiring that the heat released by granule flips is enough to produce the above ΔT:

$$q^\ell\, c_{det}^{-1}\, \Delta\psi = \Delta T \qquad \{22\}$$

where ψ is the filling factor in volume and $\Delta\psi$ the filling factor for the fraction of granules flipping at $\Delta H \le \Delta H(\Delta T)$. c_{det} is the average heat capacity per unit volume of the detector. On the other hand, $\Delta H(\Delta T)$ is related to ΔT through the expression:

$$H_{sh}(T+\Delta T)/H_{sh}(T) = H_o/(H_o+\Delta H) \qquad \{23\}$$

In the limit $\Delta H, \Delta T \ll 1$, one gets from the last two equations:

$$\delta \equiv - q^{\ell} c_{det}^{-1} T_c^{-1} H_o d\psi/dH_o H_{sh}^{-1} dH_{sh}/dt_r = 1 \qquad \{24\}$$

and the avalanche effect is expected to happen for $\delta > 1$. If $\delta < 1$, the flip of a granule can still generate other flips but the chain will stop before reaching macroscopic size. From {24} we can compute, for a given detector, the temperatures and values of H_o at which the avalanche effect can be expected to occur. q^{ℓ} can be written [31, 14] as:

$$q^{\ell} = q^{\ell}_r H_c^2/8\pi \qquad \{25\}$$

where q^{ℓ}_r is a function of $h = H_o/H_c$ and t_r. Assuming $c_{det} \simeq A T^3$, and taking [14]:

$$H_{sh}(t_r) \simeq H_{sh}(0) (1+t_r^2)^{5/12} (1-t_r^2)^{11/12} \qquad \{26\}$$

derived from approximations based on the extreme anomalous limit, equation {24} becomes:

$$\delta \simeq A^{-1} q^{\ell}_r t_r^{-2} (1 + 8/3\, t_r^2)(1-t_r^2)(1+t_r^2)^{-1} H_o d\psi/dH_o H_c(0)^2/8\pi T_c^4 \qquad \{27\}$$

Making the rough assumption that the dielectric dominates the heat capacity of the detector, we can take A to be independent of ψ, and of the superconducting material used. Then, at fixed t_r and h, the dependence on the superconducting material appears through the factor $\zeta = H_c(0)^2/8\pi T_c^4$. For Sn, one has: $\zeta = 19.3$ erg cm^{-3} K^{-4}. In the same units, one gets: $\zeta = 25.6$ for In, 73.3 for Ga and 364.1 for Cd. Such figures clearly show that the avalanche effect should occur much more easily for Cd than for Sn. Also, since superheating is lower for Sn than for Cd, q^{ℓ}_r is smaller enhancing therefore the difference in behavior of the two materials. We have attempted a rough numerical estimate for Cd, Sn, In and Ga, setting: $A \simeq 4 \times 10^{-5}$ joule cm^{-3} K^{-4}, $\psi = 10\%$ for Sn, In and Ga, and 4% for Cd., and taking the top of the superheating curve at: $H_o = H_c$ in the case of tin, $H_o = 1.6 H_c$ for cadmium and gallium, $H_o = 1.12 H_c$ for In. The value of H_o chosen for gallium must be considered as an upper limit to the quality of superheating expected for this material. The superheating curve has been approximated by a gaussian distribution of width $2\sigma = 0.16$ for Cd, In and Ga, and $2\sigma = 0.2$ for Sn. Numerical results for several temperatures are shown in Fig.10, where δ is presented as a function of h. The temperatures chosen for Cd correspond of superheating curves reported by the Garching group [29], where avalanches were observed. Our numerical results clearly explain the presence of avalanche phenomena. For tin, T and the superheating curve have been chosen to reproduce the conditions of our test, where no avalanche was found. Again, our estimate agrees with experiment but the avalanche effect may appear at slightly lower temperatures. For In, at 10% filling factor in volume, we expect avalanches to occur at $T < 600$ mK and even at higher T if a larger ψ is taken. This point should be seriously studied, since it may bring severe constraints on the design of a solar neutrino In detector. Also, Ga dark matter detectors will have to face similar phenomena, even at low filling factors. However, from {11}, superheating for Ca and Ga can be lowered adding impurities to the superconductive material.

4.2 Heat capacity of a small granule in the presence of a magnetic field

The naive expression $C = V c_s$ is only valid for a superconducting sample at $H_o = 0$. Otherwise, the perturbation of the order parameter, $f(0) < 1$ near the surface, brings an extra contribution where the normal electron specific heat (that decreases only linearly with T) plays a significant role. We expect such a term to be important at $t_T \ll 1$ for small spheres (radius $R \approx \xi$). The details of the calculation were first presented in [41] and are further developed in [31]. The heat capacity writes then:

$$C = V c_s + S c' \qquad \{28\}$$

where S is the grain surface and a simple expression for c′ can be obtained in the GL approximation with $\kappa \ll 1$, $R \gg \xi$, $H_o \simeq 2/3\, H_{sh}$ and assuming {8}:

$$S c' \simeq 0.07\, T_c^{-1} H_c^2 \xi_o t_T R^2 \qquad \{29\}$$

4.3 Numerical estimates

Assuming global heating, we have performed numerical estimates mainly for two cases: a) Solar neutrino detection with In granules; b) Galactic photino detection with Ga granules.

Results are given in terms of $(\Delta H/H)_D = 2\sigma$. Setting $H_o = H_{min}$, we require granules with $H_{sh}^{eff} = H_{max}$ to be sensitive to the deposit of energy under consideration and estimate the maximal allowed size of granules. For the In detector, granules should be sensitive to energy deposits close to minimum ionization. We thus optimize energy resolution for electrons in the range $E \leq 1$ MeV. Granules of diameter $2\,\mu m < \phi < 4\,\mu m$ turn out to be necessary for $0.5 < \alpha < 1$ if $(\Delta H/H)_D \equiv \Delta H_D/H_{max} = 0.05$, which is for the time being far from the range of any existing SSG sample. If a larger gap in H_{sh}^{eff} is allowed, let us say, $(\Delta H/H)_D = 0.1$, the required diameters would be in the range $1\,\mu m < \phi < 2\,\mu m$. The operating temperature should in both cases be $T \simeq 200$ mK,

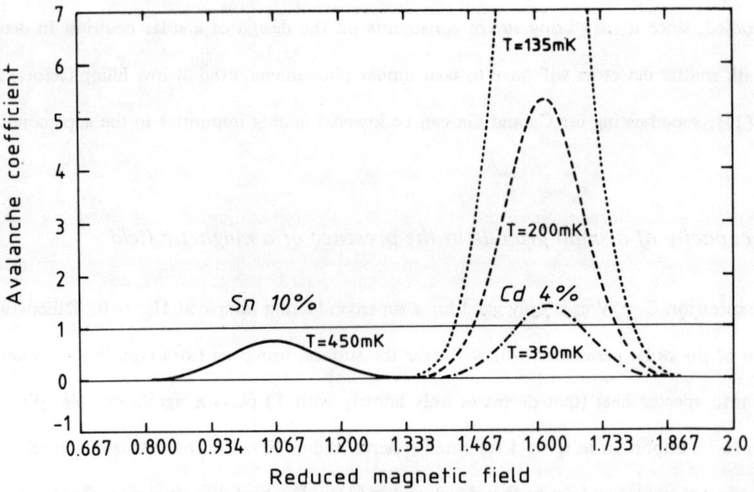

Fig.10a: Avalanche coefficient δ in terms of the reduced magnetic field $h = H_o/H_c$, for: 1) Sn granules at T = 450 mK and filling factor (volume) ψ = 10%, 2) Cd at ψ = 4% and T = 350 mK (dot-dash line), T = 200 mK (dashed line), T = 135 mK (dotted line).

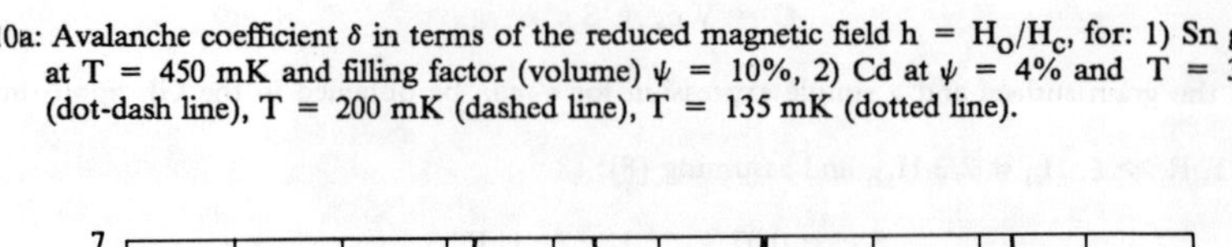

Fig.10b: Avalanche coefficient δ in terms of reduced magnetic field h for: 1) In at ψ = 10% , and T = 650 mK (full line), T = 200 mK (dot-dash line), 2) Ga at ψ = 10%, T = 100 mK (dashed line), T = 550 mK (dotted line).

but then it is not obvious that the avalanche effect can be avoided at high filling factors. Working at much higher T would, in a conventional scenario, degrade sensitivity and require smaller granules. The surface term {29} turns out not to contribute significantly to the heat capacity of In granules.

In the case of gallium, the situation is different and the surface term starts to become important if one works at $T \simeq 100$ mK. With $(\Delta H/H)_D \simeq 0.05$, about 3 μm diameter granules would be sensitive to 50 eV (approximate deposit of energy from light galactic photinos), whereas 10 μm granules would detect 1 keV energy deposits (heavy photinos, solar axions). The coherence length of gallium has been taken to be equal to 1 μm in the present estimates, which can be considered as an upper limit. Lower values of ξ would lead to less important surface terms and bring better sensitivity.

Finally, we recall that the performance of Cd SSG detectors was estimated in Seidel's Thesis [29].

5. AMPLIFICATION BY THERMAL MICRO-AVALANCHE.

Our previous study of the avalanche effect opens the way to an exciting line of research: at $0 < \delta < 1$, latent heat is released by the superconducting to normal phase transition, so that other granules can flip as heat propagates in the composite medium. Still, macroscopic avalanche is not reached and the perturbation remains localized. All existing estimates of the performance of SSG devices as particle detectors rely on the assumption that SSG response consists only in the phase transition of one or several granules crossed by the particle. The existence of the avalanche effect provides an obvious experimental counter-example to this commonly admitted scenario. We propose, here, to further study the details of heat exchanges in the composite medium, and how latent heat can improve the performance of the detector in four crucial points:

a) As previously explained, the electronic signal over noise ratio is often too weak. In a 4 ton indium solar neutrino experiment with a realistic number of electronic channels, it would be extremely difficult to see the signal produced by a 116 keV γ ray if the conventional scenario for SSG response applies. Localized thermal avalanches offer the possibility to produce extra grain flips, and as will be seen later may lead to a large increase of the electronic signal.

b) Granules size homogeneity is an usual requirement to have a reasonably reproducible SSG response to low energy particles. Otherwise, the produced signal may strongly depend on where most of the energy was deposited. Localized micro-avalanches spread heat to a slightly larger region of the detector and, under some precise conditions, we expect the dispersion in granules size to play a much less important role.

c) A major drawback of SSG experiments based on the nucleus recoil energy lies in the fact that only one granule is hit by the incident particle. In such case, SSG devices provide only a rough threshold, with no energy resolution. Again, the situation is drastically changed in the presence of thermal micro-avalanches, and we claim that good energy resolution can in principle be reached.

d) It is usually assumed that energy deposited in the dielectric material does not play a significant role in the response of SSG devices. This is obviously wrong in the new scenario, where heat exchanges are taken into account. A fundamental improvement is then the possibility of incorporating target materials that are not type I superconductors (e.g., hydrogen target for dark matter searches).

The basic phenomenon, for a point-like energy deposit ΔE and $\Delta H_{min} = 0$, can be sketched as follows. Let us assume that heat propagates as an isothermal hot sphere in the detector centered around the point where the collision took place. After a time t, the sphere of radius $\eta(t)$ will be heated by:

$$\Delta T = (\Delta E + V_{flip} q^{\ell}) / (4/3 \, \pi \, \eta^3 \, c_{det})$$

where the volume V_{flip} of granules having changed state obeys the equation:

$$dV_{flip} = 4 \, \pi \, \eta^2 \, \Delta\psi(\Delta T) \, d\eta$$

If $\Delta T \ll T$, we can write from {24}

$$\Delta\psi / \Delta T = \delta \, c_{det} / q^{\ell}$$

Therefore:

$$dV_{flip} = 4 \, \pi \, \eta^2 \, d\eta \, \delta \, c_{det} \, \Delta T / q^{\ell}$$

and:

$$q^{\ell} \, dV_{flip} / (\Delta E + V_{flip} \, q^{\ell}) = 3\delta \, d\eta \, \eta^{-1} \quad \{30\}$$

Assuming that the granule hit by the particle changes state, {30} can be integrated to:

$$\ln [(\Delta E + V_{flip} \, q^{\ell})/(\Delta E + V_o q^{\ell})] \approx 3\delta \ln \eta/\eta_o \quad \{31\}$$

where $\eta_o = \psi^{-1/3} R_o$, R_o is the mean granule size and $V_o = 4/3 \, \pi \, R_o{}^3$. Then:

$$V_{flip}(\eta) \approx [(\eta/\eta_o)^{3\delta} - 1] \, \Delta E/q^{\ell} + V_o(\eta/\eta_o)^{3\delta} \quad \{32\}$$

and, as $\eta \propto t^{-1/2}$, one gets the signal in voltage:

$$d\Phi/dt \propto t^{3\delta/2 - 1} (\Delta E/q^{\ell} + V_o) \quad \{33\}$$

The total amount of sensitive granules at $\Delta H_{min} = 0$ diverges as $\eta^{3\delta}$, but in a realistic set-up a finite small value of ΔH_{min} is always created. Then, the signal in voltage will be cut at $\eta = \eta(\Delta H_{min})$, defined as the value of η for which ΔT is too low to overcome the magnetic field threshold ΔH_{min}. If $\Delta E \gg 4/3\ \pi\ R_o^3\ q^\ell$, the last term in {32} can be neglected. The signal is then independent of the response of the first granule and proportional to the deposited energy. The best choice for granules size may then be $\phi \approx 1\mu m$. Working temperature would not be very low. For $H_o = H_c$, $q_\ell = 0$ at $T = 1/3\ T_c$. An indium detector with $\psi = 30\%$ could operate at $T \simeq 1$ K. The choice of lower cutoffs in {31} and {32} is only indicative and should not be taken literally. Further work on that point is obviously required.

A rough numerical estimate can be made for In granules as follows. Using varnish GE 7031 as the dielectric, existing data at $T \simeq 1$ K [47] give: D(dielectric) $\approx 10\ cm^2\ s^{-1}$, whereas the Kapitza resistivity is of the order: $\rho_K \approx 10^{-3}\ K^4\ m^2\ W^{-1}$. From {20}, taking $\phi = 1\ \mu m$ and $\Delta T \ll 1$, heat transfer between granules and the dielectric should happen in a time $\tau_K \approx 10^{-8}$ sec. At the beginning of the interaction, when ΔT is large, we expect exchanges to be faster. In 200 nsec, i.e., the time resolution required for the In solar neutrino experiment, heat has spread over a radius $\eta \approx 30\ \mu m$ in the composite medium. With dilution coefficient $\psi = 30\%$, $q^\ell \approx 0.3$ keV μm^{-3}, $\delta \simeq 1/3$ and $\eta/\eta_o \approx 20$, the yield in flipping volume would be of the order: $V_{flip} \approx 40\ \Delta E\ \mu m^3\ keV^{-1}$ for low energy electrons and photons. A 110 keV γ would then produce a signal in magnetic flux equivalent to a 20 μm diameter granule (with an efficiency close to 100%), which can be easily read with existing low noise fast amplifiers. Estimates for dark matter experiments based on Cd or Ga nucleus recoil also provide rather encouraging figures, but we differ their presentation to a more detailed study.

6. PROPOSED EXPERIMENTS: A CRITICAL REVIEW

None of the proposed experiments with SSG seems to us immediately feasible within the present status of the art, but recent work on the basic properties of SSG devices should open the way to fundamental improvements in the near future. Except for monopole detection, which would involve large granules, the feasibility of SSG detectors depends on two basic questions:

a) Can very small granules ($\phi < 2\ \mu m$) be produced in large quantity through a perfectly reproducible procedure? Progress made in the late years by industry is encouraging, but by now no definite answer exists.

b) Will amplification by thermal micro-avalanche provide the expected sensitivity and energy resolution? This is an open field of research.

Solar neutrino detection through coherent scattering [19] is based on the recoil energy of the scattered nucleus. Cross sections are large, but recoil energies are small (1 eV or less for pp solar neutrinos using Ga or Zn granules, about 5 eV for ^{7}Be neutrinos). No feasibility result exists for such an experiment, which appears to be the most difficult one. However, it is not excluded that with very small granules ($\phi < 1\ \mu m$) and using micro-avalanches, a detectable signal would eventually be produced.

The alternative way to detect solar neutrinos [20] uses Raghavan's reaction [21]:

$$\nu\ (E > 128\ \text{keV}) + {}^{115}\text{In} \rightarrow {}^{115}\text{Sn}^{**} + \beta^-\ (E \simeq E_\nu - 128\ \text{keV})$$

where the excited tin level decays into 2 γ's ($E \simeq 497$ keV and $E \simeq 116$ keV) with a lifetime of about 3.2 μs. The main problem here is the low event rate (a solar neutrino event per day for 4 tons of In) and the severe background due to In natural radioactivity. With very good quality granules, ($\Delta H_D/H_{max} = 0.05$, i.e. 3 to 4 times better than existing samples), grain size in the range $2\ \mu m < \phi < 4\ \mu m$ would be needed to reach full sensitivity to ≈ 1 MeV electrons, in the conventional scenario. Recent developments on SSG real time read out (LAPP Annecy, College de France, ENS Paris, INFN Torino, MPI Munchen, UBC Vancouver,...) seem to provide well suited amplifiers for the signals expected with the micro-avalanche effect. Work with $\approx$ 1cc prototypes, or even smaller ones, is still necessary before a more ambitious program can be undertaken. A reasonable step for the near future would be, in our opinion, to produce granules of $\phi < 5\ \mu m$, to optimize electronics for such granules and to test the micro-avalanche scenario with varnish GE 7031 or Epibond resin. If the basic principle works, it would then be possible to build a $\approx 1cm^3$ detector where the interactions of low energy electrons and photons could be studied in some detail.

Cold dark matter detection [24,25,26] is again based on the nucleus recoil energy, but now recoil energies are much higher (typically in the 50 eV to 1 keV range depending on the particle mass). According to our estimates [41,50], Ga granules of 3 μm diameter at $T \simeq 100$ mK, with

$\Delta H_D/H_{max} \simeq 0.05$, would be required to detect 50 eV deposit of energy, whereas 10 μm granules would be sensitive to 1 keV. A similar calculation applies o solar axion detection, where it is claimed [27] that axion conversion in atoms allows for a reasonable rate of events using comparatively small detectors. In both cases, the main problem seems to be the severe background from radioactive elements below 1 keV. The crucial problem of energy resolution should be solved by thermal micro-avalanches, and therefore the program for a feasibility study would be quite close to that proposed for In solar neutrino detection. The possibility of using the dielectric as a dark matter target looks exciting for light galactic photinos, which do not exhibit coherent scattering and may transfer a high recoil energy to a hydrogen nucleus.

Double β experiments [28] proposed with superconducting materials such as Cd (Q = 2.8 MeV) or Mo (Q = 3 MeV) have to face both the energy resolution problem and radioactive background. But space and time resolution can play a significant role in reducing the background. It remains also to be proven that small Mo granules can actually be produced.

Finally, we proposed to detect magnetic monopoles [22,23] some time ago. The case for feasibility of such an experiment has been improved by our result on the metastability of very large granules at high values of ρ, and therefore with short flipping times [23,32]. Here, the destruction of the superheated state is not due to a thermal phenomenon, but to the flux tube injected by the monopole going through the granules. As this flux tube is created for any value of β (β = v/c, v = speed of the monopole), slow monopoles can as well be detected. This is a real-time, track detector for monopoles of all velocities working with conventional electronics. Improving the surface quality of large granules and making a conclusive check of the detection principle itself seem to be the necessary steps before proposing a large scale detector.

7. CONCLUSION

We have discussed several important aspects of SSG detectors: phase transition mechanism, very low T behavior and heat exchanges in the detector. The idea of amplification by thermal micro-avalanche has been developed from a detailed study of the low temperature properties of SSGD.

Irradiation results are encouraging in the sense that sensitivity is always at least as good as that predicted by the global heating model. For the granules used, local heating phenomena seem to be at work and sensitivity is improved. It is then possible to use global heating calculations as a lower bound to the expected sensitivity in Particle Physics experiments. Even so, the conventional scenario (where heat exchanges are neglected) leads to rather pessimistic estimates for the feasibility of most proposed experiments. Only magnetic monopole detection with very large granules, as proposed by us four years ago, would have a chance of leading to a positive feasibility result. But, as stressed previously, such a scenario rests on no experimental basis at $T \leq 1K$, where latent heat can be positive.

A R&D program should be based on low temperature studies ($T \simeq 1K$ for indium) and a serious attempt to implement amplification by thermal micro-avalanche:

a) Experimental check of the proposed description of avalanche phenomena, in terms of the granule filling factor, operating magnetic field and temperature, and materials used.

b) Industrial developments for the production of very small granules are likely to be performed in several steps, the first one concerning production of $\phi < 10\ \mu m$ grains. At that stage, some micro-avalanche studies can already be performed, based on local heating and with $\Delta E \sim 1\ MeV$.

c) Development of good quality production of $\phi < 5\ \mu m$ granules and check of the micro-avalanche principle itself: thermal contact between dielectric and granules, heat propagation in the composite medium, existence of localized micro-avalanches in a realistic set-up.

d) Study of the energy resolution in a 1 cc. three-dimensional multichannel set-up. If, according to our expectations, signals turn out to be large enough, the event position can be determined by comparing the signals produced in different loops and looking at correlations between the rise-time and the signal amplitude (the upper cutoff in t will appear later for larger values of ΔE). Some information on energy resolution would then come out from irradiation experiments. We do not expect energy resolution with 5 μm granules to be good enough, but its study should lead to the development of rather accurate models, that could be extrapolated to $\phi < 1\ \mu m$ or $\phi < 2\ \mu m$.

Results obtained from $\phi < 5\ \mu m$ irradiations would then open the way to a reappraisal of SSG potentialities as a particle detector, and allow to decide on the opportunity of a more expensive and ambitious research program, based on $\phi < 1\ \mu m$ or $\phi < 2\mu m$ grains.

References

[1] See, for instance, "Superconductivity", Ed. R.D. Parks, Dekker N.Y. 1969.

[2] T.E.Faber, Proc. Royal Soc. London, Ser. A 214, 392 (1952). M.P. Garfunkel and B. Serin, Phys. Rev. 85, 834 (1952).

[3] J. Feder, S.R. Kiser and F. Rothwarf, Phys. Rev. Lett. 17, 87 (1952). J.P. Burger, J. Feder, S.R. Kiser, F. Rothwarf and C. Valette, "Superheating and supercooling in the superconducting transitions of indium and mercury", LT X, Moscow 1966. Ibid. Solid State Comm. 4, 611 (1966).

[4] Orsay Group on Superconductivity, Contribution to "Quantum Fluids", presented by P.G. de Gennes, Ed. D.F.Brewer, North Holland 1966.

[5] V.I. Ginzburg, Sov. Phys. JETP 27, 78 (1958).

[6] D. Saint-James, G. Sarma and E.J. Thomas, Type II Superconductors, Pergamon Press 1969. The original papers by V.I. Ginzburg and L.D. Landau can be found in: Collected papers by L.D. Landau, Ed. D. Ter Haar, Oxford, 1985.

[7] A. Baratoff, "Small superconducting samples" in superconductivity, Ed. R.D. Parks, Gordon and Breach 1969 ; F.W. Smith, A. Baratoff and M. Cardona, "Nonlocal effects on superheating of type I superconductors", in Proceedings of the XI International Conference on Low Temperature Physics, St Andrews 1968.

[8] H. Meissner, Phys. Rev. 109, 668 (1958) ; L. Rinderer, Helv. Phys. Acta, 29, 339 (1956) ; E. Posada and L. Rinderer, J. Low Temp. Phys. 9, 33 (1972).

[9] L.Gonzalez-Mestres and D. Perret-Gallix, Nuovo Cimento 9C 605, (1986), presented at Underground Physics 1985.

[10] J. Blot, Y. Pellan and J. Rosenblatt, J. Low Temp. Phys. 30, 729 (1978).

[11] K. Pretzl, J. Fent, P. Freund, J. Gebauer, N.Schmitz, L. Stodolsky And G. Vestzergombi, contribution to the Moriond Meeting on New and Exotic Phenomena, January 1987.

[12] J. Fent, P. Freund, J. Gebauer, K. Pretzl, N. Schmitz, A. Singsaas, L. Stodolsky and G. Vesztergombi, contribution to "Low Temperature Detectors for Neutrinos and Dark Matter", Ed. K. Pretzl, N. Schmitz and L. Stodolsky. Springer-Verlag, Heidelberg 1987.

[13] See, for instance, J.P. Burger, "Superheating and supercooling in first kind superconductors" in Superconductivity, Ed. P.R. Wallace (Gordon and Breach, N.Y. 1969)

[14] A general introduction to the use of SSG for particle detection was given in: L. Gonzalez-Mestres and D. Perret-Gallix, preprint LAPP−EXP−87−02, to be published in the Proceedings of the International School of "Astro-Particle" Physics, Erice January 1987, Ed. A. de Rujula and P. Shavers, World Scientific Pub. Singapore 1987. See also: L.Gonzalez-Mestres and D. Perret-Gallix, preprint LAPP−EXP−87−06, invited talk at the "Séminaire D. Dautreppe on Superconductivity", Lumbin (Grenoble) september 1987.

[15] H. Bernas, J.P. Burger, G. Deutscher, C. Valette and S.J. Williamson, Phys. Lett. 24A, 721(1967) ; C. Valette, Thesis 1971 .

[16] J. Blot, Y. Pellan, J.C. Pineau, J. Rosenblatt, J. Appl. Phys. 45, 1429 (1974).

[17] See, for instance C. Valette and G. Waysand, "Détecteur supraconducteur de rayons gamma à résolution intrinsèque submillimétrique", Orsay 1976, and D. Hueber, Thesis Orsay 1980

[18] A.K.Drukier, C.Valette, G.Waysand, L.C.L.Yuan and F.Peters, Lett. al Nuovo Cimento 14, 300 (1975) ; A.K.Drukier and L.C.L.Yuan, Nucl. Instrum. Methods 138, 213 (1976).

[19] A.K. Drukier and L. Stodolsky, Phys. Rev. D30, 2295 (1984)

[20] G. Waysand, Proceedings of the Moriond Meeting on Massive Neutrinos, January 1984, Ed.J. Tran Thanh Van, p.319.

[21] R. S. Raghavan, Phys. Rev. Lett. 37, 259 (1976)

[22] L. Gonzalez-Mestres and D. Perret-Gallix, Proceedings of the Moriond Meeting on massive Neutrinos, January 1984, Ed. J.Tran Thanh Van, p.461. Ibid. preprint LAPP−TH−115, July 1984, presented at LT 17 , Karlsruhe 1984.

[23] L. Gonzalez-Mestres and D. Perret-Gallix, Proceedings of the Moriond Meeting on Primordial Nucleosynthesis, March 1985, Ed. Reidel, p.580 ; L. Gonzalez-Mestres and D. Perret-Gallix, Nuovo Cimento C9, 573 (1986), presented at Underground Physics 85.

[24] M.W. Goodman and E. Witten, Phys. Rev. D30, 2295 (1984).

[25] A.K. Drukier, C. Freese and D. Spergel, Phys. Rev. D to appear.

[26] D. Spergel, Proceedings of the Moriond Meeting on Massive Neutrinos, January 1986, Ed. J. Tran Thanh Van, p.420.

[27] S. Dimopoulos, G. Starkman and B.W. Lynn, Phys. Lett. B167, 145 (1986) and Mod. Phys. Lett. A1, 491 (1986).

[28] A. Pacheco, Mod. Phys. Lett. 1, 167 (1986). P. Andreo, J.G. Esteve and A.F. Pacheco, contribution to "Low temperature detectors for Neutrinos and Dark Matter", Springer-Verlag 1987.

[29] F. Von Feilitzsch, L. Oberauer and W. Seidel, Nuovo Cimento 9C, 598 (1986), presented at Underground Physics 85. W. Seidel, Thesis (1985).

[30] F. Von Feilitzsch, T. Hertrich, H. Kraus, L. Oberauer, Th. Peterreins, F. Probst, and W. Seidel, Contribution to " Low Temperature Detectors for Neutrinos and Dark Matter", Springer-Verlag 1987.

[31] L. Gonzalez-Mestres and D. Perret-Gallix, contribution to "Low Temperature Detectors for Neutrinos and Dark Matter", Springer-Verlag 1987.

[32] L. Gonzalez-Mestres and D. Perret-Gallix, Preprint LAPP EXP−84−05 and TH−112 (Unpublished). L. Gonzalez-Mestres and D. Perret-Gallix, Proceedings of the EPS International Conference on High-Energy Physics, Bari July 1985.

[33] J. Matricon, D. Saint-James, Phys. Lett. 24A, 241 (1967)

[34] H. Parr, Phys. Rev. B14, 2849 (1976) ; G. Pettersen, H. Parr, Phys. Rev. B19, 3482 (1979) 1968.

[35] B.B. Goodman, IBM J. Res. Dev. 6, 63 (1962).

[36] T.E. Faber, Proc. Royal Soc., Ser. A 219, 75 (1953).

[37] N. Gauthier and P. Rochon, J. Low Temp. Phys. 59, 225 (1985); K. Aswathy, G. Rangarajan, R. Srinivasan and B.K. Mukherjee, Proc. of the 17th Int. Conf. on Low Temperature Physics, Karlsruhe 1984, p. 587.

[38] T. Van Duzer and C.W. Turner," Principles of superconducting devices and circuits", Ed. Elsevier North Holland, New York 1969.

[39] B.G. Turrell, A. Kotlicki, M.A. Le Gros, A.K. Drukier, "Feasibility investigation of a superheated superconducting colloid calorimeter", Smithsonian Astrophysical Observatory Order n^0 S05−06510, Final report (1986).

[40] See, for instance, N.W. Ashcroft, N.D. Mermin, Solid State Physics, Ed. Holt, Rinehart and Winston, Frome and London 1976. Pergamon, 1965.

[41] A. Baldisseri, "Comportement à très basse température d'un détecteur supraconducteur de neutrinos solaires et candidats à la matière noire galactique", rapport de stage D.E.A., Annecy (LAPP) March 1986.

[42] C.A. Bryant, P.H. Keesom, Phys. Rev. Lett. 4, 460 (1960) and Phys. Rev. 123, 491 (1961).

[43] D.H. Parkinson, Rep. Progr. Phys. 21, 226 (1958)

[44] G. Seidel, P.H. Keesom, Phys. Rev 112, 1083 (1958)

[45] See, for instance, R. Radebaugh and P.H. Keesom, Phys. Rev. 149, 209 and 217 (1966), and references therein.

[46] O. Lounasmaa, "Experimental Principles and Methods Below 1K", Academic Press 1974.

[47] R.B. Stephens, Cryogenics 15, 420 (1975).

[48] EXTRAMET, Zone Industrielle de Cranves-Sales, 74380 Bonne-sur-Menoge (Haute-Savoie), FRANCE.

[49] BILLITON WITMETAAL FRANCE, 48−58 rue Alfred Dequeant, 92000 Nanterre.

[50] L. Gonzalez-Mestres and D. Perret-Gallix, preprint LAPP-EXP−87−01, to be published in the Proceedings of the Moriond Workshop on New and Exotic Phenomena, January 1987, Ed. Frontieres, Gif-sur-Yvette 1987.

[51] L. Gonzalez-Mestres and D. Perret-Gallix, preprint LAPP−EXP−87−06, contibution to the "Rencontre sur la Masse Caché dans l'Univers et la Matiere Noire", Annecy July 1987, to be published in "Annales de Physique" (France).

[52] A. de Bellefon, D. Broskiewicz, R. Bruere-Dawson and P. Espigat, contribution to "Low Temperature Detectors for Low Energy Neutrinos and Dark Matter", Springer-Verlag 1987.

[53] J. Guets, P.L. Garbarino, M.A. Mitchell, P.G. Klemens and C.A. Reynolds, Phys. Rev. 178, 1009 (1969).

TIME AND AMPLITUDE CALIBRATION OF THE PHASE TRANSITION IN SUPERCONDUCTING GRANULES

E. MENICHETTI, N. PASTRONE, V. PENNA, G. RINAUDO

I.N.F.N. Sezione di Torino
University of Torino,
Torino - Italy

1. INTRODUCTION

Detectors based on superheated superconducting granules (SSG) have been proposed since a long time[1] to detect gamma rays and more recently[2] to detect monopoles, neutrinos, dark matter, etc. The effect, as illustrated in other talks of this workshop, is based on the disappearance of the superconducting state of a microscopic granule maintained in a metastable phase in a magnetic field larger than the critical field (H_c) but smaller than the superheated field (H_s). The transition is induced by an energy deposit in the granule, except for magnetic monopoles, where it is due to the penetration of a magnetic field. The variation of the external magnetic flux, due to the disappearance of the Meissner effect, is detected by a pick-up loop which surrounds the granule. Transitions of granules due to irradiation with alpha, beta or gamma sources have been observed [3,4].

Many problems must be solved before the implementation of a realistic detector, concerning sensitivity, efficiency and time resolution. It is important to have a reliable model of the physics of the transition, in order to predict, in particular, the risetime and the amplitude of the signal associated with the penetration of the magnetic field in the granule. In this note we present an electronic readout chain developped with the principal aim to allow a detailed study of the risetime and amplitude of the signal induced by the transition.

2. CHARACTERISTICS OF THE PHASE TRANSITION

For a description of the general features of the transition we refer to ref.2. We recall only the relevant formulae, that we will need in the following. The dipole magnetic field B due to the supercurrent in the granule repelling an external magnetic field H is:

$$B(r) = H\left(1 + \frac{R^3}{2r^3}\left(1 - 3\cos^2\vartheta\right)\right) \qquad (1)$$

where R is the granule radius and r is the distance from the center of the granule. The disappearance of the Meissner effect restores an uniform magnetic field. The variation of magnetic flux $d\Phi$ picked up by a loop of area S surrounding the granule is:

$$d\Phi = \int (B(r) - H)\, dS \qquad (3)$$

The voltage signal is the time derivative of the flux. The value of the integral in eq.3 depends on the position of the granule with respect to the loop. For a granule at the center of a loop of radius l, in the equatorial plane, we expect:

$$d\Phi = -\pi\, H\, R^3/l \qquad (4)$$

The value of the flux increases rather sharply as the granules goes off center, as illustrated in fig.1. As one can notice, the spread of the distribution is rather important for granules of the same size and it is dominating with respect to the size difference.

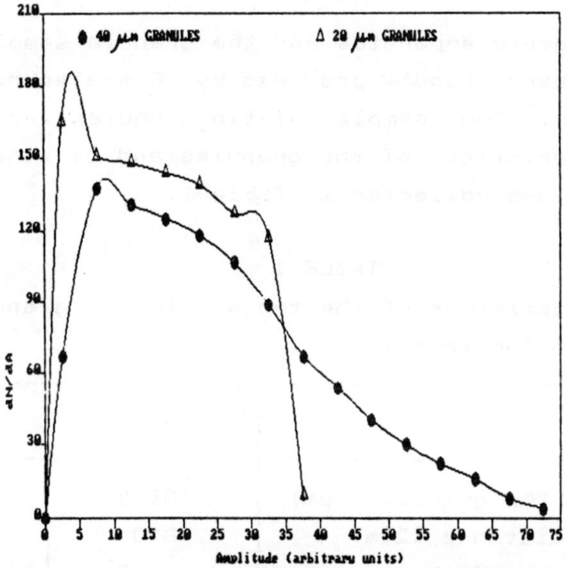

Fig.1 Amplitude distribution expected for granules with diameters of 20 and 40 μm uniformly distributed in a 2mm diameter loop.

The flipping time of the granules can be estimated on the basis of a model developped by Faber[9] and tested on macroscopic samples. Valette and Waysand[1] have shown that this model can be applied also to a microscopic granule and that one expects a flipping time t given by the following expression:

$$t = \pi \sigma R^2 H_s / (H_s - H_c) \qquad (5)$$

where σ is the conductivity. For a granule of 10μm diameter, resistivity 10^{-9} Ωcm one expects a rising time close to 1μsec. Rising times can be reduced, according to eq.5, by using higher resistivity samples or smaller granules. Going to shorter flipping times might be crucial for experiments with extremely low rates and important background.

3. DETECTOR AND RESULTS

The cryogenic apparatus and the granule samples use in the tests were kindly provided by Gonzales-Mestres and Perret-Gallix. Two samples of tin granules were used[4]: the characteristics of the granules and of the relative pickup loops are collected in Table 1.

TABLE 1

Characteristics of the two samples of granules used in the tests.

	sample a	sample b
diameter of the granules (μm)	10÷25	200÷400
residual resistance (Ωcm)	≈10^{-7}	≈10^{-8}
inner diameter of the loop (mm)	2	10
number of turns	30	15
inductance (μH)	1	10

Sample a was prepared by EXTRAMET while sample b was prepared by BILLITON (see ref.4 for details). Notice the high resistivity of the two samples.

The loops were directly connected to the entrance of the preamplifier, through a damping resistance and a capacitor. No intermediate transformer was inserted between the pickup and the transfer line (the length of the transfer line was of the order of 1m), the inductances quoted in the table are inclusive of the transfer line contribution.

As preamplifier we used a fast low-noise transimpedance amplifier (Lecroy TRA1000), which can be configured at different gain values. We adopted an intermediate gain configuration (10mV/μA), for which the expected risetime should be 24nsec and input impedance 1.5Ω. A fast amplifier (risetime around 40nsec) was connected to the preamplifier output to obtain a further gain of a factor between 10 and 40.

Since the inductive signal of the loop is far from behaving as an ideal current generator, we did not expect to obtain the nominal gain. However this fact does not affect the results, as we will show below, as far as the signal remained above the noise level. A more serious problem arised from the difficulty to adapt the relative impedances of the source, the transfer line and the amplifier entrance. Since our main concern was the possibility to calibrate in a reliable way the risetime and amplitude of the signal, we set up our working conditions (gain of the amplifier, value of the damping resistance and inductance of the pickup loop) in such a way to avoid ringing and maintain the sensitivity to the expected transition times, without trying to optimize the efficiency of the circuit.

To calibrate the time and amplitude response of the electronics, we first of all implemented a normal "testin" input, in order to check the behaviour of the preamplifier with good current signals. We then setup a calibration system consisting of a transformer, whose primary loop could be connected to a pulse generator and the secondary loop to the preamplifier (fig.2).

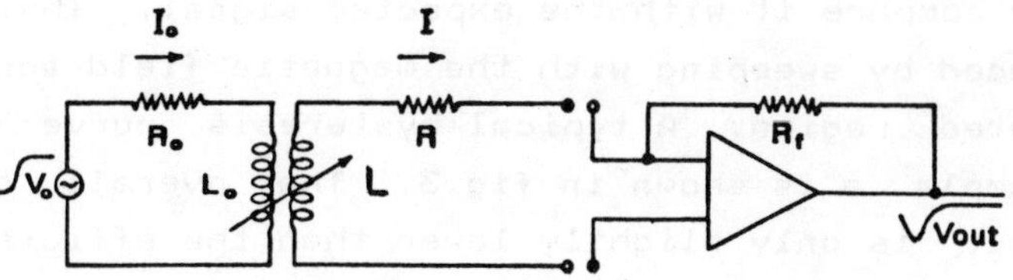

Fig.2 Setup used for electronics calibration.

The geometrical and electrical parameters of the secondary loop were similar to those of the pickup loops used in the tests with the granules. The primary loop and the geometry of the transformer were such that the value of the magnetic field in the transformer and thus the field flux seen by the secondary loop could be easily calculated. Calculations of the transinductance value were checked by direct measurements with test circuits. Signals of different risetimes and amplitudes were then sent to the primary loop and the voltage signals at the output of the preamplifier-amplifier chain were recorded.

The tests allowed to:

1- establish the correspondence between the output signal and a given variation of the magnetic field flux, thus obtaining an amplitude calibration;

2- measure the risetime response of the circuit to a given input risetime;

3- check the onset of critical conditions, which could give rise to oscillations;

4- determine the noise level.

The tests showed that the circuit could reliably reproduce risetimes longer than about 50 nsec and that the noise level was about 1μA equivalent (r.m.s.), referred to the input, with the loop connected.

Having calibrated the circuit in this way, we measured the signal produced by the flipping of the granules and tried to compare it with the expected signal. The flipping was induced by sweeping with the magnetic field across the superheated region. A typical hysteresis curve obtained with sample a is shown in fig.3. The overall detection efficiency is only slightly lower than the efficiency obtained with more optimized, slower preamplifiers.

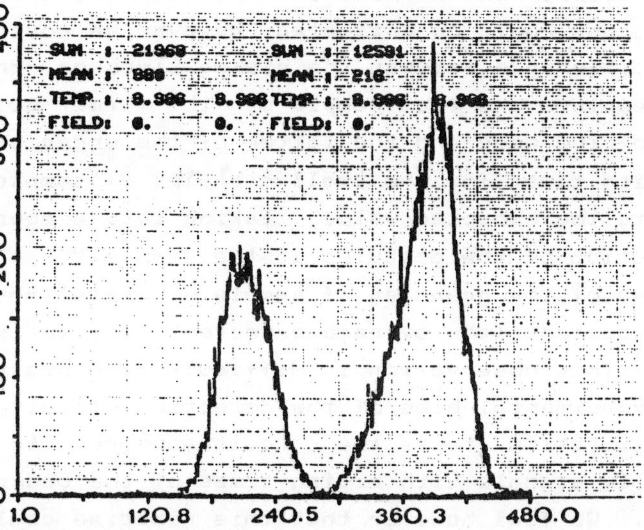

Fig.3 Hysteresis curve obtained with granules of sample a; the magnetic field is in arbitrary units.

Typical signals observed at the oscilloscope for sample a granules are shown in fig.4.

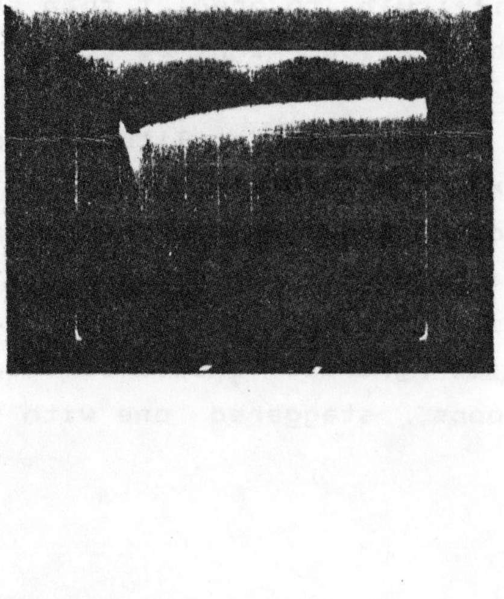

Fig.4 Signals due to the transition of granules of sample a in the superheating cycle; the horizontal scale is 200 nsec/div, the vertical scale is 200 mV/div.

Comparison of the results with expectations cannot be very precise at the present stage, since only average values based on inspection of the signals on the oscilloscope could be estimated. However we could observe a clear difference between the rising times of the signals from sample a and from sample b, the former being of the order of 100 nsec and the latter of the order of 600nsec. The rising times for the small granules of sample a is very close to the expected value, taking into account the size distribution of the granules in the sample [6]. The flipping time of the granules of sample b, is slightly shorter than expected (the average expected risetime is close to 1μsec). This observation is difficult to explain in terms of an experimental bias of the detection circuit, since in the calibration tests the circuit appeared to reproduce faithfully signals with rising times of the order of microseconds. We feel however that this risetime could still be compatible with expectations, if the uncertainties in the size of the granules and in the residual resistance are taken into account.

It was also observed that the flipping time of the granules in the supercooling cycle is much longer (of the order of 1μsec for the granules of sample a) than in the superheating cycle. Supercooling flipping time is more difficult to predict than superheating time, though this observation might be compatible with calculations by Faber[7].

As to the amplitude response, due to the wide spread expected for the signals, it is difficult to draw any firm conclusion, beyond the general statement that the order of magnitude is correct. A better geometry of the pickup loop with respect to the granules is needed and possibly the same granule should be inspected by to different oickup loops, staggered one with respect to the other, to avoid configurations with very low values of flux variation.

In conclusion, this preliminary attempt to set up a calibration chain to study the details of the transition, has shown that calibration is possible, in spite of numerous difficulties. More work needs to be done. First of all we intend to cure the matching of the source-transfer-amplifier chain in order to improve the signal. Second we will try to implement better geometries of the pickup loops, as mentioned above, with the possibility of detecting coincidence signals. Finally, we intend to use for the read-out of the amplified signals a system based on a fast flash-type ADC (SIEMENS 6020) which will allow to sample the signal with a 30MHz rate and thus reconstruct completely the single transition. A CAMAC module with the flash ADC and a fast memory has been developped in our laboratory and tested with the calibration loop described above. The response to an input step function is shown in fig.5.

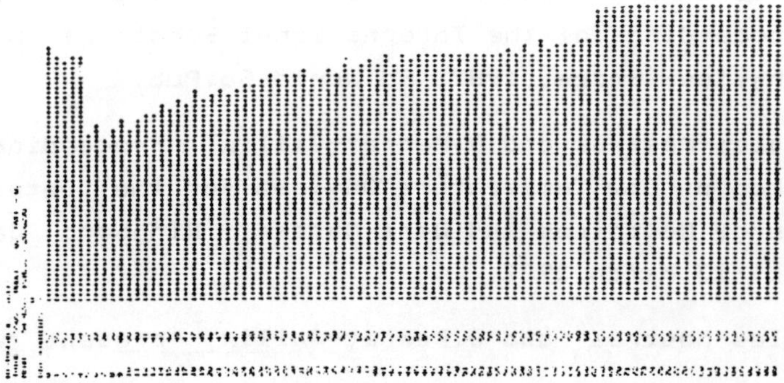

Fig.5 Response of the circuit to an input step function sent to the primary of the calibration circuit described in the text, as reconstructed by the Flash-ADC readout; each binning is 30nsec.

The use of this fast read out will allow to reconstruct completely the single flips of the granules, thus allowing a complete analysis of the transition. With such a system it will be in particular very interesting to analyze the transitions induced by radiation of ionizing particles to check if the flipping parameters are the same as in transitions induced by the change of the magnetic field.

Acknowledgments

The authors are deeply indebted to L.Gonzales-Mestres and D.Perret-Gallix for giving them access to their cryogenic apparatus and for constant help and encouragement.

References

1) See, for instance, C.Valette and G.Waysand "Detecteur supraconducteur de rayons gamma a resolution intrinseque submillimetrique", Thesis, Orsay 1976.

2) For a recent review see: L.Gonzales-Mestres and D.Perret-Gallix, Proceedings of the International School of Astro-Particle Physics, Erice, 1987, Ed.World Sc.Pub.

3) L.Gonzales-Mestres and D.Perret-Gallix, Proceedings of the Workshop on Low Temperature Devices for the Detection of Low Energy Neutrinos and Dark Matter, Ringberg Castle (Bavaria) 11-13 March 1987.

4) L.Gonzales-Mestres and D.Perret-Gallix, present workshop.

5) T.E.Faber Proc.Royal Soc. A219 (1953) 75.

6) see e.g.fig.3b of ref.4.

7) T.E.Faber Proc.Royal Soc. A223 (1954) 174.

INVESTIGATION OF THE PROPERTIES OF SUPERCONDUCTING ANTIMONY DOPED TIN GRANULES

P. FREUND, K. PRETZL, N. SCHMITZ, A. SINGSAAS, L. STODOLSKY

Max-Planck-Institut fur Physik und Astrophysik
Munchen, Germany

At a workshop on Low Temperature Detectors for Neutrinos and Dark Matter held in March 1987 at Ringberg Castle, Germany this group[1] reported on measurements with single pure tin granules. Each granule was shown to have its own characteristic superheating (H_{SH}) and supercooling (H_{SC}) field, which strongly depended on the rotational angle. The relative variations $\Delta H_{SH} / \overline{H_{SH}}$ and $\Delta H_{SC} / \overline{H_{SC}}$ of individual grains, when rotated in an external field, were measured to be as large as 30% and explained the width of the phase transition smearing observed for a group of grains[2,3]. This smearing can be explained by the presence of nucleation centers which are due to surface and internal effects (crystalline structure). To further study these effects we investigated tin granules doped with 1% antimony. It was the hope that by adding impurities in low concentrations to the superconducting metal such grains would show a more uniform behavior when rotated in the field.

Individual grains were rotated around an axis perpendicular to an external magnetic field. The grain flip was detected with a small pickup coil which measured the flux change due to the disappearence of the Meissner–effect. By cycling the magnetic field at a fixed temperature we measured H_{SH} and H_{SC} as a function of the rotation angle α (see Fig. 1). The measurements were made in the temperature range between 1.4 and 3.5 Kelvin.

As an example, the measurements of H_{SH} and H_{SC} at various temperatures for an individual Sn–Sb grain with a diameter of 250 μm are shown in Fig. 2. There is only a small variation of H_{SH} and H_{SC} for different orientations of the grain in the magnetic field. The relative variation $\frac{\Delta H_{SH}}{H_{SH}}$ is only a few percent. This behavior was also found for many other grains with the same diameter.

In Fig. 3a the mean values of H_{SC} and H_{SH} are plotted versus T. The reduction of metastability as expected from the increase of the Ginsburg–Landau parameter κ caused by the added impurity (1% antimony) is clearly seen. For comparison, in Fig. 3b the mean values of H_{SC} and H_{SH} for a 56 μm diameter pure tin granule are shown as a function of T. The bars in Fig. 3a and Fig. 3b represent the variations of H_{SC} and H_{SH} as shown above in Fig. 2 and in fig. 5 of Ref. 1. The dashed curves in both Fig. 3a,b marked with H_{SC}, H_C, H_{SH} show the supercooling, critical and superheating fields calculated for pure tin using κ values taken from Ref. 4.

For smaller size Sn–Sb granules (diameter 24 – 50 μm), this rather flat angular dependence of H_{SC} and H_{SH} on the rotation angle was not observed. As an example, H_{SH} for an 24 μm Sn–Sb grain at 1.4 k is shown in Fig. 4a. The relative variation is 14 %. In table 1 a summary of $\Delta H_{SH} / \overline{H_{SH}}$ is given for various grain diameters. With exception of the 250 μm Sn–Sb grain there is no difference in the relative variation between pure Sn grains and small size Sn–Sb grains.

TABLE 1. The relative variation $\frac{\Delta H_{SH}}{H_{SH}}$ is shown for various grain diameters and different temperatures

	Sn (99%) – Sb (1%)			pure Sn		
T[K]	250 μm	40 μm	24 μm	112 μm	56 μm	20 μm
3.3	0.07	–	–	0.14	0.33	–
3.0	0.06	–	–	0.17	0.22	0.10
2.2	0.04	–	–	0.16	0.13	0.10
1.4	0.03	0.25	0.14	0.17	0.12	0.10

With the concept of nucleation centers at or near the surface of a grain, it is possible to explain the observed variation of the superheating field with the rotation angle. The value of the magnetic field at the surface of a superconducting granule is given by

$$H = \frac{3}{2} H_{app} \cdot sin\vartheta \tag{1}$$

ϑ being the polar angle with respect to the direction of the applied field H_{app}. A nucleation center on or near the surface of a granule (Fig. 5) will see different values of H, when the granule is rotated by an angle α around an axis perpendicular to the direction of H_{app}.

$$H = \frac{3}{2} H_{app} \cdot \sqrt{1 - A \cdot cos^2\alpha} \tag{2}$$

The constant A depends on the location of the nucleation center and is given by $sin^2\vartheta'$, with ϑ' the polar angle with respect to the rotation axis. To flip the granule from the superconducting to the normal state a certain H_{SH} is necessary; the applied field H_{app} will therefore depend on the rotation angle α:

$$H_{app} \sim H_{SH} / \sqrt{1 - A \cdot cos^2\alpha}. \tag{3}$$

Each nucleation center has its own α–dependence with a certain amplitude and phase. Fig. 4b shows the superposition of 3 nucleation centers, which reproduce the data plotted in Fig. 4a. The more uniform α–dependences, as shown in Fig. 2, find their explanation in the superposition of many nucleation centers. Up to now it is unclear whether surface defects (as shown in Fig. 6) or crystalline structures act as the dominant part of the nucleation centers.

In order to construct a low energy neutrino detector with uniform homogeneity of response and good energy resolution, it is of utmost importance to overcome the problem of the smearing of the phase transition.

REFERENCES

1. K. Pretzl et al.: Proceedings of the workshop on Low Temperature Detectors for Neutrinos and Dark Matter, at Ringberg Castle 1987, edited by K. Pretzl, N. Schmitz, L. Stodolsky (published in Springer Verlag Heidelberg), page 30.

2. L. Gonzales–Mestres and D. Perret–Gallix: Proceedings of the workshop on Low Temperature Detectors for Neutrinos and Dark Matter, at Ringberg Castle 1987, edited by K. Pretzl, N. Schmitz, L. Stodolsky (published in Springer Verlag Heidelberg), page 9;
F.v. Feilitzsch et al., Proceedings of the workshop on Low Temperature Detectors for Neutrinos and Dark Matter, at Ringberg Castle 1987, edited by K. Pretzl, N. Schmitz, L. Stodolsky (published in Springer Verlag Heidelberg), page 94.

3. F.v. Feilitzsch et al.: Il Nuovo Cimento, Vol. 9C, N2 (1986) 598.

4. J. Feder and D.S.Mc Lachlan, Phys.Rev. **177** (1969) 763, fig. 10.

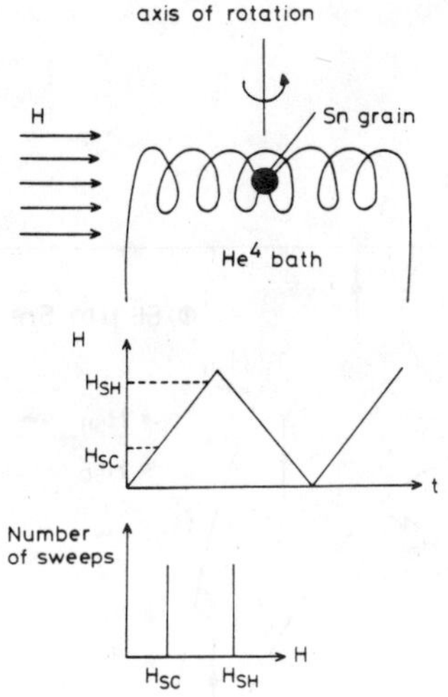

Fig. 1 : Schematic view of the experimental method.

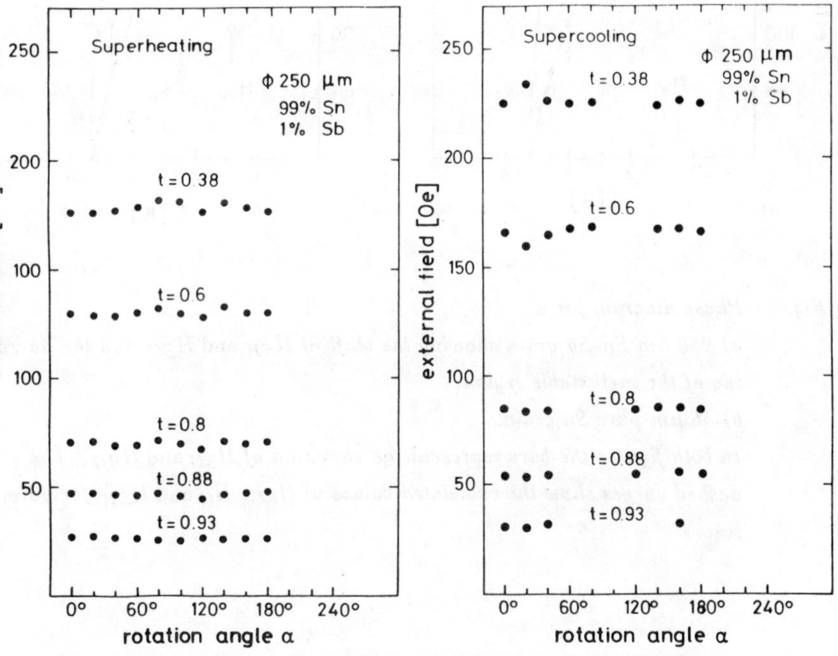

Fig. 2 : Superheating and supercooling fields for a Sn (99%) Sb (1%) grain of 250 μm diameter at various temperatures and rotation angles α.

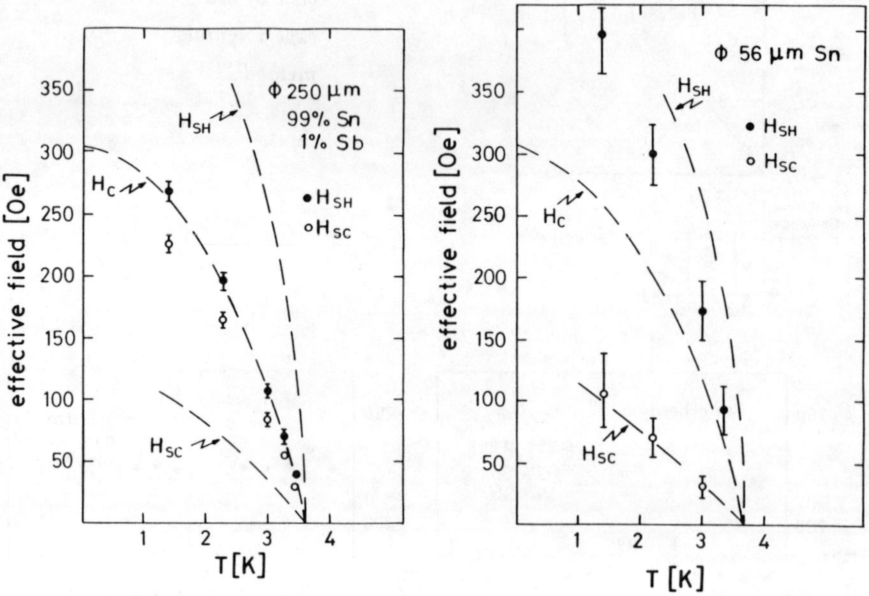

Fig. 3 : Phase diagram for a

a) 250 μm Sn–Sb grain showing the shift of H_{SH} and H_{SC} *and the narrowing of the metastable region.*

b) 56 μm pure Sn grain.

In both figures the bars represent the variation of H_{SC} and H_{SH}. *The dashed curves show the calculated values of* H_{SC}, H_C and H_{SH} *for pure tin.*

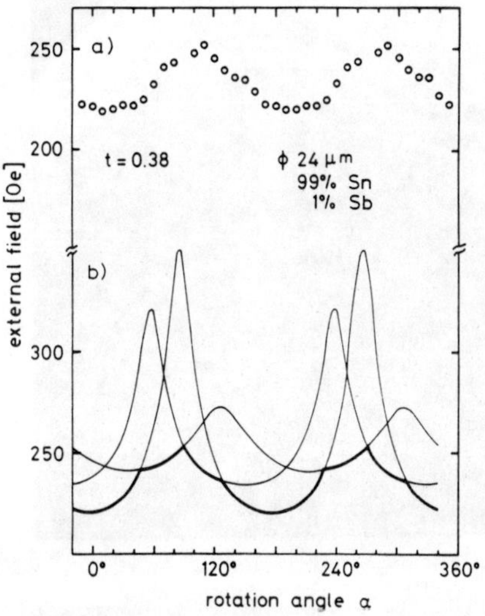

Fig. 4 : a.) Superheating field H_{SH} as a function of rotation angle α of a 24 μm Sn–Sb grain at 1.4 Kelvin.
b.) As indicated by the fat line, three nucleation centers are sufficient to fit the data in a).

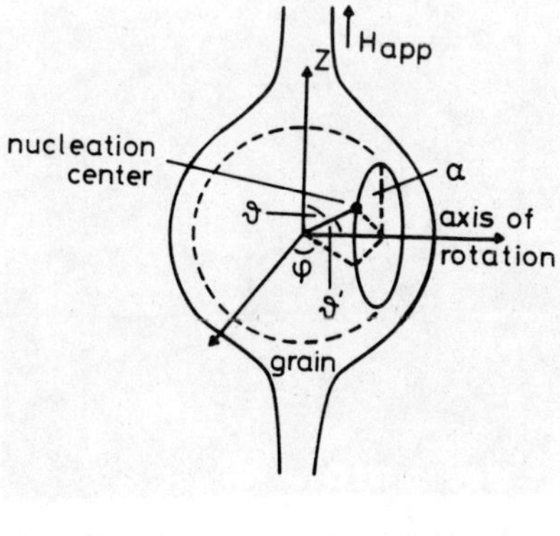

Fig. 5 : Schematic view of a superconducting grain in an applied field H_{app}. A nucleation center with the polar angle ϑ and the rotation angle α is shown.

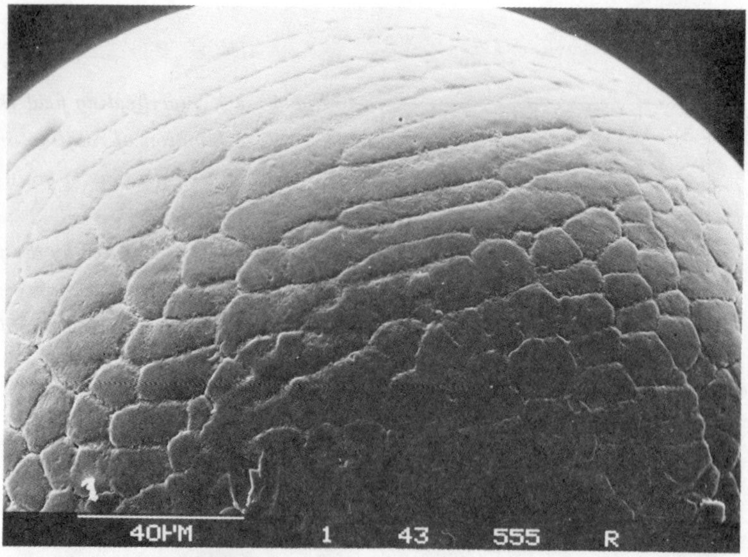

a) grain diameter 240 μm

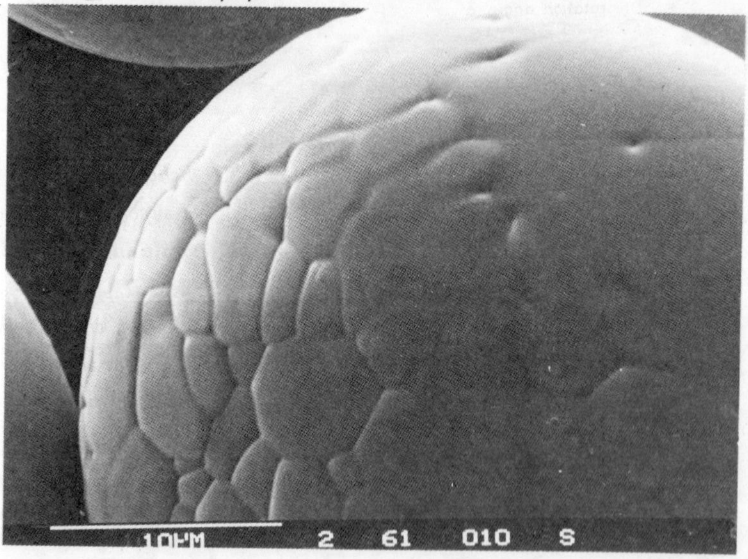

b) grain diameter 40 μm.

Fig. 6 : Photographs of Sn (99%) Sb(1%) grains taken with an electron–microscope.

DETECTING SOLAR COSMIONS WITH SUPERCONDUCTING COLLOID DETECTORS

ANDRZEJ K. DRUKIER
Applied Research Corporation
8201 Corporate Drive, Suite 920
Landover, MD 20785

and

Harvard-Smithsonian Center for Astrophysics
60 Garden Street, Cambridge, MA 02138

The growing synergy between astrophysics, particle physics, and low background experiments strengthens the possibility of detecting astrophysical non-baryonic matter. The idea of direct detection is that an incident, massive weakly interacting particle could collide with a nucleus and transfer an energy that could be measured. The present low levels of background achieved by the PNL/USC Ge detector represent a new technology which yields interesting bounds on galactic cold dark matter and on light bosons emitted from the Sun. Further improvements require the development of cryogenic detectors. We analyse the practicality of a superheated superconducting colloid detector using a SQUID readout system, and discuss the possibility of building a low radioactive background dilution refrigerator necessary for its operation.

1. INTRODUCTION

One of the outstanding questions in astrophysics is understanding what makes up 90% of the mass of the universe. We observe the gravitational effects [1] of this "missing mass", but see neither stars, nor gas, nor dust. A variety of theoretical arguments [2] suggest that this "missing mass" may not be composed of ordinary protons and neutrons, but rather exotic non-baryonic particles.

The early universe was a tremendous particle accelerator. Its high temperatures and densities were the perfect environment for producing forms of matter inaccessable at laboratory energies. If any of the particles produced in the big bang are stable, then they could survive to the present epoch and account for the "missing mass". The candidates for this "missing mass" range in size from 10^{39} gram black holes through very low mass stars down to 10^{-39} gram axions. In the middle of this range lie the GeV mass "sparticles" offered by supersymmetric theories. The effect of all these "cold dark matter" candidates on galaxy formation is very similar, so we must turn to the laboratory to elucidate the mysteries of the heavens.

When our galaxy was formed, it would have accreted not only the baryons that formed stars, but also the non-baryonic matter. Since the non-baryonic matter is not dissipative, it would not have cooled to form a disc, but rather be distributed in a spheroidal halo. We do observe the gravitational effects of this dark matter halo. If weakly interacting massive particles (WIMPs) comprise the halo dark matter, then millions of these particles are streaming through a square centimeter every second. Ultralow background experiments may soon be capable of detecting this incident flux. Actually, some limits on non-baryonic matter in our Galaxy have been already obtained using data from an underground, ultralow background Ge-semiconducting spectrometer.[3] Several cryogenic detectors are being developed for WIMPs searches.[4]

If WIMPs comprise the galactic halo, they will scatter off the baryons in the stars in the galaxy and some of the WIMPs will become bound to stars. In stars, WIMPs can alter energy transport and produce observable effects. "Recently, astrophysicists (5) have explored the possibility that WIMPs could solve the solar neutrino problem and the appropriate particles have been proposed.(6)

The solar neutrino mystery ranks with the missing mass problem as one of the more important issues at the interface between particle physics and astrophysics. One recent attempt at an explanation for both the deficit in the ^{8}B neutrino flux from the sun and the missing mass in the galaxy invokes a solar population of weakly interacting particles (that have been captured from a galactic reservoir of such particles) which can improve energy transport from the center of the sun (5). This would decrease the central temperature of the sun and would therefore reduce the ^{8}B neutrino flux. The hypothetical particle has been dubbed the cosmion. Its mass must be greater than 5 GeV to prevent evaporation and less than 15 GeV to prevent its orbits from being too small to produce efficient thermal transport. The elastic cross section for the cosmion with hydrogen must be about 10^{-36} cm^2 to insure trapping from the galactic reservoir and good energy transport within the sun. Explicit particle physics models for cosmions have been given (6) which are acceptable in terms of avoiding excessive annihilations of cosmions within the sun. In one of the models, the cosmion is a Majorana fermion which would have a scattering rate in an appropriate detector of several thousand per kg per day. In two other models, the cosmion is a Dirac fermion having a scattering rate up to 10^9 per kg per day in a suitable detector! These are enormous rates, which will enable us to probe for the cosmion existence without substantial improvements in existing technology. The primary focus of this contribution is to study experimental questions associated with direct detection of these solar cosmions.

2. METHODOLOGY OF EXPERIMENTAL SEARCHES FOR WIMPS, E.G., SOLAR COSMIONS

It is clearly important to cover the principal theoretically-motivated particles, but at the same time design the experiments to cover as wide a range as possible of alternative types of particles. This can be achieved by means of two rather general classes of experiments (see Ref. 7 for review articles):

(a) searches based on the principle that many such particles can be converted to photons by interaction with electrons or electromagnetic fields; and

(b) searches based on the detection of low energy nuclear recoils.

The second method is clearly of considerable generality: although some particle types would not interact directly with electrons (e.g., the photino), all conjectured heavy particles would interact with nucleons and produce a nuclear recoil, the energy of which is governed solely by kinematics. Furthermore, atomic nuclei happen to be in the ideal mass range for optimum sensitivity to incident particles of mass 1-1000 GeV.

These particles scatter elastically from nuclei producing nuclear recoils of a few keV. Since the typical WIMP has a cross-section of order 10^{-37} cm^2, a few thousands such recoils are produced in a kilogram of detector per day. The experimental challenge is to detect these rare weak recoils amidst the background from internal radioactivity. The task is complicated by the need of using very sensitive detectors able to detect recoiling nuclei of a few keV. Unfortunately, the ionization detectors, (scintillators and gas detectors, even semiconductors) are unable to detect recoiling nuclei with $E_{recoil} \leq 10$ keV.

For the detection of WIMPs with $M \leq 10$ GeV, the optimum approach is to use ultralow temperature detectors, which has already attracted the interest of many experimental groups. The recoiling nucleus moves only a short distance, giving up the majority of its energy to the lattice, and the resulting temperature increase can be observed at very low temperature. For any of the detection schemes, a crucial problem is, of course, the reduction of background to below the signal levels estimated in the above quoted references.

We will focus on the particular detection scheme of superheated superconducting colloids, first proposed by Drukier and Stodolsky [8] as a detector for neutrinos. Goodman and Witten [4] then noticed that it could be used as a dark matter detector. In a paper by Drukier, Freese and Spergel[4], the actual count rates one could expect for various particles using realistic models of the halo were estimated. The total mass of the detector is below 1 kg, with micron-size grains in a dielectric filling material. The superconducting grains are placed in a magnetic field and maintained in a superheated state. When a cosmion passes through the detector, it interacts with a nucleus inside a grain: the nucleus recoils, energy is deposited in the grain, the temperature of the grain increases, and the grain goes normal. This change of state can be detected by the Meissner effect. Penetration of the magnetic flux into the grains produces an observable signal in the readout electronics. The flipping of a grain a few microns in size produces a flux change of typically 0.1 ϕ_0 (flu quanta) in a few centimeter diameter superconducting loop encircling the detector. SQUIDs can detect this flux change. Demanding a signal-to-noise ratio of at least 10, one expects 10^4 counts/day for neutrinos and Dirac neutrinos and 10^6-10^9 counts/day for solar cosmions.

Advantages over conventional particle detectors for GeV mass particles are the sensitivity to heavy nucleus recoils with energy < 1 keV and good background rejection. The detector is small ($\leq$ 1 kg) so one can use anticoincidence or shielding effectively to eliminate cosmic rays. Also the signal from such small energy deposits will be the flip of one and only one grain. Most charged background particle will flip many grains: for example a 0.5 MeV electron with range 0.5 gm/cm^2 must pass through 100 grains to lose its energy. The worst background for high energy thresholds is natural radioactivity (^{40}K, U, Th); for low energy thresholds it is solar neutrinos. The recent experimental tests of SSCD are described in the next section (see also ref. 9).

3. THE SUPERHEATED SUPERCONDUCTING COLLOID DETECTOR

The superheated superconducting colloid detector (SSCD) placed in very low temperature (20 mK) satisfies all requirements for a highly sensitive, low background detector of solar cosmions. The basic idea of the SSCD is quite simple. One prepares a colloidal suspension of spherical grains of a Type I superconducting material, which constitutes the detecting medium.

To enable the detection of energy deposited in the grains, one first brings them into the superconducting state by cooling them below the critical temperature (using pumped helium cryostats or dilution refrigerators). It is well known that if a sufficiently large magnetic field (H) is applied to such grains, they will revert to the normal state:

$$H = H_C = (2/3)H_O(1 - T^2/T_C^2),$$

where T is the grain temperature, T_C is the transition temperature (at zero magnetic field) at which the superconducting state is reached, and H_O is the "critical field." H_O is defined as the field longitudinal to the axis of a long cylinder of a superconductor which will cause the material to revert to the nonsuperconducting state at T = 0. For spheres, the transition occurs for $H = (2/3)H_O$. In small grains however, one observes superheating, i.e., if one fixes H and increases T beyond the point for which $H = H_C(T)$, transition to the normal state will not occur until $H = H_{sh}(T) > H_C(t)$, where H_{sh} defines the superheating transition line. A grain entering the normal state will not return to the superconducting state if it now returns to its initial temperature. Rather it must be cooled to the point for which $H = H_{sc}(T)$, which defines the supercooling line.

To take advantage of the above principles in producing a radiation detector, one first establishes an ambient field H slightly smaller than $H_{sh}(T)$. If radiation is absorbed by the grain, this will eventually be converted to heat. This is true whether the energy is initially deposited as electronic excitation or in the form of nuclear recoils. The heat which is deposited in the grain will increase the temperature of the grain by the amount

$$\Delta T = (3\Delta E)/(4\pi r^3 C_V(T)),$$

where ΔE is the deposited energy in excess of that which goes toward dislocation production, and r is the grain radius. If $H > H_{sh}(T + \Delta T)$, then the grain will revert to the normal state.

The change of state of the grain can be monitored with a superconducting curcuit which consists of an interconnected pickup and output coil (so that whatever current flows in the pickup coil also flows through the output coil). The grain is placed within the pickup coil, the the output coil is placed near a SQUID. When the grain changes to the normal state (or "flips"), the magnetic field (which had been excluded by the Meissner effect) now penetrates freely. Effectively, a magnetic dipole has been removed from the interior of the colloid, and this changes the externally-produced field which threads the pickup coil. Due to quantization and conservation of magnetic flux within a closed loop of a superconductor, a supercurrent is induced in the superconducting circuit. The supercurrent sets up a small magnetic field at the SQUID which is detected by well known procedures. The flux induced at the SQUID ($\Delta\phi_{SQUID}$) by the flip of a grain located at the center of the pickup coil is given by

$$\Delta\phi_{SQUID} = F\pi^3 H/R,$$

where R is the pickup coil radius, and $F \approx 0.2 - 0.3$. For $r = 1$ micron, $R = 10$ cm, and $H = 100$ G, $\Delta\phi_{SQUID} \approx 10^{-12}$ Gcm^2. This is a small fraction of a flux quantum (ϕ_o, 2.07×10^{-7} Gcm^2) but is detectable by modern, low noise SQUID systems. In preliminary experiments, we reached a good signal to noise ratio ($S/N \geq 10$) for grains with $R \leq 2\mu m$ in about a 1 cm diameter loop. The noise level was however, considerable ($\delta\phi \approx 5 \times 10^{-2}\phi_o$). In forthcoming experiments we hope to have $S/N \geq 10$ for $= 1$ μm and a few cubic centimeter detector.

The ultimate sensitivity of SSCD's depends on the detailed mechanism of the grain heating and flipping process. The range of a recoiling nucleus of energy less than the order of 10 keV is much smaller than a micron. Furthermore, any inner shell excitation processes will be unlikely at the small velocities involved so that the energy of the recoil nucleus will almost always be localized initially to a small fraction of the volume of a single grain. To determine the subsequent temperature increase of the grain, we adopt the "uniform-heating picture" where it is assumed that the deposited energy spreads uniformly as heat through the grain whose temperature increase is determined by the heat capacity of the grain in the nonsuperconducting state. This is reasonable since the heating of the initial volume containing deposited energy is much more than is necessary to drive this small volume out of the superconducting state. In certain instances, the uniform heating picture may underestimate the sensitivity of the grain flipping process. This would be the case if it is possible for the grain to flip, if the normal zone created by the initial energy deposition reaches the surface without the whole grain heating up. However, the uniform heating picture is in rough agreement with experimental observations, and in any case, the sensitivity of the grains will be at least as good as that obtained by assuming the validity of this simple model. The heat capacity of the grain in the normal state is dominated at low temperatures by the electronic component, given by

$$C_V(T) = \gamma T,$$

where γ is of the order 1 mJ/mole-deg^2 for soft superconductors. In the case of tin at 50 mK, which is easily reached with ^3He-^4He dilution refrigerators, the specific heat is only 34 eV/K-μm^3. Since the temperature stability of commercially available cryostats is better than 0.05 mK, a highly stable operating point would exist if the magnetic field was established such that the cryostat temperature was about 1 mK below the superheating transition line. For these conditions, we can heat a grain with a radius of 1 micron sufficiently to cause it to flip in the magnetic field by depositing an energy of only 15 eV. We saw above that such a grain flip would produce an observable signal in a SQUID. Thus, the SSCD has an intrinsic energy sensitivity which is much better than 100 eV.

The idea of direct detection is that an incident, weakly interacting, particle could collide with a nucleus and transfer an energy that could be measured [4,9]. The maximum energy transfer in a collision of a projectile particle, with velocity v, with a target of mass M at rest is

$$T_{max} = 2\beta^2 \times Mm^2/(m + M)^2 \qquad (\beta = v/c)$$

Typical velocities of dark matter particles in the halo should be $v \approx 10^{-3}c$, and about 10% of particles have $v \geq 600$ km/sec, i.e., $v \geq 2.0 \times 10^{-3}c$. For example, a particle with m = 5 GeV interacting with nuclei of tin (M ≈ 116 GeV, have $T_{max} \approx 1.6$ KeV. Thus the energy deposited is very much higher than threshold, ($T_{max}/E_{th} \geq 16$) and we expect a good quantum detection efficiency.

Finally, let us estimate the expected count rate. We have

$$M_{detector} = F \times \alpha \times V_{pickup} \times \rho_{Sn} \approx 0.36 \text{ gram}$$

where the geometrical filling factor of SSCD in the pickup loop is F=0.5; the grains occupy α = 10% of the volume of the colloid, $V_{pickup} = 1\ cm^3$ and $\rho_{Sn} = 7.3 g/cm^3$. For a typical interaction cross section of $10^{-35}\ cm^2$ for cosmions with $V \geq 600$ km/sec, we expect $\geq 10^6$ counts/kg x day and with QDE≈50% we have

CR ≥ 180 counts/day.

4. LOW BACKGROUND DILUTION REFRIGERATORS

Detecting of solar cosmions requires a detector with low radioactive background. Fortunately, SSCD may be based on two materials known to have very low radioactive contamination: very pure (preferably zone refined) Sn and GeO powder.

Cryogenic detectors should operate at a very low temperature. For reasonable mass detectors (≤ 0.5 kg) this is not too difficult: for example, the dilution refrigerator which our collaboration operates in NBS can cool masses of 500 g to a temperature of 10 millikelvin. This dilution refrigerator was developed for the purpose of highest precision thermometry and possesses an excellent long term temperature stability of about 10 microkelvin. Actually, in similar cryostats at NBS, a temperature stability of one microkelvin was obtained.

Thus, the requirements of a stable, very low temperature environment are not the main challenge, but rather the development of an <u>ultralow</u> <u>radioactive</u> <u>background</u> <u>dilution</u> <u>refrigerator</u>. Several sources of radioactive background are anticipated:

A) The typical construction material is nonmagnetic stainless steel, which is contaminated with radioactive cobalt;

B) Commonly-used aluminium and fiberglass dewars are contaminated with radioactive Al^{26} and K^{41} (however, "old petrole" epoxy armed with quartz fiber can be used: it contributes only soft electrons from C^{14} decay which are easily shielded);

C) Superinsulation used in dewars is highly contaminated with radioactive thorium (recently, the PNL/USC group developed a technique permitting a decrease in the radioactivity of superinsulation by at least one order of magnitude);

D) The mixing chambers of the dilution refrigerator contain sintered silver, which is a strong neutron absorber (62 barn) and becomes radioactive, emitting both β^- and γ's ($E\gamma(max) \approx$ 1.5 MeV);

E) The indium O-rings used in typical dilution refrigerators are highly radioactive (In - 115, $t_{1/2} = 6 \times 10^{14}$y, $\beta^- = 0.63$ MeV): for example, in the PNL/USC ultralow background spectrometer, a <u>single</u> indium O-ring was responsible for the majority of background;

F) Hard soldering rather than electron welding is used in existing dilution refrigerators (the experience of the PNL/USC group shows that even a milligram of solder contributes significantly to the radioactive background).

There is, however, also good news: first, liquid helium (LHe) is the purest material known; secondly, quartz or germanium oxide (GeO) glass can be used as a low radioactive background material; third, the CHFC copper used extensively in cryogenic systems (e.g., as cold fingers between mixing chamber and sample) is one of the lowest background materials. Finally, the experience of the PNL/USC groups shows that radioactively pure, 400 year old lead can be used instead of indium in O-rings. Furthermore, it can be used as solder.

On the base of this data, we have designed a low radioactive background dilution refrigerator. Note that we postulate the use of a commercially available mixing chamber which is separated from the detector by a very long copper "cold finger" and a series of "old lead" radiation shields. We call this design a "long neck" dilution refrigerator. The construction and testing (thermal performance and radioactive background) of this dilution refrigerator is one of the main tasks of our current research.

CONCLUSIONS

It is apparent that direct experimental searches for low energy galactic neutral particle fluxes, covering all of the currently favoured "cold dark matter" candidates, should now be feasible with a few year development period. More specifically, the use of existing superheated superconducting colloid detectors with SQUID read-out permits detection of solar cosmions. This offers an exciting opportunity for astrophysicists and particle physicists to collaborate on some challenging problems of common interest, with the possibility of creating a new area of observational physics. Advancing this goal requires collaboration between workers in these fields and low temperature physicists.

More specifically, we expect the count-rate of about 200 counts/day in a gram SSCD detector. This suggest, that with appropriate low radioactive background detector, the search for solar cosmions can be performed on the earth surface.

REFERENCES

1. For a recent review see:
"Dark Matter in the Universe," eds. J. Knapp and J. Kormendy (Reidel, 1986), proceedings of the International Astronomical Union Symposium No. 117, Princeton, June 24-28, 1985; S. M. Fall, "Dark Matter in Early-Type Galaxies", preprint STSI 149, Dec. 1986; K. A. Olive, "Dark Matter as a Probe of the Early Universe", preprint UMN-TH-561/86, May 1986; V. C. Rubin et al., Ap. J. 255 (1978) L107; D. Burstein, V. C. Rubin, Ap. J. 297 (1985) 423; J.

N. Bahcall, S. Casertano, "Some Possible Regularities of the Missing Mass Problem", preprint IAS, Princeton 1985.

See also:

V. C. Rubin et al., Ap. J. 255 (1978) L107; S. M. Faber, T. S. Gallagher, Ann. Rev. Astr. Ap. 17, 135 (1979).

D. Fabricant, P. Gorenstein, Ap. J. 267, 535 (1983) W. Forman, C. Jones, W. Tucker, Ap. J. 293, 102 (1985).

2. D. J. Hegyi, K. A. Olive, Phys, Lett. 126B (1983) L8 ibid. Ap. J. 303 (1986) 56; J. Primack, SLAC preprint 3387 (1984); C. J. Hogan et al., preprint Fermilab-Conf-85/57-A, April 1985; N. Kaiser, J. Silk, Nature 324 p. 539-537, 11 Dec. 1986.
3. S. P. Ahlen, et al., preprint CFA No. 2292, March 1986. (See also Physics Lett. 195 (603-608) 1987.
4. M. Goodman, E. Witten, Phys. Rev. D31 (1985), 3059; I. Wasserman. Phys. Rev. D. 33, 2071 (1986).

   A. K. Drukier, Acta. Physica. Polonica, B17 (1986), 229.

   A. K. Drukier, K. Freese, D. N. Spergel, Phys. Rev., D33 (1986), 3495.
5. D. N. Spergel, W. H. Press, Ap. J. 294 (1985) 663; W. H. Press, D. N. Spergel, Ap. J. 296 (1985) 679; J. Faulkner, R. Gilliland, Ap. J. 299 (1985) 663; K. Freese et al., Ap. J. 299 (1985) 1001; R. L. Gilliland, J. Faulkner, W. M. Press, D. N. Spergel, Ap. J. in print.
6. G. B Gelmini, L. J. Hall, M. J. Lin, "What is Cosmion", preprint Harvard HUTP-86/A042.

   S. Rabi, G. West, Los Alamos preprint LA-UR-87-1734.
7. P. F. Smith, "Possible Experiments for Direct Detection of Particle Candidates for the Galactic Dark Matter", preprint RAL-86-029.

   A. K. Drukier et al., "Neutrino Masses and Neutrino Astrophysics" Ashland, Wisconsin, 1987 p. 470-487 ed. World Scientific.
8. A. K. Drukier, L. Stodolsky, Phys. Rev. D30 (1984), 2295.
9. See Proc. Workshop on "Low Temperature Detectors for Neutrinos and Dark Matter", Ringberg, May 12-13, 1987, ed. K. Pretzel et al., Springer Verlag.

IMPACT OF HIGH Tc SUPERCONDUTCORS ON THE DETECTION OF SMALL AND "EXOTIC" EFFECTS

A. GRASSI, G. STRINI

Dipartimento di Fisica
Università di Milano
Via Celoria 16,
20133 Milano, Italy

1. INTRODUCTION

Following the discovery by Bednorz and Müller [1] of superconductivity at 30K, a lot of work everywhere did push the critical temperature at least in excess of 90K.

Suspects of superconductivity also at 160K and 240K have been reported [2,3,4], and in a recent review Ginzburg [5] did suggest maximum theoretical possible value of Tc as high as 500K.

Here we limit ourselves to 90K superconductivity, in the search for applications to various experiments.

Surprisingly enough, almost every experiment in a wide class may be improved by a large scale, but simple application of the new superconducting materials.

The first application may be to build cheap large high field magnets for the search of very small but important effects, as the vacuum polarization by macroscopic magnetic fields [6,7], or effects of nearly massless, spin-zero particles on light propagation in a magnetic field [8].

The most useful property of the new mterials, even if apparently trivial, is to provide large volumes with superconducting shields without the restriction to use very low temperatures.

This has two main advantages: namely the possibility of effectively shield large volumes cheaply, as it is not necessary to use very low temperatures, and the possibility of use superconducting shields of systems at high temperature.

The former may be useful in sensitive and well known "null" experiments, as the search of an anisotropy of space, inertia, etc., the latter may be useful in the search of less well known but very important small temperature effects.

To have an idea of the possible improvements on "null experiments", it is enough to consider the Michelson experiment on the search for an anisotropy of the speed of ligth.

In a recent repetition of this experiment, Brillet and Hall[9] did reach a sensitivity of $dl/l=(1.5 \pm 2.5)*10^{-15}$ for the expected signature. In [10] it has been pointed out the possibility of reach a corresponding sensitivity of 10^{-18} for a multipass interferometer having: laser power of 1mW, arm length of 0.1 m, hundred reflections on mirrors and integration time of 10^{6} s.

2. VACUUM POLARIZATION BY MACROSCOPIC FIELDS

The vacuum polarization by a magnetic field is due to

diagrams shown in fig. 1

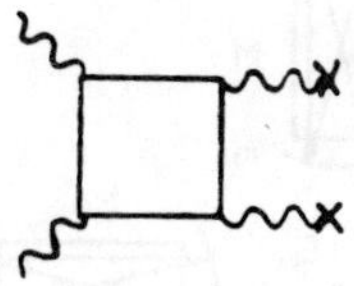

Fig. 1: Diagram for photon-magnetic field scattering.

As a result of this interaction, a computation dating back to Euler[11], gives a slowing down of the speed of electromagnetic radiation in a magnetic field ortogonal to the direction of propagation. For the incident field polarization with the electric field parallel to the polarizing magnetic field B_o, the variation is[6]:

$$\frac{\delta c}{c} = -3.5\gamma B_o^2 = -2.1*10^{-21}\left(\frac{B_o}{15\ T}\right)^2 \qquad (1)$$

and for the electric field ortogonal to B :

$$\frac{\delta c}{c} = -2.0\gamma B_o^2 = -1.2*10^{-21}\left(\frac{B_o}{15\ T}\right)^2 \qquad (2)$$

with:

$$\gamma = \frac{\hbar e^4}{45\pi m^4 c^7} \qquad (3)$$

These values require an high sensitivity, but not impossible to achieve. In fact an interferometer as in fig. 2 may reach a sensitivity of $\delta c/c \cong 10^{-23}$ with[6]: laser power of 15 W, arm length of 10m, reflectivity of mirrors of 0.995, 400 reflections and integration time of 10^6 s.

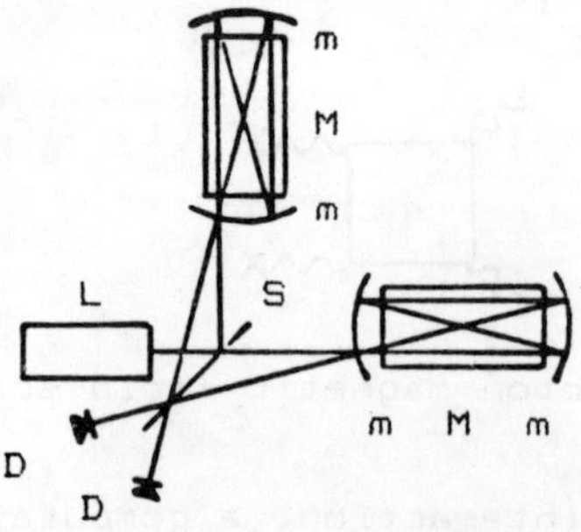

Fig. 2: Experimental lay-out for photon-magnetic field scattering: L = Laser, D = Detectors, M = Magnets, m = High reflectance mirrors, S = Semitransparent mirror.

Interferometers with similar characteristics are currently under active development in various laboratories as gravitational wave detectors.

3. CASIMIR EFFECT AT T>O

The Casimir force is due to an alteration of the vacuum fluctuations of the electromagnetic field due to boundary conditions effects (for a recent review, see [12]).

The effect is well known, and also experimentally investigated by Sparnaay [13]. Here we explore the possibility of experimentally study the temperature dependence of such an effect. The system is shown in fig. 3.

The standard Casimir effect is when a<<b and of the order of some 10^-6 m. Standard and well known computations [12] give as a result of the boundary conditions an attractive

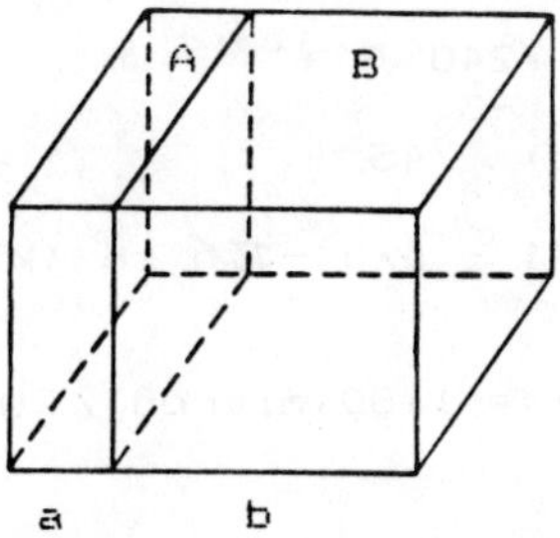

Fig. 3: Geometry for the computation of the Casimir effect: the regions A and B have dimensions a and b respectively, and infinite in the ortogonal directions.

force between the plates at distance a:

$$F/S = (\hbar c \pi^2) * a^{-4} / 240 \qquad (4)$$

where S is the area of the plate.

Turn now to the case of T>0. In the case a>>10^-6 m, the Casimir force in not more effective, due to the dependence a^{-4} in (4), and the forces on the central slab are due to the blackbody pressure given by:

$$F/S = (\pi^2) * (kT)^4 * (\hbar c)^{-3} / 45 \qquad (5)$$

on both sides of the slab; so there is no net force.

In the case T>0 and a<<b and a of the order of 10^-6 m, the blackboby pressure on the left side is almost freezed as the levels for the a degree of freedom are more and more spaced as the dimension a decreases.

A computation of the total force is given in [14,15,16]:

$$F/S = (\hbar c \pi^2) * a^{-4}/240 + \pi^2 * (kT)^4 * (\hbar c)^{-3}/45 + \pi kT a^{-3} * \sum_{n=1}^{\infty} n^2 * \ln(1 - \exp(-\pi \hbar cn/(kTa))) \qquad (6)$$

As $a > \hbar c \pi^3/(60 kT) = 1180(\text{micron})/T(K)$, the (6) simplifies in

$$F/S = (1.202/4\pi) * kT * a^{-3} \qquad (7)$$

Surprisingly the Planck constant drops out in this limit. The Casimir effect depends only on the quantum properties and boundary conditions of the electromagnetic field, but not on the electron charge and mass at lower order.

The results are plotted in fig. 4.

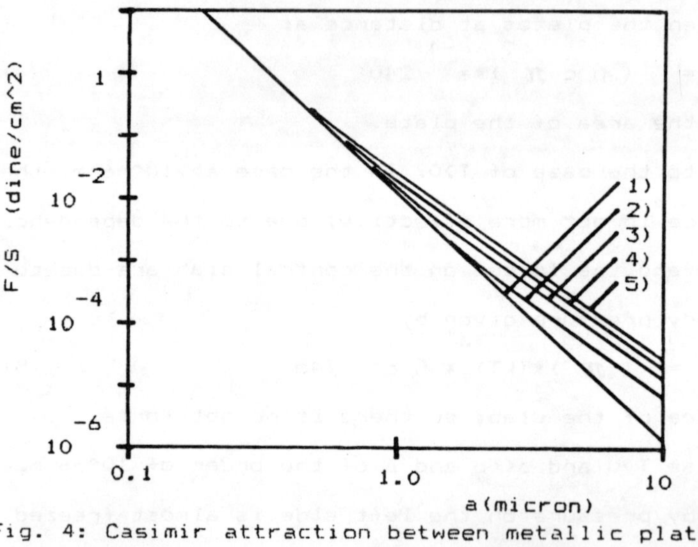

Fig. 4: Casimir attraction between metallic plates distant a microns and at various temperatures: 1) T = 0K, 2) T = 500K, 3) T = 1000K, 4) T = 1500K, 5) T = 2000K.

At higher order, radiative corrections depend obviously on the properties of the electron, but for the practical range of the parameters, these corrections can be shown to be less than 1 ppm [17,18].

4. QED AT T>O

Various effects of QED at T>O have been searched for. First, it is important to ceck which effects are null or too small to be detected.

4.1 Thermal effects on the speed of light

This effect is due to photon-photon interaction, and has been computed by Tarrach [19], with the result:

$$\frac{\delta c}{c} = - \frac{106}{2835}\left(\frac{\alpha}{\pi}\right)^2\left(\frac{kT}{m c^2}\right)^4 = - 2.65*10^{-31}\left(\frac{T}{2000K}\right)^4 \tag{8}$$

and so it is too small to be detected.

4.2 Thermal correction on the electron inertial mass

This effect at the one loop level in the radiative corrections has been computed various times. As by Peressutti and Skagerstam [20], amounts to:

$$\frac{\delta m}{m} = \frac{\pi \alpha}{3}\left(\frac{kT}{m c^2}\right)^2 = 8.09*10^{-16}\left(\frac{T}{2000K}\right)^2 \tag{9}$$

This value is small, but not too far to be seen.

4.3 Thermal correction to the electron charge

To the one loop level QED corrections, as shown by Yee[21], the electron electric charge is not affected by the temperature, so the effect in 4.2 can be searched for by measuring the ratio e/m.

4.4 Thermal corrections to the electron a=(g - 2)/2

This effect has been evaluated by Peressutti and Skagerstam[20], with the result:

$$\frac{\delta a}{a} = \left(\frac{2\pi}{3}\right)^2 \left(\frac{kT}{m\,c^2}\right)^2 = 5.1*10^{-13} \left(\frac{T}{2000K}\right)^2 \qquad (10)$$

This effect seems to be the most sensitive for an experimental search. In fact the present precision is of the order of 10^{-12}.

5. PRINCIPLE OF EQUIVALENCE AT T>O

In finite temperature quantum field theory, the variation of the inertial mass has already been given in (9). In general relativity the gravitational field couples to the energy-momentum tensor of matter, and so, by defining the gravitational mass by means of the energy-momentum tensor, it has been shown by Donoghue et al.[22,23] that temperature gives a contribution to the gravitational mass opposite to the corrisponding contribution to the inertial mass. The resulting violation of the equivalence principle is:

$$\frac{\delta m}{m} = -\frac{4\alpha}{3}\left(\frac{kT}{mc^2}\right)^2 = -3.55*10^{-15}\left(\frac{T}{2000K}\right)^2 \tag{11}$$

The present sensitivity of the verification of the equivalence principle is:

$$\frac{\delta m}{m} = (-0.3 \pm 0.9)\ 10^{-12} \tag{12}$$

by Braginskii and Panov[24], in an experiment more then 15 years old.

Recently various proposals have been made to push the sensitivity to the 10^{-15} and even to the 10^{-17} level. so the sensitivity requested in (11) does not seem too far to be reached.

6. REFERENCES

1) Bednorz, J. G. and Müller, K. A., Z. Phys. B64, 189 (1986).

2) Cai, X., Joynt, R. and Larbalestier, D. C., Phys. Rev. Lett. 58, 2798 (1987).

3) Huang, C. Y., Dries, L. J., Hor, P. H., Meng, R. L., Chu, C. W. and Frankel, R. B., Nature 328, 403 (1987) Erratum, Nature 329, 181 (1987).

4) Chen, J. T., Wenger, L. E. and McEwan, C. J., Phys. Rev. Lett. 58, 1972 (1987).

5) Ginzburg, V. L., Preprint: Miramare-Trieste IC/87/164.

6) Grassi Strini, A. M., Strini, G. and Tagliaferri, G., Phys. Rev. D19, 2330 (1979).

7) Iacopini, E. and Zavattini, E., Phys. Lett. 85B, 151 (1979).

8) Maiani, L., Petronzio, R. and Zavattini, E., Phys. Lett. B175, 359 (1986).

9) Brillet, A. and Hall, J. L., Phys. Rev. Lett. 42, 549 (1979).

10) Grassi Strini, A. M., Strini, G. and Tagliaferri, G., Lett. Nuovo Cim. 24, 212 (1979).

11) Euler, H., Ann. Phys. 26, 398 (1936).

12) Plunien, G., Müller, B. and Greiner, W., Phys. Rep. 134, 87 (1986).

13) Sparnaay, M. J., Physica XXIV, 751 (1958).

14) González, A. E., Physica 131A, 228 (1985).

15) Scharnhorst, K., Robaschik, D. and Wieczorek, E., Annalen der Phys. 7,44, 351 (1987).

16) Plunien. G., Müller, B. and Greiner, W., Physica 145A, 202 (1987).

17) Wieczorek, E., Robaschik, D. and Scharnhorst, K., Sov. J. Nucl. Phys. 44, 665 (1987).

18) Robaschik, D., Scharnhorst, K. and Wieczorek, E., Ann. Phys. 174, 401 (1987).

19) Tarrach, R., Phys. Lett. 133B, 259 (1983).

20) Peressutti, G. and Skagerstam, B. S., Phys. Lett. 110B 406 (1982).

21) Yee, J. H., Phys. Lett. 141B, 411 (1984).

22) Donoghue, J. F., Holstein, B. R. and Robinett, R. W., Phys. Rev. D34, 1208 (1986).

23) Donoghue, J. F. and Holstein, B. R., Eur. J. Phys. 8, 105 (1987).

24) Braginskiĭ, V. B. and Panov, V. I., Sov. Phys. JETP 34, 463 (1972).

LOW COST, COMPACT DILUTION REFRIGERATOR FOR PARTICLE DETECTOR APPLICATIONS

K.E. GRAY, P.R. ROACH and W.D. GUSCHWAN

Materials Science Division
Argonne National Laboratory,
Argonne, Illinois 60439

We have developed a one-shot ^3He-^4He dilution refrigerator using charcoal pumps that is of a very compact and simple design (essentially portable) and eliminates many of the problem areas of continuously operating refrigerators. It will cool below 0.015 K and run at this temperature for more than 10 hours. It is likely that a typical user would make whatever measurements were desired during the 10 hour running time at low temperature and then let the system recycle automatically overnight. For operation at higher temperatures, the running time is expected to be considerably longer. For example, extrapolating the T^2 dependence of the cooling power to 0.1 K implies a running time well in excess of 100 hours, or 40 hours with an order-of-magnitude larger heat leak of 1 microwatt[1]. We have also developed a complete computer control for the operation and recycling of the refrigerator. This was simplified by the fact that there are no moving parts and only heaters have to be controlled in response to measurements of resistance thermometers and thermocouple pressure gauges.

This refrigerator would also be ideally suited for particle detector applications requiring temperatures below 300 mK. With computer control, the refrigerator can be run relatively unattended, except for replenishing the liquid ^4He bath, for indefinite periods of time by persons with little or no expertise with dilution refrigerators, and at considerably less expense (and maintainence) than conventional systems. Thus the inherent low noise accompanying very low temperature operation can be available to everyone.

CONTINUOUS REFRIGERATORS

^3He-^4He dilution refrigerators of the continuously operating type have been used for many years to provide temperatures from 1.0 K down to as low as 0.002 K. These refrigerators operate on the principle of cooling that is achieved in a mixing chamber when ^3He crosses a phase boundary separating almost pure ^3He from a dilute mixture of 6% ^3He in ^4He. The flow of ^3He across the phase boundary is driven by the removal of ^3He from the dilute phase in a separate container, the still, which is thermally isolated from the mixing chamber but connected to the liquid He in it by a thin tube. The temperature of the still is maintained by a heater at a value such that the vapor above the dilute liquid is almost pure ^3He. An external pump then removes this ^3He from the still, leaving the ^4He behind as a stationary background component. In a continuously operating system, the ^3He is re-compressed, cooled to 1 K, whereupon it re-liquefies, and is returned to the mixing chamber. In order to minimize the heat load on the mixing chamber from this returning ^3He, it is precooled by the cold, dilute ^3He exiting the mixing chamber in a series of sophisticated heat exchangers.

Such dilution refrigerators include 1 K pots in which liquid ^4He is cooled using an external pump. Outside the cryostat will be, in addition to the pumps mentioned, storage tanks for the ^3He-^4He gas mixture, various traps for cleaning the ^3He before it is returned to the refrigerator, large diameter pumping lines to the cryostat, and numerous valves and gauges for operating the system. Such systems require appreciable training or experience to run and to keep them running. Although they can operate relatively unattended except for replenishment of liquid ^4He, the external mechanical pumps and gas handling equipment must be periodically monitored and maintained to avoid contamination of the returning gas. Contamination with either air or pump oil vapor can cause catastrophic blockages of low temperature capillaries. Such blockages may only require warming the whole system to room temperature or, in more serious cases, surgical intervention into the sealed ^3He-^4He mixture system. Given operators with sufficient low temperature expertise and dilution refrigerator experience, these systems can run for many days at their lowest temperature and some have been kept cold for many months at a stretch.

ONE-SHOT DILUTION REFRIGERATOR

Principle of Operation. The one-shot refrigerator we have developed operates on the same principle as the continuously operating refrigerator except that ^{3}He is not returned to the mixing chamber after it leaves the still. In addition, the external pumps are replaced by charcoal adsorption pumps operating in the 4.2 K bath of the cryostat. A charcoal pump removes ^{3}He from the still until the concentrated ^{3}He above the phase boundary in the mixing chamber is depleted. Then cooling stops and the system must be recycled by heating the charcoal to desorb the gas and to recondense the ^{3}He into the refrigerator.

Figure 1 shows a schematic of the lowest temperature parts of our refrigerator. There are 3 systems in our design: a ^{4}He system for the 1 K pot which is used to condense ^{3}He into the refrigerator and to provide thermal shielding; a ^{3}He system which cools a pot attached to the mixing chamber in order to set up the initial phase boundary in the mixing chamber; and the ^{3}He-^{4}He mixture system for the dilution refrigerator itself. At room temperature the amount of ^{3}He for the ^{3}He pot is small enough that it is stored in its charcoal pump at elevated pressure (~1.6 MPa). The ^{3}He-^{4}He mixture is stored in internal storage tanks just above the bath at ~0.6 MPa. The amount of ^{4}He used is large enough that it must be stored in an external tank (at ~0.6 MPa). Sintered copper powder in the bottom of the mixing chamber provides a large surface for good thermal contact between the liquid helium and the copper walls.

Running the System. The refrigerator is started at 4.2 K by heating the charcoal pump containing the ^{4}He. When the pressure of the ^{4}He reaches 1 atm, it condenses where it runs through the 4.2 K bath and drops into and fills the ^{4}He pot. The heating is stopped and exchange gas is admitted to the vacuum space of the pump by heating a small exchange gas charcoal pump. The ^{4}He exchange gas cools the pump towards the bath temperature, whereupon it begins to pump and the ^{4}He pot cools to about 1 K. Then the pure ^{3}He and the ^{3}He-^{4}He mixture are condensed in using the low temperature of the ^{4}He pot. When all the systems have reached 1 K, the ^{3}He pot is pumped to about 0.3 K by turning its pump heater off and exchange gas pump heater on, thereby admitting ^{4}He into its vacuum space. This cooling sets up a phase boundary in the mixing chamber and draws the maximum amount of concentrated ^{3}He into the mixing chamber above the boundary. The still is then pumped and heat is supplied to it to keep it at ~0.6K. The temperature of the mixing chamber is controlled by varying the pumping rate of ^{3}He (determined by the still temperature) or by adding heat to the mixing chamber.

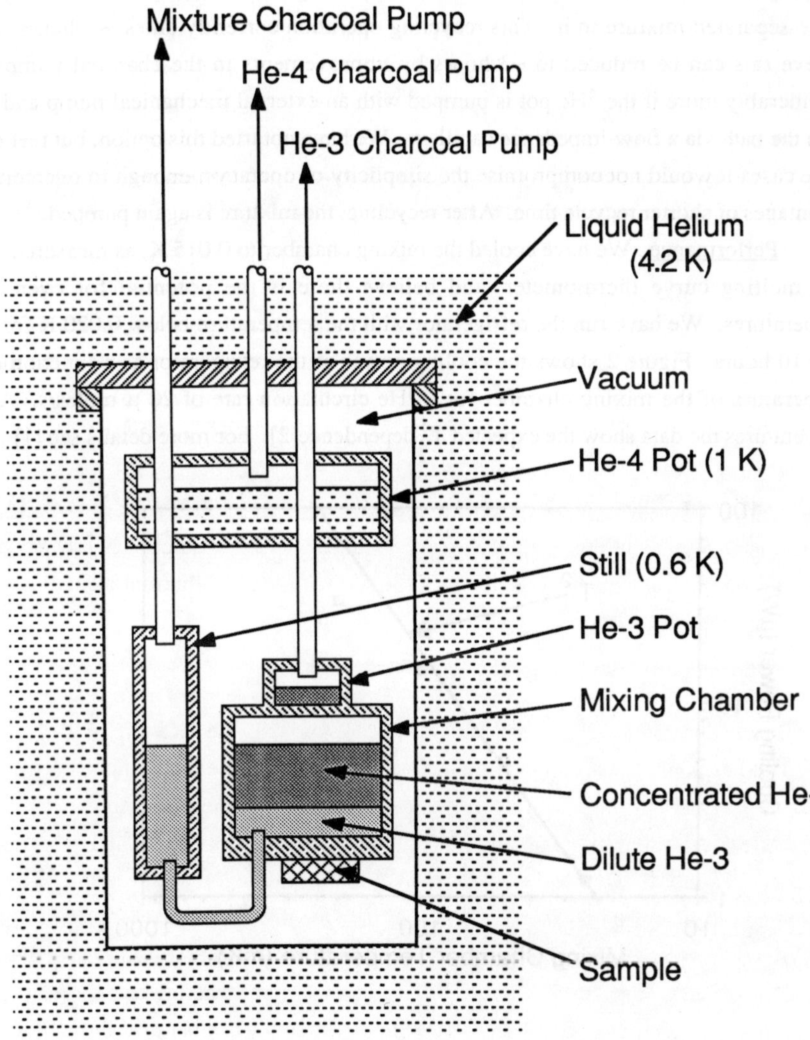

Fig. 1. Schematic diagram of the low temperature part of the refrigerator.

When the concentrated ^{3}He in the mixing chamber is depleted and cooling stops, it is necessary to recycle the system. The sequence of condensing ^{4}He into the ^{4}He pot, pumping it to 1 K, etc., is then repeated until the mixing chamber is again at 0.3 K with the phase-separated mixture in it. This recycling operation currently takes ~5 hours but we believe this can be reduced to ~3 hours by improvements in the charcoal pumps, and considerably more if the ^{4}He pot is pumped with an external mechanical pump and filled from the bath via a flow-impedance capillary. We have not tried this option, but feel that in some cases it would not compromise the simplicity of operation enough to overcome the advantages of shorter recycle time. After recycling, the mixture is again pumped.

Performance. We have cooled the mixing chamber to 0.015 K, as measured with a ^{3}He melting curve thermometer, and believe there is the potential for even lower temperatures. We have run the refrigerator with the temperature below 0.020 K for more than 10 hours. Figure 2 shows the cooling power of the refrigerator as a function of the temperature of the mixing chamber for a ^{3}He circulation rate of 20 μ moles/s. At low temperatures the data show the expected T^2 dependence[2]. For more details see [1]

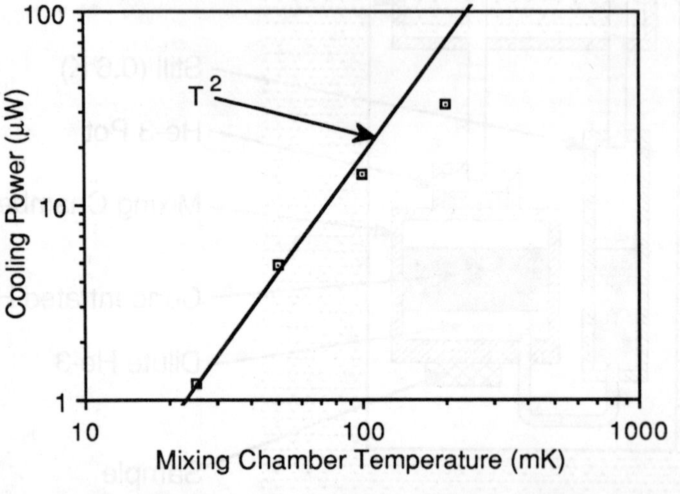

Fig. 2. Cooling power of refrigerator vs. mixing chamber temperature for a circulation rate of 20 μ moles/s.

Design Advantages. One of the important features of our design is that the still is at the same level as the mixing chamber and is of much smaller cross-sectional area. This means there will be a free liquid surface in the mixing chamber at approximately the same level as the one in the still. Then, as the ^3He in the mixing chamber is depleted, the level in the mixing chamber and still will both drop until the ^3He is completely gone from the mixing chamber at which point the mixing chamber will be nearly empty of liquid and the level in the still will be rather low. Since the total amount of dilute phase in the system is approximately constant, the large drop in level of dilute phase in the still can be accommodated by a rather small rise in the level of dilute phase in the mixing chamber because they have very different cross-sectional areas. The advantage of this feature is that the limited cooling power of the refrigerator is not wasted cooling a large amount of dilute phase. In the standard configuration with the still above and of comparable volume to the mixing chamber, dilute phase would enter the mixing chamber to replace the ^3He leaving it and the end point of the operation would be with the mixing chamber full of dilute phase. This greatly reduces the available cooling capacity and running time.

Because the ^3He is not returned to the mixing chamber, the need for complex heat exchangers is eliminated. Because the external pumps are replaced by internal charcoal pumps, the external gas handling system is eliminated. Because the exchange gas is controlled by small charcoal pumps in the system, there is no need of valves or external pumps for this function. These simplifications mean the present system is much smaller and, because everything is contained in the cryostat insert, rather portable. The low-temperature part of the system is much simpler without the heat exchangers. Because the heat load on the mixing chamber from returning ^3He is eliminated, this design should easily reach temperatures of 0.010K or below. Because the He gas mixture never leaves the low-temperature region during operation, the chance of contamination of the gas is eliminated and the reliability of the system is much improved. Reliability is further enhanced because there are no mechanical pumps nor valves needed for operation of the refrigerator. An additional benefit of this is that the operation is completely silent and vibrationless.

COMPUTER CONTROL

The operation of the refrigerator can be divided into eight distinct stages. The first three stages represent cooling the cryostat from room temperature to 4.2 K and at present involve no control functions by the computer, only measurements of temperatures and pressures. In stage 4, the ^{4}He is condensed into its pot (see Fig. 1 or 5) at 4.2 K and then pumped by its charcoal pump to about 1 K. Next, in stage 6, the ^{3}He and ^{3}He-^{4}He mixture are condensed into their pot and mixing chamber/still respectively at a temperature of 1 K by passing the gases through the ^{4}He pot. If the ^{4}He pot runs out before this is accomplished, or at any other ensuing stage of operation, the program switches to stage 5. Stage 5 recondenses the ^{4}He and pumps it in a similar fashion to stage 4, except that the other pumps are maintained in an appropiate holding condition (e.g., neither pumping nor desorbing gas). After the ^{3}He and mixture are condensed, the ^{3}He is pumped in stage 7 to lower the temperature of the mixing chamber to about 0.3 K and so ensure that the phase separation occurs there and that the ^{3}He-rich component of the mixture remains there. When this temperature is reached, stage 8 begins by pumping on the still and supplying the appropiate still heater power to maintain a still temperature of about 0.6 K so that ^{3}He is preferentially pumped. During each of the stages 4-8, the computer controls the power in each of the heaters, and may vary it during that stage in response to measured temperatures and pressures.

Hardware. Several components besides the computer are required to control the heaters and read the resistance thermometers and pressure gauges. For the resistance thermometers the need to minimize self-heating requires an ac bridge and the desire for automation requires it to be self-balancing. The BioTechnology, Inc. (BTi) Model 1000 Conductance Bridge satisfies these requirements, although it is very slow when operating in the auto-ranging mode and does not have the IEEE interface. Therefore the BTi range was fixed at 4 mS and its output, together with the thermocouple pressure gauges, were measured with a Keithley Model 175 digital voltmeter (DVM) which has the IEEE interface. For switching the BTi input between the various resistance thermometers and the DVM input between the various thermocouple gauges and the BTi output, we chose a Keithley Model 706 Scanner, with an IEEE interface. We required two low-voltage scanner cards for the above operations. Two additional general purpose scanner cards were used to control the heater power to the three pump heaters (three levels plus off), the three exchange gas pump heaters (on or off) and the still heater (seven levels plus off).

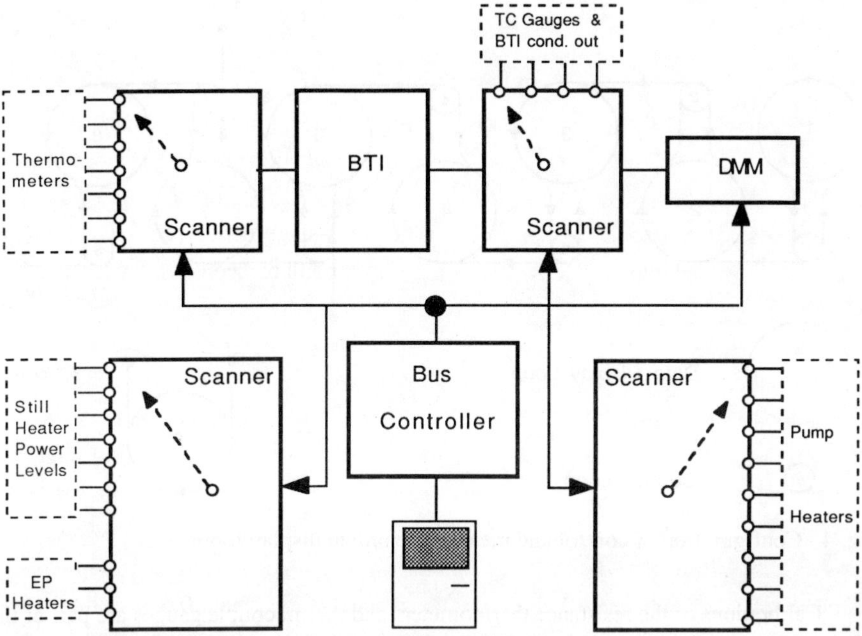

Fig. 3. Schematic of the electronic hardware for the computer control.

These are all illustrated schematically in Fig. 3, together with the Macintosh computer which offers a simple system for an interactive visual display using Microsoft Basic. Because of the serial output of the Macintosh, a bus controller (IOTech Mac488A) was used to communicate with the IEEE bus.

Software. All the source code, from the display to the control decisions are in Basic. The decision of passing from one stage of the refrigerator operation to the next is based on several measurements of temperature and/or pressure, which are different for each of the various stages. Therefore in each stage there is a measurement loop and a control loop in the program. The time required to measure all the parameters is sufficiently long that for each measurement loop only the critical parameters are sampled. The measurement loop sequences the Scanner, DVM and BTi to make the appropiate measurements, after which the control loop determines whether to go to another stage or return to the measurement loop. These are shown schematically in Fig. 4.

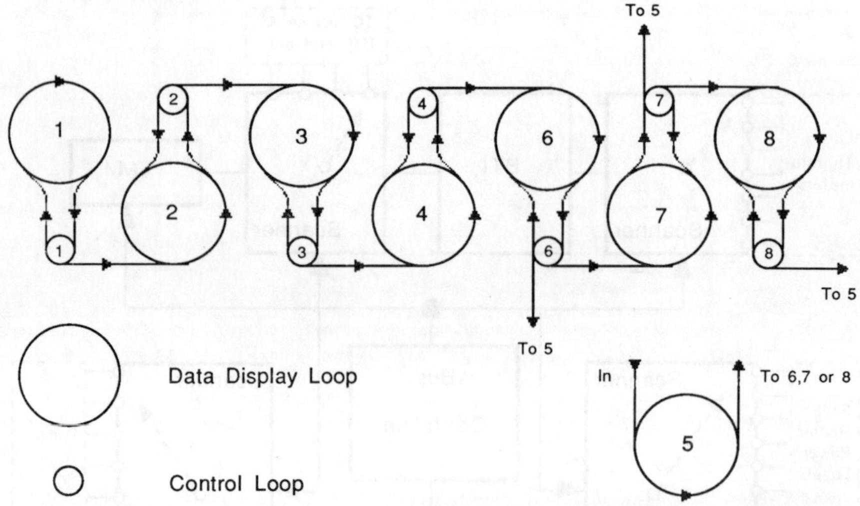

Fig. 4. Configuration of control and measurement/data display loops.

Calibrations of the resistance thermometers and thermocouple gauges are part of the software since only the resistance (conductance) and thermocouple voltages are measured. A simple circuit for operating a number of thermocouple vacuum gauges simultaneously, without uncertainties in the contact resistance of the connectors, was used and will be described elsewhere[3]. For the resistance thermometers, we have fit the measured temperatures to a power series in log(R-R'), where R' and the coefficients of the power series are determined individually for each thermometer.

The interactive visual data display is an important part of the program because we wanted the user to be able to evaluate the proper functioning of the refrigerator and interrupt or change the operation if deemed necessary. The information, mainly temperatures and pressures, is displayed on a schematic diagram of the refrigerator shown in Fig. 5. Although only the most recent data are displayed, older data may be retrieved by using one of the "pull-down" menus shown at the top of the display. Another menu allows the user to add extra parameters to be measured during each measurement loop. Note that the display shows the stage number on the upper left; that too can be changed manually. Finally, the heaters can be controlled manually by using the A/M (automatic/manual) "pushbuttons" located by the heaters in Fig. 5, and using the "hand" to move the slide switches to a new power level.

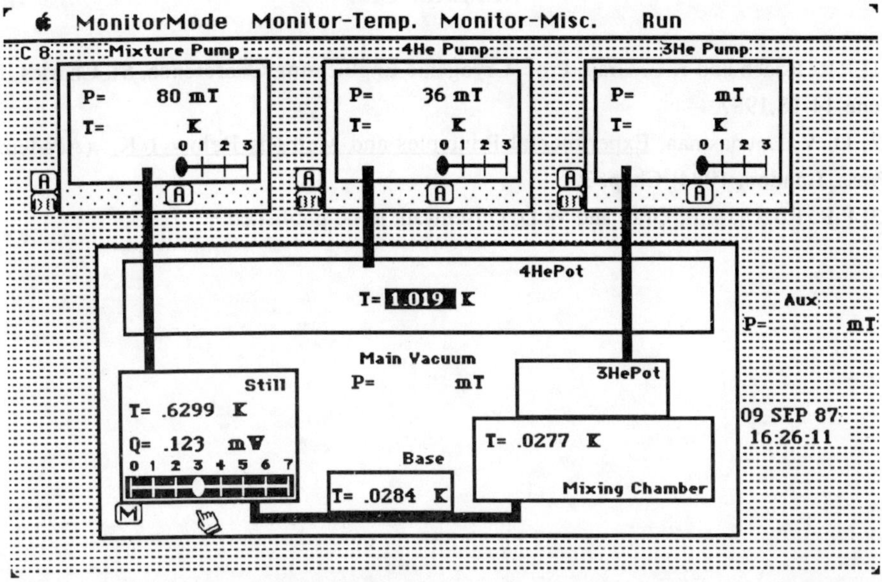

Fig. 5. Visual data display shown on the Macintosh during the various stages of operation. Note the manual heater controls and the "pull-down" menus at the top.

Note that when the refrigerator is at low temperatures, i.e., stage 8, very little time is required for the control functions and the computer can be used for data acquisition or control of the experiment on a time-share basis.

Performance. Our prototype compact dilution refrigerator has been run under computer control on several occasions with no complications. All the functions described above have been tested except the time-share operation when the refrigerator is at low temperatures.

ACKNOWLEDGEMENTS

We would like to thank Charles Schaeffer, Paul Morrison, Zenon Sungaila,Yong Jia Qian and Russ Hilleke for their assistance in the construction and testing of various stages of this refrigerator. Work supported by the U.S. Department of Energy, Basic Energy Sciences, Materials Sciences, under contract #W-31-109-ENG-38.

REFERENCES

[1] P.R. Roach and K.E. Gray, Proc. Cryogenic Engineering Conference, St. Charles, Il., June 14-18,1987.

[2] O. V. Lounasmaa, Experimental Principles and Methods Below 1 K. (Academic Press, London, 1974) Chap. 3.

[3] P.R. Roach, K.E. Gray and W.D. Guschwan, to be published.